SIR WILLIAM JONES

A READER

SIR WILLIAM JONES

A READER

Edited with Introduction and Notes by

SATYA S. PACHORI

Foreword by Rosane Rocher
and a Preface by Peter H. Salus

DELHI
OXFORD UNIVERSITY PRESS
OXFORD   NEW YORK
1993

Oxford University Press, Walton Street, Oxford OX2 6DP

New York Toronto
Delhi Bombay Calcutta Madras Karachi
Kuala Lumpur Singapore Hong Kong Tokyo
Nairobi Dar es Salaam
Melbourne Auckland
and associates in
Berlin Ibadan

SBN 0 19 562928 0

Computerset by Rastrixi, New Delhi 110030
Printed in India at Rekha Printers Pvt. Ltd., New Delhi 110020
and published by Neil O'Brien, Oxford University Press
YMCA Library Building, Jai Singh Road, New Delhi 110001

To my wife
Kamal Rani

Contents

ACKNOWLEDGMENTS

I am grateful to the University of North Florida (UNF) for awarding me a sabbatical, as well as for time off from teaching a full load in order to facilitate my work on William Jones. For travel grants to research libraries I am beholden to, especially, the University of Florida, the University of Texas at Austin, Texas A&M University, and Emory University. For her diligence and perseverance in typing the manuscript I thank Miss Fran Rigsby, a research assistant at UNF. For their timely research funding and other support, I am grateful to Professor E. Allen Tilley, former chairperson, Department of Language and Literature, and to former Dean Peter H. Salus, College of Arts and Sciences, UNF.

I am most grateful to Professor Garland Cannon for offering me many valuable suggestions in the preparation of this manuscript and for providing a few holographs of previously unpublished works by Jones. Cannon's biography of Jones and his edition of Jones's *Letters* are among the prime sources of information regarding the dates and the circumstances surrounding the composition of Jones's works included here. Cannon's studies are indispensable to any future Jones researcher.

While most of Jones's writings are in the public domain, I wish to thank the India Office Library and Records (of the British Library) for letting me reprint Sir Joshua Reynold's portrait of Jones. As always, the India Office Library and Records, the British Library, and the Bodleian Library were magnificently helpful.

Jacksonville, Florida   S.S.P.
November 1991

INVOCATION OF GANESA[1]

May success attend the actions of good men, by the favour of that mighty God, on whose head a portion of the moon appears written with the froth of the Gangā!

This *amicable instruction*, exquisitely wrought in *Sanscrit* phrases, exhibits continually, when heard, a prodigy of wisdom and the true knowledge of morals.

The learned man may fix his thoughts on science and wealth, as if he were never to grow old or to die; but when death seizes him by the locks, he must then practise virtue.

Knowledge produces mildness of speech; mildness a good character; a good character wealth; wealth, if virtuous actions attend it, happiness.

Among all possessions knowledge appears eminent; the wise call it supreme riches; because it can never be lost, has no price, and can at no time be destroyed.

Knowledge acquired by a man of low degree places him on a level with a prince, as a small river attains the irremeable ocean; and his fortune is then exalted.

The science of arms, and the science of books, are both causes of celebrity; but the first is ridiculous in an old man, and the second is in all ages respectable.

[1] [In honour of the Hindu god of wisdom and learning, the poem appears in Jones's translation *Hitōpadēsa of Vishnusarman*, the famed collection of animal fables illustrating the general morals that Vishnu-Sarman was assigned to teach his king's erring sons. Translated in 1786, *Hitōpadēsa* was published posthumously in *Works* (1807) 13: 1-210. All Hindu creative works begin with such an invocation.]

FOREWORD

The publication of Edward W. Said's controversial book, *Orientalism* (1978), has kindled an interest in the dynamics of the East–West intellectual encounter among a larger public in the past decade. The same period also saw, in 1984, the bicentenary of the founding of the Asiatic Society, a society 'instituted for the purpose of inquiring into the history and antiquities, the arts sciences, and literature of Asia'. This momentous step in the internationalization of culture was commemorated not only in Calcutta, the seat of the first Asiatic Society, but around the world. In the USA the Modern Language Association, the American Historical Association, and international conferences in Princeton on the historiography of linguistics and in Philadelphia on Sanskrit studies, underscored the significance of this event. It is good that this new publication presents, in a manner accessible to a general educated readership, some of the major achievements of the Asiatic Society's founder, Sir William Jones.

The range of Jones's scholarly interests was vast. It spanned concerns from literature to botany, from music to epigraphy, from Persian and Sanskrit to law. The expertise of modern scholars is, of necessity, more narrowly circumscribed. The author of *Sir William Jones : A Reader* is a Professor of English, and in a position to highlight the features of Jones's works which are of the greatest relevance for the general reader and the college student today. He is a lifelong student of Jones, and is familiar with the entire range of Jones's oeuvre.

By far the largest section in the *Reader* is devoted to Jones's contribution to literature. From his early studies of Persian poetry to his discovery of Sanskrit theatre in later life, Jones's goal was to impart to his contemporaries an appreciation of Oriental letters. He argued the point in essays, and illustrated it with poetical translations and imitations of his own. Though his translations have been superseded by more recent scholarship, their literary quality has given them an enduring value. It was not possible for Jones to convince all his contemporaries. His friend, Samuel Parr, quipped that 'when Jones dabbled in metaphysics, he forgot his logic; and when he meddled with Oriental literature, he lost his taste'. Yet this eccentric Latin scholar should be rated a less credible critic than the poet Goethe, whom the German translation of Jones's translation of *Shakuntala* moved to pen immortal lines, the vibrancy of which withstands translation:

Wouldst thou the young year's blossoms and the
fruits of its decline,
And all by which the soul is charmed,
enraptured, feasted, fed,
Wouldst thou the earth and Heaven itself in one
sole name combine?
I name thee, O Sakontala! and all at once is said.

Today's readers would no longer doubt the genuineness of newly discovered Oriental literary treasures, as too many eighteenth-century literary forgeries led some of Jones's contemporaries to do. And no one would, we trust, be narrow-minded enough to censure him any longer, as the Serampore missionary William Ward did, for writing hymns to Hindu gods such as are included in the *Reader*.

A shorter section on languages and linguistics follows Jones's path from his early and pattern-setting grammar of Persian to his later Sanskrit studies in India. Contrary to Charles (later Sir Charles) Wilkins, who asked to be stationed in Banaras for the benefit of his Sanskrit studies, Jones learned the ancient language in the seat of the British government, Calcutta. With tact, perseverance, and unfeigned respect, he convinced pandits, the learned guardians of the Hindu tradition, of his genuine and unselfish desire to learn their sacred language. Their instruction was a *sine qua non* for steady progress. Jones's knowledge of Sanskrit led to his statement of 1786 on the kinship of the languages now known as Indo-European, a statement which became so famous that it has often been quoted in isolation and taken as the charter of modern comparative linguistics. Historians of linguistics have shown that, viewed in its context, it is steeped in speculations on the origin of nations which were current in the eighteenth century. The Third Anniversary Discourse, in which this statement is found, is printed in full in the *Reader*.

A reader must organize its materials under discrete headings. Yet the section devoted to religion, mythology, and metaphysics should be read together with that on languages and linguistics. For Jones, a man of the eighteenth century, speculations on the origin and kinship of nations encompassed all aspects of their culture, language, religion, philosophy, and the arts. Jones's conjectures on the origin of the Indian zodiac make particularly interesting reading in this respect. They illustrate that it is possible to argue a point intelligently, even when wrong. More important, Jones's study of the Indian zodiac, read in conjunction with his comparisons between the gods of Greece, Italy, and India, and his statement on the affinity of Greek, Latin, and Sanskrit, brings vividly to the fore the difficulty of assessing and mapping kinship patterns at a time when scholars were still groping for a basic taxonomy.

The *Reader* includes, under the heading of religion, mythology, and metaphysics, an extract of Jones's translation of the Laws of Manu. From Jones's point of view, of course, the text was an ancient lawbook, and his interest in it stemmed from his duties as a Puisne Judge of the Supreme Court in Calcutta. Jones's wish to have unmediated access to the sources of Hindu law was the very reason why he undertook to study Sanskrit. The Laws of Manu, therefore, continue and expand the legal scholarship he had conducted on British common law and on the Greek orator Isaeus prior to his long-sought appointment to the bench in India. Legal scholarship, indeed, runs as a thread through most of his adult life.

To all who read it *Sir William Jones : A Reader* should be a tantalizing introduction to one of the most accomplished personalities in eighteenth-century Britain, a man superbly described as 'harmonious Jones'. It should foster the interest in interdisciplinary eighteenth-century studies which is gathering strength in the United States and elsewhere. Its success will be measured by the incentive it gives generalists and college instructors and their students to probe further into the many aspects of Jones's wide intellectual world. Above all it is hoped that the mission which 'Oriental Jones' took upon himself, to inculcate in the West an appreciation of Oriental literatures and cultures, will be given a new life. Today as in Jones's times, the college curriculum and the information of the general public remain disturbingly ethnocentric and increasingly out of step with a multinational, multicultural world. It is time, two hundred years after the founding of the Asiatic Society, for the West not just to learn about Oriental cultures, but to approach them, as any culture deserves to be approached, with respect and sympathy, and above all without the intellectual arrogance which is the hallmark of the uninformed. It is not enough that there are now in the West a number of distinguished Orientalists, learned societies and professional associations such as the American Oriental Society and the Association for Asian Studies, and a number of famed graduate departments which provide training in these still recondite subjects. In this, it is hoped, Jones may still play a role, for he provides a model and a counterpoint to the voices of Empire, such as that of Rudyard Kipling. East need not cease to be East, nor West cease to be West, yet, assuredly, if we are to grow stronger, the twain must meet.

University of Pennsylvania   ROSANE ROCHER
August 1989

PREFACE

Little more than a single printed page is devoted to Sir William Jones in *The Mid-Eighteenth Century* volume of the *Oxford History of English Literature*. About three-fourths of a page is lavished upon him in the bibliographical section. By way of comparison, Joseph Wharton, a decidedly minor poet, gets nearly double the space.

I begin with this rather unfortunate and melancholy statement because it illustrates, in my view, the importance of the present *Reader*. Virtually none of Jones's voluminous works is in print (two editions of his works appeared in six volumes in 1799, and then in 13 volumes in 1807) and even major cis- and trans-Atlantic libraries lack any but the most sketchy materials. This is tragic, for not only was Jones of major importance himself, he was of lasting influence in the fields of literature, linguistics, mythology and metaphysics, law, and the history of science.

One cannot open a work on historical or comparative linguistics, on Arthur Schopenhauer, on Goethe, without tripping over Jones's influential work. This anthology contains works which illustrate these points. At this point I shall only attempt to sketch Jones's all-too-brief life and works, as well as his influence upon posterity, and comment briefly on the organization of this volume.

Life and Works

William Jones was born in London, 28 September 1746, the second son (third child) of William Jones, a native of Anglesey, whose mathematical talents had brought to him the esteem of both Isaac Newton and Edmund Halley and the vice-presidency of the Royal Society. He was relatively old when his third child was born and died before William's third birthday. William was brought up by his mother, the daughter of George Nix (Chippendale's rival in cabinet-making), who had learnt algebra and other aspects of mathematics from her husband. Maria Jones taught William to read by age four. He was especially brought up on Shakespeare.

Michaelmas 1753, when he was seven, Jones was sent to Harrow. In the lower school he appears to have been sickly and made little impression; by the time he was twelve (1758), he had a reputation for precocity. His memory was phenomenal (he is said to have once written out *The Tempest* from memory), and he was both tasteful and elegant in expression. He apparently wrote and translated much, but most of his juvenilia has not survived.

(An exception is a poem, 'Saul and David', written when he was fourteen which Mrs Thrale recorded in her diary.) When he was seventeen Jones put together a small volume of verse which he presented to his friend John Parnell. His short epic 'Caissa' (360 verses), on the invention of chess, was included here. Jones became a chess addict when in his teens and remained one throughout his life. At Harrow until 1764, he made a distinguished record, winning all the prizes offered in composition and language study. There he learned Greek, Latin, Hebrew, modern languages like French and Italian, and even the Arabic writing system.

Jones was admitted to University College, Oxford, in 1764, and elected Bennett Scholar the following Michaelmas. He had taught himself Hebrew while at Harrow, and now added Arabic and Persian to his normal studies, filling in his time by reading Italian, Spanish, and Portuguese.

Because of financial difficulties, Jones became the private tutor of Lord Althorp, the heir of Earl Spencer, in the summer of 1766. That August 7, Jones was elected to a fellowship at University College, and was invited by the Duke of Grafton (at the Treasury) to become interpreter for the Eastern Languages. Jones modestly declined the offer, for his monetary needs were well met by the tutorship. That summer he also met Anna Maria Shipley, daughter of Dr Jonathan Shipley, then Dean of Winchester and later Bishop of St Asaph, many years later to become his wife.

In the summer of 1767 Jones accompanied the Spencer family on their trip to Spa, his first visit to Europe, where he greatly improved his German. The following winter was spent at the famed library at Althorp, where Jones read a great deal and added Chinese to his list of 'hobbies'.

1768 brought Jones an invitation—he was (at 22!) already known as an Orientalist—from Christian VII of Denmark to translate the history of Nader Shah from Persian into French. Jones (again modest) suggested that Alexander Dow might be better suited, but when Dow declined, he readily accepted. Though Jones stated that he did not like the work, he executed his commission faithfully and it was published at his own expense, entitled *L'Histoire de Nader Chah*, in 1770, together with an essay on Oriental poetry and thirteen verses of Hafiz in French. The *Histoire* became his first published book, clearly establishing his reputation as a Persian scholar. During this time he engaged in a steady correspondence about his Persian studies with the translator-diplomat Count Charles Reviczky, a Hungarian-born bibliophile, and about his Arabic studies with the Dutch Orientalist Henry Albert Schultens.

Though his pupil, George John Spencer, entered Harrow in 1769, Jones continued under the Spencer employment thereafter. He spent the winter of 1769 in the Midi, perfecting his French (Louis XVI is said to have remarked on meeting him: 'He is a most extraordinary man! He understands the language of my people better than I do myself!'), and planning a number of works. He also collected materials for a grammar of Persian. The *Grammar of the Persian Language* was published in 1771, following the return of the Spencer party to England and Jones's resignation of his tutorship to read for the Bar. He was admitted to the Temple on 19 September 1770, and was called in January 1774. The Persian grammar set a model both for language learning and for major grammars in Persian, Arabic, Hindustani, and Bengali.

On 30 April 1772, Jones was elected a Fellow of the Royal Society. As has been noted, perhaps the best indicator of the high esteem in which he was held at twenty-seven is the fact that on 2 April 1773 he was elected a member of The Literary Club—Samuel Johnson's coterie—four weeks prior to Boswell receiving that honor. As part of The Club, Jones associated with Sir Joshua Reynolds, Edward Gibbon, Richard Sheridan, Edmund Burke, David Garrick, Thomas Percy, and (of course) Dr Johnson. Jones was President of The Club in 1780.

Later in 1772 appeared his famous *Poems, Consisting Chiefly of Translations from the Asiatick Languages.* Four of the nine poems in the volume have classical sources, with pastoral subjects from Spenser, Addison, and Petrarch. Jones's title was thus somewhat misleading. The classical poems are: 'Caissa; or, The Game at Chess', 'Arcadia, a Pastoral Poem', and two poems inspired by Petrarch. Oriental richness dominates such poems as a pseudo-Indian tale, the allegorical 'The Seven Fountains', and 'A Turkish Ode of Mesihi', the only poem in the volume to be a literal translation of an Asiatic one. The most famous poem in the book is, of course, an expanded translation of 'A Persian Song of Hafiz', which had already become renowned from its appearance in the Persian grammar. Neither Jones's name nor his initials appear in this publication, though it was apparently well known that Jones was the author. It is worth noting here that Jones's *Grammar* and the Hafiz poem directly inspired Edward Fitzgerald, as noted by Kejariwal.[1]

In 1774 Jones published his *Poeseos Asiaticae Commentariorum (Commentaries on Asiatic Poetry,* in six

[1] O.P. Kejariwal, *The Asiatic Society of Bengal and the Discovery of India's Past 1784-1838* (Delhi : OUP, 1988).

books, in Latin). The 542 pages focus on Jones's implicit thesis that Asians have a fondness for poetry. The work also suggests his critical views on Middle Eastern metrics, imagery, subject matter, diction, and on a few individual poets. There are many comparisons with Latin, Greek, and Hebrew literature, thus opening the door to the field of comparative literature and its many unique perspectives and methodologies. Retrospectively, the volume tells us much about Jones and how little he knew about himself. In *Poeseos,* Jones bids a (premature) adieu to Oriental pursuits. His iambics, which were translated anonymously (Calcutta: 1800), were highly praised by his contemporaries:

> Vale, Camena, blanda cultrix ingeni,
> Virtutis altrix, mater eloquentiae,
> Linquenda alumno est laurus et chelys tuo.
> At, O Dearum dulcium dulcissima,
> Seu Suada mavis, siva Pitho dicier,
> A te receptus in tua vivam fide:
> Mihi sit, oro, non inutilis toga,
> Nec indiserta lingua, nec turpis manus!

> Farewell, O muse! sweet former of the mind!
> Parent of Eloquence and thought refin'd!
> Your pupil now desires his lov'd pursuit,
> Nor wears the laurel more, nor strikes the lute!
> Supreme of the sweet denizens in Heaven!
> Whether it be to your fond votary given,
> To gain applause by fair Persuasion's speech,
> Or should strong Eloquence his words enrich,
> Receiv'd in youth by you, he lives in you,
> Beneath his auspices the stripling grew.
> Hence aiming at professional renown,
> Let him with decency assume the Gown,
> Appropriate language give him to command,
> And spirit firm, without a venal hand.

Notwithstanding this public withdrawal from Oriental studies, Jones's ardent love of languages led him to translate ten legal orations and some fragments of Isaeus, the legendary teacher of Demosthenes, which appeared in 1779 as *The Speeches of Isaeus in Causes Concerning the Law of Succession to Property at Athens.* Its preface makes a case for the study of comparative law. The work thus established Jones as a solid legal scholar, capable of applying other nations' legal systems to that of the British.

Called to the Bar, Jones's correspondence now shows him increasingly interested in a parliamentary seat, ever more preoccupied with politics. He failed where politics was concerned, but he redirected his parliamentary interest towards a judicial appointment—and, as one might expect, found time to plan other writing projects. In a letter of 4 September 1780, he wrote to Edmund Cartwright that he was embarked upon:

two little works . . . . The first is a *Treatise On the Maritime Jurisprudence of the Athenians,* illustrated by five speeches of Demosthenes in *commercial* causes; and the second, a dissertation *On the Manners of the Arabians before the Time of Mahomet,* illustrated by the seven poems, which were written in letters of gold, and suspended in the temple at Mecca, about the beginning of the sixth century.

The second of these 'little works' bore fruit in 1783, when *The Moallakát* (of which one poem appears in this volume) was published. This work introduced important Arab literature to European readers and Jones became immediately known as a pioneer here.

In 1781 Lord Althorp had wed Miss Bingham, daughter of Lord Lucan, and Jones penned a nuptial ode, 'The Muse Recalled', which Horace Walpole liked so well that he printed it at his Strawberry Hill Press and lauded it to the Earl of Strafford on 31 August 1781. The ode is included in this volume.

In the same year, Jones produced his *Essay on the Law of Bailments* (concerning the holding of property in trust), which was considered a classic legal treatise at the time and was studied for many decades on both sides of the Atlantic. He soon turned to Shafite (Arabic) law of inheritance, writing an English translation of Ibn al-Mulaqqin's verse *Bughyat al-bahith.* In his Preface, Jones noted: 'It appears indubitable, that a knowledge of *Mahomedan* jurisprudence (I say nothing of the *Hindu* learning), and consequently of the *languages* used by *Mahomedan* writers, are essential to a complete administration of justice in our *Asiatick* territories'. Jones's legislative experience while working for Burke on several Indian bills in 1781-2 further convinced him that the Indians should be governed by their own laws and customs. To fill that need, he rendered his pioneering translation of *The Mahomedan Law of Succession* in 1782. It helped the British legislators in the implementation of the new Judicature Act.

Jones was quite eager to get an appointment to the colonial administration, and even his patience was beginning to wear a bit by the end of 1781. On 1 March 1782 he wrote Lord Althorp that 'I *feel* and know myself

capable of doing some good in Asia . . . and I should like to see the country, and give the finishing stroke to my oriental knowledge'. Jones had also received an invitation to travel in the opposite direction, to visit America, and he wrote Althorp, 'I am very ready (as I need not repeat) to traverse immense seas and burning sands . . .' (letter of 6 April 1782). Three weeks later (30 April), Jones wrote to the Attorney-General, Lloyd Kenyon, about 'being appointed to fill the vacant seat on the bench at Calcutta—which I have been four years hoping to attain . . .' Jones had come to know about the death of the puisne judge of the Bengal Supreme Court in 1778, a post that paid 6,000 pounds a year. This could be a cure for his financial ills—he may have surmised—as well as an aid to his ambition to launch a parliamentary career. Oddly enough, Jones seems to have been unaware that his liberal views on the American struggle for independence and his close ties to such political rebels as Benjamin Franklin, Dr. Shipley (his father-in-law to be), and Arthur Lee were obstacles in realizing his Indian ambitions.

During the course of these political frustrations, however, Jones kept his scholarly pursuits alive by working constantly on his Arabic translation of the *Moallakát,* which appeared late in 1782. This work introduced the concept of *kismet* and other hedonistic notions of life to the West. Poets like Tennyson (e.g. 'Locksley Hall') and Fitzgerald were influenced by this work. His prose translations of the camel descriptions, ruby wine, and desert fighting long haunted the European imagination and constituted another landmark for Jones in world literature. John Drew's *India and the Romantic Imagination* elaborates at length on both the extensive influences that Jones's introduction of the orient had on later British literature and on the many poets who drew upon him: Blake, Coleridge, and Shelley, to name but a few.[2]

It is unfortunate that India and things Indian appear to have fallen in Britain's esteem from 1785 to 1815. While the *Puráṇas,* for example, inspired the eighteenth-century imagination, they were cited by Lord Macaulay in introducing his education bill—which devastated traditional Indian learning.

The summer of 1782 saw Jones on the Continent again, complaining to Lord Althorp that his intended journey to the United States had resulted in an 'excursion to the United *Provinces* of Holland' (letter of 13 September 1782). Jones's hopes rose again, for he wrote to his favourite correspondent at the end of October in a more

[2] Delhi : OUP, 1986

vigorous and optimistic vein. But the matter was, unfortunately, not resolved, and Jones wrote to Lloyd Kenyon again on 27 January 1783. This time, the gods (of some pantheon) smiled, and on 3 March 1783 Lord Ashburton conveyed the news to Jones for which he had waited for five years: 'When I consider this appointment [as judge of the High Court in Bengal] as securing to you at once, two of the first objects of human pursuit, those of ambition and love, I feel it a subject of very serious and cordial congratulation, which I desire you to accept'.

Only a fortnight later, Benjamin Franklin wrote to Jones from Passy: 'I just now learn from Mr Hodgson, that you are appointed to an honourable and profitable place in the Indies; so I expect now soon to hear of the wedding . . . .'

The appointment was made public on 4 March 1783, and a knighthood was conferred on Jones to mark the occasion. His marriage to Anna Maria Shipley was solemnized on 8 April. On the 11th, Sir William and Lady Jones set sail from Portsmouth on the frigate 'Crocodile', on the long voyage to Calcutta. Jones now had what he most wanted in his life: a happy marriage and financial independence—ironically, both acquired by means of a lucrative colonial salary.

Looking ahead, it is difficult to believe that little more than a decade remained to Jones: in a letter to Lord Althorp from aboard the 'Crocodile' towards the end of April he says he would devote the next 'six years' to 'Persian and Law, and whatever relates to India' and he lists twenty-six items for future 'Enquiry':

the laws of the Hindus and Mohammedans, Modern Politics and Geography of Hindustan, Best Mode of Governing Bengal, Arithmetic and Geometry, and Mixed Sciences of the Asiatics, Medicine, Chemistry, Surgery, and Anatomy of the Indians, Natural Productions of India, Poetry, Rhetoric, and Morality of Asia, Music of the Eastern Nations, Trade, Manufacture, Agriculture, and Commerce of India, . . .

To Jones this was no joke, no impossible goal. In a public discourse delivered in early 1784 he remarked that he had (while on board the 'Crocodile') reflected upon 'how important and extensive a field was yet unexplored'. And in this, of course, Jones was correct.

For nearly a millennium Europe had benefited from its slight contacts with Asia: spices and silk; printing and paper money; the notion of a postal system; 'Arabic' numerals and the sternpost rudder; these, and many other objects and concepts, had found their way West through contacts in the Levant and through isolated traders like the Polos. But with the British conquest of India there was wholesale contact with a non-classical ancient civilization. And, naturally, there was culture shock on both sides.

Militarily and politically, we look to individuals like Warren Hastings and Robert Clive. Scientifically and culturally, the impact began when Sir Elijah Impey welcomed Sir William on behalf of the other judges of the Court at the end of September 1783. Jones sat on the Bench for the first time in early December—eight months after leaving Portsmouth. But, in the meantime, Sir William had busied himself with an enterprise close to his heart: the creation of a learned society in Calcutta, modelled along the lines of the Royal Society.

The response to his proposal was enthusiastic. He quickly sought out scholars like Governor-General Warren Hastings, Judge Robert Chambers, Charles Wilkins, Richard Johnson, and others who knew or professed interest in Persian, Sanskrit, and other Asian languages, and who were eager to investigate the unexplored treasures of Indian culture, literature, and the sciences. The Asiatick (now the Asiatic) Society of Bengal held its first meeting on 15 January 1784 in the company of twenty-nine British administrators and scholars. Jones asked Hastings to accept the Presidency, but when this was refused, he accepted the unanimous election to the chair. Jones remained President till his death, and under his aegis the Asiatick Society made tremendous progress towards its goals of 'enquiry into the history and antiquities, arts, sciences, and literature of Asia'. In his Preliminary Discourse, Jones outlined the Society's objectives, principles, and procedures. He raised the issue of including learned Indians among its members—something which fell onto the deaf ears of his ethnocentric fellow Britons.

The Society's business, however, kept him very busy, along with his judicial duties. Over the ensuing decade, Jones read twenty-seven of his linguistic, anthropological, historical, mythological, literary, and scientific (botanical) papers to the members. In 1784 alone, for instance, he held twenty-three meetings of the Society; there were about ten each year from 1785 to 1794.

During his brief vacation in Benares (now called Varanasi) in the summer of 1784, after recovering from a severe fever, Jones learned of the *Manava-Dharma-sastra*, an ancient Hindu law-book, which he very much wanted to translate. He felt committed to seeing Hindus and Muslims governed in accordance with their respective traditional laws and customs. He was unhappy with the few Persian translations of the Hindu law-books, for they were grossly inaccurate in their renderings. To his disgust, the Indian pandits who assisted the Calcutta Supreme Court had to undergo a system which was so cumbersome and fraught with difficulties that they were prone to make errors in their citation of the relevant laws in Sanskrit.

Jones immediately recognized the importance of the study of Sanskrit and, with his linguistic background, duly mastered it. Actually, his repeatedly unsuccessful requests to Wilkins to execute the needed translations from the Sanskrit law-books further impelled him to acquire the language of the Aryans. His letters to Patrick Russel (8 September 1785)—'I am now at the ancient university of Nadeya, where I hope to learn the rudiments of that venerable and interesting language which was once vernacular in all India', and to Charles Chapman (27 September 1785)—'I am proceeding slowly, but surely... in the study of Sanscrit', reveal the beginnings of his study to us. But already in October, Jones was able to write to John Macpherson of 'the Sanscrit mine which I have just opened'.

Despite recurring illness, which plagued Jones throughout his residence in India, by his Third Anniversary Discourse to the Asiatick Society on 2 February 1786 he had clearly made enormous strides in his Sanskrit, and he was able to make an announcement which gave rise to the entire field of comparative and historical linguistics of the nineteenth and twentieth centuries. The entire discourse is included in this volume, but it is vital for the reader of these pages to recognize just how insightful and revolutionary Jones's claims for Sanskrit were.

For European scholarship toward the end of the eighteenth century, the Classical languages were the basis of all non-Biblical wisdom. And it was upon the grammars of the two Classical languages that Renaissance linguistics was founded. For Jones to declare, as he did, that Sanskrit was on a level with Greek and Latin, more ancient than either, and certainly kin to both—and probably to Germanic and Celtic as well—was a definite show-stopper. Indeed, the Third Discourse laid down the foundation of modern comparative and historical linguistics not by mere assertion, but by defining language similarities on the basis of their descent from some earlier common language that might no longer exist. Jones's lucid formulation suggests the necessary concept of language families, the necessity of analysis of quantities of data in order to arrive at logical conclusions, and the possibility that the ancestral language, possibly an Indo-European language, no longer existed. This was his lasting contribution to linguistic scholarship.

As a consequence, it is impossible to examine the works of the fathers of comparative linguistics—August Schleicher, Jacob and Wilhelm Grimm, Rasmus K. Rask, August Pott, Franz Bopp, and others—without hearing the echoes of Jones's words. Revisionist linguistic history has sought to deprive Jones of his role, but Jones was unhesitating and unstinting in his praise of the earlier Sanskrit studies of Wilkins and the Bengali studies of Halhed, and it is Jones who is the 'onlie begetter' of the formulation in the discourse.

But Jones was not content with introducing European scholarship to the language of the Hindus. On 6 May 1786 he wrote to Macpherson that he was labouring on a digest of 'Hindu and Musulman law ... and I have begun a translation of Menu into English'. By 28 September he was involved (as he wrote to Russel) with the 'Hitopadesa, which is a charming book ... ' It is a rendering of two recensions of the major tales of the *Pañcatantra*. And in November he wrote to Macpherson that he spent an hour before dawn every day reading Sanskrit. (Wilkins published his translation of the *Hitōpadēśa* the following year.)

Jones loved India, even though he was not healthy there; his Anna Maria was frequently ill, too. But Jones was full of energy and vigour. He became interested in botany and in completing Wilkin's translation of the *Institutes of Hindu Law* (which finally appeared in 1794). On 19 March 1788 Jones wrote to Lord Cornwallis, Hasting's successor, a long memorandum containing his plan for a compilation of 'a complete digest of Hindu and Mohammedan laws ...' Jones did not live to execute the charge placed upon him by Cornwallis. The digest was completed by Henry Thomas Colebrook, Jones's successor, in 1797.

Jones's prose translation of Jayadeva's *Gīta Govinda* (1789), a lyrical drama included in this volume, presents in a most delicate fashion the love story of Krishna (a Hindu incarnation of Vishnu) and Radha. Jones's puritanic bent prevented him from including Sarga XII, which describes the marriage bower in highly sensual imagery. It is possible that Jones feared that his British readers might be offended by the passage, which would have the effect of jeopardizing his mission of bringing Indian literature in translation to the European reader. Here Jones exercised his translator's privilege, though we might quibble with this. However, the contemporary reader must attempt to view Jones in his own milieu.

In 1788 Jones undertook the editing of the *Asiatick*

Researches of the Society and brought out the first volume that year. Further volumes of the *Researches* appeared in 1790, 1792 and 1794. These volumes, containing the monumental essays of Jones and his colleagues, stirred up a wave of intellectual curiosity in Europe and in America about the existence and latest discoveries concerning the rich culture of India. Jones's essay 'On the Literature of the Hindus' delivers a lengthy commentary on a wide variety of Sanskrit writings. His 'On the Chronology of the Hindus' identifies a sequence of major Indian events and their dates, suggestive of the Western style of sequential recording of historical phenomena. The interpretation of the 'Inscriptions on the Staff of Firus Shah', as provided by Pandit Radhakanta Sarman, indicates Jones's early attempts at epigraphic analysis and translation of Indian inscriptions. His single most important essay, 'On the Gods of Greece, Italy, and India'—with its focus on the Hindu pantheon—paved the way for the entire discipline of comparative mythology. Previous to Jones, mythology was confined to the Classical world of Greece and Rome. Similarly, 'On the Antiquity of Indian Zodiac' supports the notion of the originality of Indian astronomy, thereby opening a dialogue about the originality of the Greek zodiac and a comparative study of the two systems.

In 1788, too, appeared Jones's translation of Kālidāsa's drama *Shakuntalā* (London edition : 1790), which had as great an impact on Western literature as the Third Discourse had on linguistics. Jones's Preface to the play gave Kalidasa the lasting epithet 'The Indian Shakespeare' and seated the Sanskrit playwright in the world's literary pantheon.

Shakuntalā was an immediate success in England and was translated into German (from Jones's English) by Georg Forster in 1791. It was retranslated many times, but the Forster translation was the one read by Herder, who recommended it to Goethe. Without Jones's translation of Kālidāsa's famous play, *Faust* would lack its 'Prologue on the Stage', for it is derived directly from the 'Prologue' of *Shakuntalā*.

> Do you desire the blooms of springtime, the
>   fruits of autumn;
> Do you desire that which charms and allures,
>   which satisfies and nourishes;
> Do you desire heaven and earth, encompassed in
>   one name?
> I name you Shakuntala, and all is said.
>   (J.W. von Goethe, transl. by P.H.S.)

Along with the *Shakuntalā*, some of Jones's poetic compositions which received a warm welcome in both Indian and European circles were the nine hymns to the Hindu divinities. Captivated by the Hindu mythologies found in Sanskrit literature, Jones culled from his reading the images and personalities of gods and goddesses: Kamadeva, Narayana, Saraswati, Ganga, Indra, Surya, Lakshmi, Durga, and Bhavani. His scholarly introductions to these hymns, published in European journals, aided his Western readers in their understanding of the rich fabric of the Hindu pantheon, conspicuously absent from the monotheistic Judaeo-Christian scriptures. Jones brilliantly dealt with the cosmic conception of the divinities. Most prominent of these hymns are the ones addressed to Kamadeva, the god of love, and Narayana, the Supreme Being, which influenced English Romantics like Southey, Coleridge, Shelley, and Byron. By treating Hindu mythology in such an aesthetic fashion, Jones again was a pioneering influence on English literature.

Under the auspices of the Asiatick Society, Jones was able to make another major contribution to the study of different national cultures, origins, languages, and literatures through the medium of his eleven Anniversary Discourses delivered to the members and other researchers. Five of these Discourses were devoted to individual cultures: the third to the Indians, the fourth to the Arabians, the fifth to the Tartars, the sixth to the Persians, and the seventh to the Chinese. Each Discourse takes up language, letters, philosophy, religion, arts, and the architecture of these various peoples. The ninth Discourse takes up the ethnic origins of these five peoples, while the tenth provides us with a basis for scholarly advantage in possessing this kind of knowledge.

Originally, Jones had planned to stay in India till about 1800, but Anna Maria's failing health and his own frequent illnesses caused him to contemplate a much earlier return. Towards the outset of 1792 he writes of it; in October he wrote to George Harding that Lady Jones had promised to sail for Europe in January 1793. She finally did set sail from Calcutta at the beginning of December of that year.

On 20 February 1794 Jones delivered his eleventh Discourse to the Asiatick Society: 'On the Philosophy of the Asiaticks'. This being his last Discourse, and possibly sensing that his own end was near, Jones formulated his views on the definition of Indian philosophy, a subject that was very close to him.

I yield the final biographical words to Lord Teignmouth:

on the evening of the 20th of April, or nearly about that date, after prolonging his walk to a late hour, during which he had imprudently remained in conversation, in an unwholesome situation, he called upon the writer of these sheets, and complained of anguish symptoms . . . He had no suspicion at the time of the real nature of his indisposition, which proved in fact to be a complaint common in Bengal, an inflammation of the liver . . . The progress of the complaint was uncommonly rapid, and terminated fatally on the 27th of April, 1794. On the morning of that day, his attendants, alarmed at the evident symptoms of approaching dissolution, came precipitatedly to call the friend who has now the melancholy task of recording the mournful event. Not a moment was lost in repairing to his house. He was lying on his bed in *a posture of meditation*; and the only symptom of remaining life was a small degree of motion in the heart, which after a few seconds ceased, and he expired without a pang or groan.

Sir William Jones, aged forty-eight, was buried the following morning. Lord Teignmouth adds a coda:

The pundits who were in the habit of attending him, when I saw them at a public *durbar*, a few days after that melancholy event, could neither restrain their tears for his loss, nor find terms to express their admiration at the wonderful progress which he had made, in the sciences which they professed.

Scope and Influence

It is hard to place Jones. The *Oxford History of English Literature* remarks:

Jones is admittedly of more importance to the historian of ideas than to the literary reader, although it is characteristic of him that he should have been one of the earliest exponents of the view that poetry and the arts derive not from imitation, but from 'a strong and animated expression of the human passions'. (vol. VIII, p.199)

At the outset of this Preface, I enumerated a few places where one might encounter Jones: among these was in Schopenhauer's magnum opus. While the poetic, linguistic, legal, dramatic, and critical writings (both original and translation) of Jones are generally nodded towards, his influence on nineteenth-century German (and hence American) philosophy is little noted. Yet, at the very

beginning of the *The World as Will and Representation*, it is Jones who is cited. And from several important pieces of verse, it is clear that Ralph Waldo Emerson, thousands of miles away, paid fealty to Jones's translations.

Jones published translations from the Arabic, the Persian, the Turkish, and the Greek , as well as from the Sanskrit. His versions of Hafiz are still worth examination. His *The Speeches of Isāeus* (1779) was nearly as acclaimed for its legal scholarship as was his *Essay on the Law of Bailments* (1781). His *The Principles of Government* (1782) was extremely liberal for his age, and may have been the principal reason for the delay in his judicial appointment. I have no doubt that his 'Caissa' was the source of Benjamin Franklin's delightful essay on chess (Passy, June 1779; Jones's *Poems* was originally published in 1772, pirated in Germany in 1774, appeared in a second edition in 1777, and was reprinted the following year).

Moreover, the Asiatick Society of Bengal became the mother of all other Oriental learned societies (by comparison, the American Oriental Society was founded in 1842). Because of his great interest in non-Western languages, Jones became concerned with the problems of the transliteration of alphabets—what we would call phonetic transcription. The result of this was Jones's essay on orthography: 'Dissertation on the Orthography of Asiatick Words in Roman Letters' (an excerpt from which appears below). This essay, as I have writtern elsewhere, may well be considered the beginning of the International Phonetic Alphabet, the 'Dissertation on the Orthography' starting us on the path to linguistically-based transcription.

It would be inappropriate of me to devote a great deal of space here to the Asiatick Society. Kejariwal's excellent monograph covers much of the territory and the works of Ludo Rocher and Rosane Rocher have illuminated the lives and efforts of Halhed, Hamilton, Colebrook, and several other members of the Society. Half-a-dozen and more years after Jones's death the Asiatick Society continued what he had begun with the founding of Fort William College and the press at Serampore. The importance two centuries later of what Sir William and his associates set in motion in 1784 is great.

Jones's translations from the Sanskrit: the *Hitōpadēsa* (1786), *Gītagovinda*(1789), *Shakuntalā* (also 1789), and *The Ordinances of Menu* (published in the last weeks of his life, 1794), did more than convey the glories of the Indian subcontinent and its past to Europe, they also stimulated Indian pride in their literary heritage. The *Ordinances*, for example, not only established Jones's reputation as a linguist, translator and jurist, but helped Indians to be accepted by Europeans as respectable inheritors of a rich and ancient culture. Jones succeeded in changing attitudes towards the whole subcontinent.

Jones's other unique contribution to the governance of India was what came to be known as the Cornwallis Code, within which the native law of custom was earnestly observed by British administrators. Jones formally presented to the Governor-General his proposal for an English digest of Hindu and Muslim law (intended primarily as a compendium of Hindu law). Many learned pandits under the supervision of Pandit Jagannātha Tarkapañcānana were engaged to write the compendium (the *Vivādabhaṅgārṇava*). Colebrooke later translated it in its entirety, and it remains the modern text for the digest.

In 1968, S.N. Mukherjee published his fascinating monograph *Sir William Jones : A Study in Eighteenth-Century British Attitudes to India*. The 1987 second edition notes that 'no significant work on Jones has appeared since the first edition of this book'. This anthology may start the snowball of Jones inquiry rolling.

Mukherjee states:

Jones was a key figure in the development of both the oriental studies and the British policies of the eighteenth century . . . He was involved in politics; in England he was an ardent supporter of America and the movement for parliamentary reform, and in India he developed a definite theory of law and government for the British Raj in Bengal. He was constantly in touch with men in power and often consulted by them. Thus his 'Digest' of Indian laws was considered complementary to Cornwallis's Permanent Settlement . . . . Jones occupies a large place in the history of British India, than has hitherto been recognized. (pp. 3-4)

Prior to Jones, the West (as can be seen in d'Herbelot's *Bibliothèque Orientale* and Diderot's *Encyclopédie*) saw India as disorderly, chaotic, superstitious. But though there were contacts with India from about the twelfth century on, until the Hastings period there was little regard for Indian thought or government. Clive, Hastings and their followers were intimately involved in Bengal power politics, and Britain had a pragmatic need for a workable administrative policy. But all their information was second (and third) hand, and they needed knowledge: they needed a scientific approach to an alien culture. Jones brought reason, study and analysis to the subcontinent.

It is difficult to praise Jones's influence too highly: when one looks at his six charges to the grand jury in Calcutta, one sees a model that British imperialism failed to sustain for very long. A member of Johnson's Literary Club; a friend of Franklin, Burke, Percy, Reynolds, Sheridan, and other prominent figures—Jones was one of the great British scholars of his time. Certainly none of his contemporaries achieved eminence in so many diverse areas. And certainly, no one else in the eighteenth century widened Western intellectual horizons in the generous spirit of universal brotherhood and tolerance for non-European cultures. Jones's living legacy, the Asiatick Society of Bengal, celebrated its bicentenary in January 1984 with a resplendent international conference in Calcutta.

The Reader

The volume at hand may seem enormous, yet it leaves untouched vast quantities of Jones's work: his translation of Isaeus, for example; 'On the Musical Modes of the Hindus', *Principles of Government*; the *History of the Life of Nader Shah;* the various botanical works; his speeches—none of these are included. Yet there is enough here to give the interested reader a good taste, and from that taste one can be led to Garland Cannon's superb two-volume edition of Jones's letters. Little else is easily accessible. If the 'trunkless legs of stone' were all that remained of Ozymandias, then calf-bound volumes of Sir William's *Works* and the *Asiatick Researches* in academic libraries are all that remain of Jones. Yet this may not be so bad, for it is in the Fourth Volume of the *Researches* (1794) that we find both Jones's Eleventh Discourse ('On the Philosophy of the Asiatics') and his obituary. And it is a copy of this volume that was owned by Schopenhauer and which is cited in Book I, Article I of *The World as Will and Representation*.

In his address to the Asiatick Society of 22 May 1794, Sir John Shore, elected to succeed Jones as President, said:

It was lately our boast to possess a President, whose name, talents, and character, would have been honourable to any institution; it is now our misfortune to lament, that *Sir William Jones* exists but in the affections of his friends, and in the esteem, veneration, and regret of all . . . .

To define, with accuracy, the variety, value, and extent of his literary attainments, requires more learning than I pretend to possess.

With nearly two centuries' hindsight, I, too, find the extent of Jones's attainments difficult to define. However, his achievements and their effects on posterity, as suggested by the editorial notes and introductions, are manifest in the works printed in this *Reader*, many for the first time in two centuries.

The editor has provided an introductory headnote to each selection in this anthology, providing a context for that work's composition (when known) and intimating its value to modern scholarship. Only obscure references in the texts have been annotated in an attempt to eschew heavy scholarly documentation, which might impede the reader. The editor's notes are placed in square brackets, differentiating them from Jones's own notes. Jones's diacritics and orthography are preserved, though they frequently deviate from modern transliterations.

Beginning with an invocation to Ganesha, the Hindu god of wisdom and learning, this *Reader* is organized in three major sections, each with its own introduction: Literature; Language and Linguistics; and Religion, Mythology and Metaphysics. As Jones's major writings were in literature, the first section is the longest.

It begins with Jones's own poetry, containing 'A Persian Song of Hafiz', 'The Palace of Fortune, An Indian Tale', and 'The Seven Fountains, An Eastern Allegory' (both influential allegories), 'Caissa, or, The Game at Chess', 'An Ode in Imitation of Alcaeus' and 'The Muse Recalled' (acclaimed political poems), 'Damsels of Cardigan', 'The Poem of Amriolkais', 'Lines from the Arabic' (addressed to Anna Maria), 'On Parent Knees' (a famous quatrain), 'The Concluding Sentence of Berkeley's Siris', 'Written after a Perusal of Barrow', and the nine hymns to Hindu deities. The section proper also includes the entire *Gīta Govinda* of Jayadeva. This is followed by a sub-section on drama which features the complete text of Kālidāsa's *Shakuntalā*. Finally, there is a subsection on criticism containing five essays: the two which appeared in the early *Poems* ('Essay on the Arts, Called Imitative' and 'An Essay on the Poetry of the Eastern Nations'), two on Persian and Sanskrit literature

('Mystical Poetry of the Persians and Hindus' and 'Literature of the Hindus from the Sanscrit'), and his 'Preface to the *Ṛitusaṃhāra*' of Kālidāsa, which rationalizes the need of publication in Sanskrit.

The next section, on Language and Linguistics, consists of five notable essays or extracts from essays: the Preface to Sir William's *Grammar of the Persian Language*, his 'Proposal to Edit Meninski's *Thesaurus*', an extract from the 'Dissertation on the Orthography of Asiatick Words', 'A Discourse on the Institution of a Society' (the First Anniversary Discourse), and the revolutionary 'Third Anniversary Discourse'. These works alone are sufficient to cement Jones's stature in the history of linguistics.

The final section of this *Reader* deals with religion, mythology and metaphysics. It contains an extract from the lengthy 'On the Gods of Greece, Italy, and India', an extract from 'On the Antiquity of the Indian Zodiac', Jones's last Anniversary Discourse on Indian Philosophy, two sections from the pioneering translation of the *Institutes of Hindu Law* of Manu, some extracts from the *Vedas* and didactic *Mohamudgara*, attributed to Shankaracharya. The Vedas were incorrectly published in the *Works*, and appear here transcribed from holographs, generously provided to the editor by Professor Garland Cannon. The *Mohamudgara* appears here for the first time, also transcribed from Cannon's holographs.

This anthology thus not only brings to the contemporary reader many works by Jones which have been unavailable for nearly 200 years, it also publishes for the first time some of Jones's translations.

Lady Anna Maria Jones quoted her illustrious husband in her Preface to the 1799 *Works*: 'The best monument that can be erected to a man of literary talents, is a good edition of his works'. Satya S. Pachori, in this *Reader*, has erected such a monument.

Boston, Massachusetts   P.H. SALUS
February 1991

I. LITERATURE

Jones was not only one of the most learned men of the eighteenth century, but also his literary prolificacy was such that a reprinting of all of his works would require several volumes. We have concentrated on those representative items which have made their imprint on English and Oriental literature, while conveying the essential meaning of Eastern culture to the West. His interest in the literature of India, Iran, and Arabia richly permeates these works, with fine examples from poetry, a translated Sanskrit drama, and criticism.

The sequence is generally chronological, since this also reveals his intellectual and artistic development. First comes his poetry in England, especially the works in his innovational *Poems; Consisting Chiefly of Translations from the Asiatick Languages. To which are added Two Essays* (Oxford : 1772). His lovely 'A Persian Song of Hafiz' was one of his most frequently reprinted and best poems. His devotion to allegories shows through in 'The Palace of Fortune, An Indian Tale' and 'The Seven Fountains, An Eastern Allegory'. 'Caissa, or, the Game at Chess' is an imaginative poem showing his capacity for invention and use of classical sources. 'An Ode in Imitation of Alcaeus' and 'The Muse Recalled' reflect his strong, independent political principles that were to impede his career in an age of patronage. 'Damsels of Cardigan' is one of several light poems showing his élan amid a tedious law-circuit, as well as his capacity to compose striking light verse apart from intellectual topics. The first translation of the pre-Islamic *Mu 'allaqát* is illustrated here with the first of the seven poems, 'The Poem of Amriolkais'. Besides the famous quatrain from the Persian, 'On Parent Knees', three other poems are included here preparatory to his Indian materials. Next we reprint the comic verse epic taken from a story in the *Mahābhārata*, 'The Enchanted Fruit; or, the Hindu Wife' and nine innovative hymns to Hindu divinities, of which those to Kama and Narayana are the best. Jones's smooth prose translation of Jayadeva's drama *Gīta Govinda* concludes the poetic section. This is followed by his translation of the *Śakuntalā*, which revolutionized Western thought about ancient Indian culture and literature and gave Kalidasa his status in world literature.

Jones's acumen as a literary critic is demonstrated in his 'Essay on the Arts, Commonly Called Imitative' and 'An Essay on the Poetry of the Eastern Nations'. Then two essays on the poetry and aesthetics of India, Iran, and Arabia further reflect his competent, humanistic interpretation of cultures and minds widely separated in time

and geography from his European heritage. The brief Advertisement to his pioneering publication of the *Ritusaṃhāra* explains his profound, intellectual motives in making this Sanskrit work available to Indian readers.

A Persian Song of Hafiz[1]

Sweet maid, if thou would'st charm my sight,

And bid these arms thy neck infold;
That rosy cheek, that lily hand,
Would give thy poet more delight
Than all Bocara's vaunted gold,
Than all the gems of Samarcand.

Boy, let yon liquid ruby flow,
And bid thy pensive heart be glad,
Whate'er the frowning zealots say:
Tell them, their Eden cannot show
A stream so clear as Rocnabad,
A bower so sweet as Mosellay.

O! when these fair perfidious maids,
Whose eyes our secret haunts infest,
Their dear destructive charms display;
Each glance my tender breast invades,
And robs my wounded soul of rest,
As Tartars seize their destin'd prey.

In vain with love our bosoms glow:

---

[1] [Jones's love for Hafiz, a fourteenth-century Persian lyricist, is expressed in a letter of April 1768 to Count Reviczky (1737-93): 'Our Hafiz is most assuredly a poet worthy to sup with the god; every day I take pleasure in his work' (see *Letters* 1 : 9). He expanded each of the nine rubaiyats of 'Shirazi Turk', employing an innovational rhyme of a b c a b c and sensually communicating Hafiz's mosaic of sounds and symbols. By including his ode in *A Grammar of the the Persian Language* (London, 1771), Jones assured the immediate success of that famous grammar. When he reprinted the ode in *Poems* in 1772, he became one of the most famous living poets. Today it is still perhaps the third most famous English translation from the Persian, surpassed by the *Rubaiyat* and *Sohrab and Rustum*. See V. de Sola Pinto, 'Sir William Jones and English Literature', and A.J. Arberry, 'Orient Pearls at Random Strung', *Bulletin of the School of Oriental and African Studies* 14 (1946) : 687, 699-712, respectively.]

Can all our tears, can all our sighs,
New lustre to those charms impart?
Can cheeks, where living roses blow,
Where nature spreads her richest dyes,
Require the borrow'd gloss of art?

Speak not of fate:—ah! change the theme,
And talk of odours, talk of wine,
Talk of the flowers that round us bloom:
'Tis all a cloud, 'tis all a dream;
To love and joy thy thoughts confine,
Nor hope to pierce the sacred gloom.

Beauty has such resistless power,
That even the chaste Egyptian dame
Sigh'd for the blooming Hebrew boy:
For her how fatal was the hour,
When to the banks of Nilus came
A youth so lovely and so coy!

But ah! sweet maid, my counsel hear
(Youth should attend when those advise
Whom long experience renders sage):
While musick charms the ravish'd ear;
While sparkling cups delight our eyes,
Be gay; and scorn the frowns of age.

What cruel answer have I heard!
And yet, by heaven, I love thee still:
Can aught be cruel from thy lip?
Yet say, how fell that bitter word
From lips which streams of sweetness fill,
Which nought but drops of honey sip?

Go boldly forth, my simple lay,
Whose accents flow with artless ease,
Like orient pearls at random strung:
Thy notes are sweet, the damsels say;
But O! far sweeter, if they please
Thy nymph for whom these notes are sung.

The Palace of Fortune, an Indian Tale[1]

Written in the Year 1769

Mild was the vernal gale, and calm the day,
When Maia near a crystal fountain lay,
Young Maia, fairest of the blue-eyed maids,
That rov'd at noon in Tibet's musky shades:
But, haply, wandering through the fields of air,
Some fiend had whisper'd—Maia, thou art fair!
Hence swelling pride had fill'd her simple breast,
And rising passions robb'd her mind of rest;
In courts and glittering towers she wish'd to dwell,
And scorn'd her labouring parent's lowly cell.
And now, as gazing o'er the glassy stream,[2]
She saw her blooming cheek's reflected beam,
Her tresses brighter than the morning sky,
And the mild radiance of her sparkling eye,
Low sighs and trickling tears by turns she stole,
And thus dischar'd the anguish of her soul:
'Why glow those cheeks, if unadmir'd they glow?
Why flow those tresses, if unprais'd they flow?
Why dart those eyes their liquid ray serene,
Unfelt their influence, and their light unseen?
Ye heavens! was that love-breathing bosom made
To warm dull groves, and cheer the lonely glade?

[1] [After extensively polishing 'The Palace of Fortune' and reducing the length from 550 to 253 closed couplets, Jones included it in *Poems* (1772). Basing it on a story in Alexander Dow's *Tales Translated from the Persian of Inatulla of Delhi* (London, 1768), Jones added several descriptions and episodes from other Oriental sources and modified its moral in accord with the polished, artificial style of the day. A discontented, ambitious maiden sees a series of visions, in which Pleasure (the youth), Glory (the knight), Riches (the aged sire), and Knowledge (the sage) are granted their wishes, but then are destroyed by the fruits of these wishes. Thus, in the Oriental fable tradition, she is taught about the vanity of human wishes. See Sir H. Sharp, 'Anglo-Indian Verse', *Essays by Divers Hands* 16 (1937) : 98-99; and Emil Koeppel, 'Shelley's Queen Mab and Sir William Jones's "Palace of Fortune"', '*Englishce Studien* 28 (1900) : 45-53.]

[2] [Perhaps an allusion to the Narcissus motif. Maya is the Hindu goddess of illusion.]

Ah, no: those blushes, that enchanting face,
Some tap'stried hall, or gilded bower, might grace;
Might deck the scenes, where love and pleasure reign,
And fire with amorous flames the youthful train'.

While thus she spoke, a sudden blaze of light
Shot through the clouds, and struck her dazzled sight.
She rais'd her head, astonish'd, to the skies,
And veil'd with trembling hands her aching eyes;
When through the yielding air she saw from far
A goddess gliding in a golden car,
That soon descended on the flowery lawn,
By two fair yokes of starry peacocks drawn:
A thousand nymphs with many a sprightly glance
Form'd round the radiant wheels an airy dance,
Celestial shapes! in fluid light array'd;
Like twinkling stars their beamy sandals play'd;
Their lucid mantles glitter'd in the sun,
(Webs half so bright the silkworm never spun)
Transparent robes, that bore the rainbow's hue,
And finer than the nets of pearly dew
That morning spreads o'er every opening flower,
When sportive summer decks his bridal bower.

The queen herself, too fair for mortal sight,
Sat in the centre of encircling light.
Soon with soft touch she rais'd the trembling maid,
And by her side in silent slumber laid:
Straight the gay birds display'd their spangled train,
And flew refulgent through th' aerial plain;
The fairy band their shining pinions spread,
And, as they rose, fresh gales of sweetness shed;
Fann'd with their flowing skirts, the sky was mild;
And heaven's blue fields with brighter radiance smil'd.

Now in a garden deck'd with verdant bowers
The glittering car descends on bending flowers:
The goddess still with looks divinely fair
Surveys the sleeping object of her care;
Then o'er her cheek her magick finger lays,
Soft as the gale that o'er a violet plays,
And thus in sounds, that favour'd mortals hear,
She gently whispers in her ravish'd ear:
'Awake, sweet maid, and view this charming scene
For ever beauteous, and for ever green;
Here living rills of purest nectar flow
O'er meads that with unfading flowerets glow;
Here amorous gales their scented wings display,
Mov'd by the breath of ever-blooming May;
Here in the lap of pleasure shalt thou rest,
Our lov'd companion, and our honour'd guest'.

The damsel hears the heavenly notes distil,
Like melting snow, or like a vernal rill.

She lifts her head, and, on her arm reclin'd,
Drinks the sweet accents in her grateful mind:
On all around she turns her roving eyes,
And views the splendid scene with glad surprize;
Fresh lawns, and sunny banks, and roseate bowers,
Hills white with flocks, and meadows gemm'd with
    flowers;
Cool shades, a sure defence from summer's ray,
And silver brooks, where wanton damsels play,
Which with soft notes their dimpled crystal roll'd
O'er colour'd shells and sands of native gold;
A rising fountain play'd from every stream,
Smil'd as it rose, and cast a transient gleam,
Then, gently falling in a vocal shower,
Bath'd every shrub, and sprinkled every flower,
That on the banks, like many a lovely bride,
View'd in the liquid glass their blushing pride;
Whilst on each branch, with purple blossoms hung,
The sportful birds their joyous descant sung.

While Maia, thus entranc'd in sweet delight,
With each gay object fed her eager sight,
The goddess mildly caught her willing hand,
And led her trembling o'er the flowery land,
Soon she beheld, where through an opening glade
A spacious lake its clear expanse display'd;
In mazy curls the flowing jasper wav'd
O'er its smooth bed with polish'd agate pav'd;
And on a rock of ice, by magick rais'd,
High in the midst a gorgeous palace blaz'd;
The sunbeams on the gilded portals glanc'd,
Play'd on the spires, and on the turrets danc'd;
To four bright gates four ivory bridges led,
With pearls illumin'd, and with roses spread:
And now, more radiant than the morning sun,
Her easy way the gliding goddess won;
Still by her hand she held the fearful maid,
And, as she pass'd, the fairies homage paid:
They enter'd straight the sumptuous palace-hall,
Where silken tapestry emblaz'd the wall,
Refulgent tissue, of an heavenly woof;
And gems unnumber'd sparkled on the roof,
On whose blue arch the flaming diamonds play'd,
As on a sky with living stars inlay'd;
Of precious diadems a regal store,
With globes and sceptres, strew'd the porphyry floor;
Rich vests of eastern kings around were spread,
And glittering zones a starry lustre shed:
But Maia most admir'd the pearly strings,
Gay bracelets, golden chains, and sparkling rings.

High in the centre of the palace shone.

Suspended in mid-air, an opal throne:
To this the queen ascends with royal pride,
And sets the favour'd damsel by her side.
Around the throne in mystick order stand
The fairy train, and wait her high command;
When thus she speaks: (the maid attentive sips
Each word that flows, like nectar, from her lips.)
'Favourite of heaven, my much-lov'd Maia, know,
From me all joys, all earthly blessings, flow:
Me suppliant men imperial Fortune call,
The mighty empress of yon rolling ball:
(She rais'd her finger, and the wondering maid
At distance hung the dusky globe survey'd,
Saw the round earth with foaming oceans vein'd,
And labouring clouds on mountain-tops sustain'd.)
To me has fate the pleasing task assign'd
To rule the various thoughts of humankind;
To catch each rising wish, each ardent prayer,
And some to grant, and some to waste in air.
Know farther; as I rang'd the crystal sky,
I saw thee near the murmuring fountain lie;
Mark'd the rough storm that gather'd in thy breast,
And knew what care thy joyless soul opprest.
Straight I resolved to bring thee quick relief,
Ease every weight, and soften every grief;
If in this court contented thou canst live,
And taste the joys these happy gardens give:
But fill thy mind with vain desires no more,
And view without a wish yon shining store:
Soon shall a numerous train before me bend,
And kneeling votaries my shrine attend;
Warn'd by their empty vanities beware,
And scorn the folly of each human prayer'.
　　She said; and straight a damsel of her train
With tender fingers touch'd a golden chain.
Now a soft bell delighted Maia hears,
That sweetly trembles on her listening ears;
Through the calm air the melting numbers float,
And wanton echo lengthens every note.
Soon through the dome a mingled hum arose,
Like the swift stream that o'er a valley flows;
Now louder still it grew, and still more loud,
As distant thunder breaks the bursting cloud:
Through the four portals rush'd a various throng,
That like a wintry torrent pour'd along:
A croud of every tongue, and every hue,
Toward the bright throne with eager rapture flew.
A lovely stripling[3] stepp'd before the rest

[3] Pleasure.

With hasty pace, and tow'rd the goddess prest;
His mien was graceful, and his looks were mild,
And in his eye celestial sweetness smil'd:
Youth's purple glow, and beauty's rosy beam,
O'er his smooth cheeks diffus'd a lively gleam;
The floating ringlets of his musky hair
Wav'd on the bosom of the wanton air:
With modest grace the goddess he addrest,
And thoughtless thus preferr'd his fond request.
　　'Queen of the world, whose wide-extended sway,
Gay youth, firm manhood, and cold age obey,
Grant me, while life's fresh blooming roses smile,
The day with varied pleasures to beguile;
Let me on beds of dewy flowers recline,
And quaff with glowing lips the sparkling wine;
Grant me to feed on beauty's rifled charms,
And clasp a willing damsel in my arms;
Her bosom fairer than a hill of snow,
And gently bounding like a playful roe;
Her lips more fragrant than the summer air,
And sweet as Scythian musk her hyacinthine hair;
Let new delights each dancing hour employ,
Sport follow sport, and joy succeed to joy'.
　　The goddess grants the simple youth's request,
And mildly thus accosts her lovely guest:
'On that smooth mirror, full of magick light,
Awhile, dear Maia, fix thy wandering sight'.
She looks; and in th' enchanted crystal sees
A bower o'er-canopied with tufted trees:
The wanton stripling lies beneath the shade,
And by his side reclines a blooming maid;
O'er her fair limbs a silken mantle flows,
Through which her youthful beauty softly glows,
And part conceal'd, and part disclos'd to sight,
Through the thin texture casts a ruddy light,
As the ripe clusters of the mantling vine
Beneath the verdant foliage faintly shine,
And, fearing to be view'd by envious day,
Their glowing tints unwillingly display.
　　The youth, while joy sits sparkling in his eyes,
Pants on her neck, and on her bosom dies;
From her smooth cheek nectareous dew he sips,
And all his soul comes breathing to his lips.
But Maia turns her modest eyes away,
And blushes to behold their amorous play.
　　She looks again, and sees with sad surprize
On the clear glass far different scenes arise:
The bower, which late outshone the rosy morn,
O'erhung with weeds she saw, and rough with thorn;
With stings of asps the leafless plants were wreath'd,

And curling adders gales of venom breath'd:
Low sat the stripling on the faded ground,
And in a mournful knot his arms were bound;
His eyes, that shot before a sunny beam,
Now scarcely shed a saddening, dying gleam;
Faint as a glimmering taper's wasted light,
Or a dull ray that streaks the cloudy night:
His crystal vase was on the pavement roll'd,
And from the bank was fall'n his cup of gold;
From which th' envenom'd dregs of deadly hue
Flow'd on the ground in streams of baleful dew,
And, slowly stealing through the wither'd bower,
Poison'd each plant, and blasted every flower:
Fled were his slaves, and fled his yielding fair,
And each gay phantom was dissolv'd in air;
Whilst in their place was left a ruthless train,
Despair, and grief, remorse, and raging pain.

Aside the damsel turns her weeping eyes,
And sad reflections in her bosom rise;
To whom thus mildly speaks the radiant queen:
'Take sage example from this moral scene;
See, how vain pleasures sting the lips they kiss,
How asps are hid beneath the bowers of bliss!
Whilst ever fair the flower of temperance blows,
Unchang'd her leaf, and without thorn her rose;
Smiling she darts her glittering branch on high,
And spreads her fragrant blossoms to the sky'.

Next tow'rd the throne she saw a knight[4] advance;
Erect he stood, and shook a quivering lance;
A fiery dragon on his helmet shone;
And on his buckler beam'd a golden sun;
O'er his broad bosom blaz'd his jointed mail
With many a gem, and many a shining scale;
He trod the sounding floor with princely mien,
And thus with haughty words address'd the queen:
'Let falling kings beneath my javelin bleed,
And bind my temples with a victor's meed;
Let every realm that feels the solar ray,
Shrink at my frown, and own my regal sway:
Let Ind's rich banks declare my deathless fame,
And trambling Ganges dread my potent name'.

The queen consented to the warrior's pray'r,
And his bright banners floated in the air:
He bade his darts in steely tempests fly,
Flames burst the clouds, and thunder shake the sky;
Death aim'd his lance, earth trembled at his nod,
And crimson conquest glow'd where'er he trod.

And now the damsel, fix'd in deep amaze,

Th' enchanted glass with eager look surveys:
She sees the hero in his dusky tent,
His guards retir'd, his glimmering taper spent;
His spear, vain instrument of dying praise,
On the rich floor with idle state he lays;
His gory falchion near his pillow stood,
And stain'd the ground with drops of purple blood;
A busy page his nodding helm unlac'd,
And on the couch his scaly hauberk plac'd.
Now on the bed his weary limbs he throws,
Bath'd in the balmy dew of soft repose:
In dreams he rushes o'er the gloomy field,
He sees new armies fly, new heroes yield;
Warm with the vigorous conflict he appears,
And ev'n in slumber seems to move the spheres.
But lo! the faithless page, with stealing tread,
Advances to the champion's naked head;
With his sharp dagger wounds his bleeding breast,
And steeps his eyelids in eternal rest:
Then cries (and waves the steel that drops with gore),
'The tyrant dies; oppression is no more'.

Now came an aged sire[5] with trembling pace;
Sunk were his eyes, and pale his ghastly face;
A ragged weed of dusky hue he wore,
And on his back a ponderous coffer bore.
The queen with faltering speech he thus addrest:
'O, fill with gold thy true adorer's chest!'
'Behold', said she, and wav'd her powerful hand,
'Where yon rich hills in glittering order stand:
There load thy coffer with the golden store;
Then bear it full away, and ask no more'.

With eager steps he took his hasty way,
Where the bright coin in heaps unnumber'd lay;
There hung enamour'd o'er the gleaming spoil,
Scoop'd the gay dross, and bent beneath the toil.
But bitter was his anguish, to behold
The coffer widen, and its sides unfold:
And every time he heap'd the darling ore,
His greedy chest grew larger than before:
Till, spent with pain, and falling o'er his hoard,
With his sharp steel his maddening breast he gor'd:
On the lov'd heap he cast his closing eye,
Contented on a golden couch to die.

A stripling, with the fair adventure pleas'd,
Stepp'd forward, and the massy coffer seiz'd;
But with surprize he saw the stores decay,
And all the long-sought treasures melt away:
In winding streams the liquid metal roll'd,

---

[4] Glory.

[5] Riches.

And through the palace ran a flood of gold.
    Next to the shrine advanc'd a reverend sage,[6]
Whose beard was hoary with the frost of age;
His few gray locks a sable fillet bound,
And his dark mantle flow'd along the ground:
Grave was his port, yet show'd a bold neglect,
And fill'd the young beholder with respect;
Time's envious hand had plough'd his wrinkled face,
Yet on those wrinkles sat superiour grace;
Still full of fire appear'd his vivid eye,
Darted quick beams, and seem'd to pierce the sky.
At length, with gentle voice and look serene,
He wav'd his hand, and thus address'd the queen:
'Twice forty winters tip my beard with snow,
And age's chilling gusts around me blow:
In early youth, by contemplation led,
With high pursuits my flatter'd thoughts were fed;
To nature first my labours were confin'd,
And all her charms were open'd to my mind,
Each flower that glisten'd in the morning dew,
And every shrub that in the forest grew:
From earth to heaven I cast my wondering eyes,
Saw suns unnumber'd sparkle in the skies,
Mark'd the just progress of each rolling sphere,
Describ'd the seasons, and reform'd the year.
At length sublimer studies I began,
And fix'd my level'd telescope on man;
Knew all his powers, and all his passions trac'd,
What virtue rais'd him, and what vice debas'd:
But when I saw his knowledge so confin'd,
So vain his wishes, and so weak his mind,
His soul, a bright obscurity at best,
And rough with tempests his afflicted breast,
His life, a flower, ere evening sure to fade,
His highest joys, the shadow of a shade;
To thy fair court I took my weary way,
Bewail my folly, and heaven's laws obey,
Confess my feeble mind for prayers unfit,
And to my Maker's will my soul submit:
Great empress of yon orb that rolls below,
On me the last best gift of heaven bestow'.
    He spoke: a sudden cloud his senses stole,
And thickening darkness swam o'er all his soul;
His vital spark her earthly cell forsook,
And into air her fleeting progress took.
    Now from the throng a deafening sound was heard,
And all at once their various prayers preferr'd; .
The goddess, wearied with the noisy croud,

6 Knowledge.

Thrice wav'd her silver wand, and spoke aloud:
'Our ears no more with vain petitions tire,
But take unheard whate'er you first desire'.
She said: each wish'd, and what he wish'd obtain'd;
And wild confusion in the palace reign'd.
    But Maia, now grown senseless with delight,
Cast on an emerald ring her roving sight;
And, ere she could survey the rest with care,
Wish'd on her hand the precious gem to wear.
    Sudden the palace vanish'd from her sight,
And the gay fabrick melted into night;
But, in its place, she view'd with weeping eyes
Huge rocks around her, and sharp cliffs arise:
She sat deserted on the naked shore,
Saw the curl'd waves, and heard the tempest roar;
Whilst on her finger shone the fatal ring,
A weak defence from hunger's pointed sting,
From sad remorse, from comfortless despair,
And all the painful family of care!
Frantick with grief her rosy cheek she tore,
And rent her locks, her darling charge no more:
But when the night his raven wing had spread,
And hung with sable every mountain's head,
Her tender limbs were numb'd with biting cold,
And round her feet the curling billows roll'd;
With trembling arms a rifted crag she grasp'd,
And the rough rock with hard embraces clasp'd.
    While thus she stood, and made a piercing moan,
By chance her emerald touch'd the rugged stone;
That moment gleam'd from heaven a golden ray,
And taught the gloom to counterfeit the day:
A winged youth, for mortal eyes too fair,
Shot like a meteor through the dusky air;
His heavenly charms o'ercame her dazzled sight,
And drown'd her senses in a flood of light;
His sunny plumes descending he display'd,
And softly thus address'd the mournful maid:
'Say, thou, who dost yon wondrous ring possess,
What cares disturb thee, or what wants oppress;
To faithful ears disclose thy secret grief,
And hope (so heaven ordains) a quick relief'.
    The maid replied, 'Ah, sacred genius, bear
A hopeless damsel from this land of care;
Waft me to softer climes and lovelier plains,
Where nature smiles, and spring eternal reigns'.
She spoke; and swifter than the glance of thought
To a fair isle his sleeping charge he brought.
    Now morning breath'd: the scented air was mild,
Each meadow blossom'd, and each valley smil'd;
On every shrub the pearly dewdrops hung,

On every branch a feather'd warbler sung;
The cheerful spring her flowery chaplets wove,
And incense-breathing gales perfum'd the grove.
   The damsel rose; and, lost in glad surprize,
Cast round the gay expanse her opening eyes,
That shone with pleasure like a starry beam,
Or moonlight sparkling on a silver stream.
She thought some nymph must haunt that lovely scene,
Some woodland goddess, or some fairy queen;
At least she hop'd in some sequester'd vale
To hear the shepherd tell his amorous tale:
Led by these flattering hopes from glade to glade,
From lawn to lawn with hasty steps she stray'd;
But not a nymph by stream or fountain stood,
And not a fairy glided through the wood;
No damsel wanton'd o'er the dewy flowers,
No shepherd sung beneath the rosy bowers:
On every side she saw vast mountains rise,
That thrust their daring foreheads in the skies;
The rocks of polish'd alabaster seem'd,
And in the sun their lofty summits gleam'd.
She call'd aloud, but not a voice replied,
Save echo babbling from the mountain's side.
   By this had night o'ercast the gloomy scene,
And twinkling stars emblaz'd the blue serene,
Yet on she wander'd till with grief opprest
She fell; and, falling, smote her snowy breast:
Now to the heavens her guilty head she rears,
And pours her bursting sorrow into tears;
Then plaintive speaks, 'Ah! fond mistaken maid,
How was thy mind by gilded hopes betray'd!
Why didst thou wish for bowers and flowery hills,
For smiling meadows, and for purling rills;
Since on those hills no youth or damsel roves,
No shepherd haunts the solitary groves?
Ye meads that glow with intermingled dyes,
Ye flowering palms that from yon hillocks rise,
Ye quivering brooks that softly murmur by,
Ye panting gales that on the branches die;
Ah! why has Nature through her gay domain
Display'd your beauties, yet display'd in vain?
In vain, ye flowers, you boast your vernal bloom,
And waste in barren air your fresh perfume.
Ah! leave, ye wanton birds, yon lonely spray;
Unheard you warble, and unseen you play:
Yet stay till fate has fix'd my early doom,
And strow with leaves a hapless damsel's tomb.
Some grot or grassy bank shall be my bier,
My maiden herse unwater'd with a tear'.

Thus while she mourns, o'erwhelm'd in deep
      despair,
She rends her silken robes, and golden hair:
Her fatal ring, the cause of all her woes,
On a hard rock with maddening rage she throws;
The gem, rebounding from the stone, displays
Its verdant hue, and sheds refreshing rays:
Sudden descends the genius of the ring,
And drops celestial fragrance from his wing;
Then speaks, 'Who calls me from the realms of day?
Ask, and I grant; command, and I obey'.
   She drank his melting words with ravish'd ears,
And stopp'd the gushing current of her tears;
Then kiss'd his skirts, that like a ruby glow'd,
And said, 'O bear me to my sire's abode'.
   Straight o'er her eyes a shady veil arose,
And all her soul was lull'd in still repose.
   By this with flowers the rosy-finger'd dawn
Had spread each dewy hill and verdurous lawn;
She wak'd, and saw a new-built tomb that stood
In the dark bosom of a solemn wood,
While these sad sounds her trembling ears invade:
'Beneath yon marble sleeps thy father's shade'.
She sigh'd, she wept; she struck her pensive breast,
And bade his urn in peaceful slumber rest.
   And now in silence o'er the gloomy land
She saw advance a slowly-winding band;
Their cheeks were veil'd, their robes of mournful hue
Flow'd o'er the lawn, and swept the pearly dew;
O'er the fresh turf they sprinkled sweet perfume,
And strow'd with flowers the venerable tomb.[7]
A graceful matron walk'd before the train,
And tun'd in notes of wo the funeral strain:
When from her face her silken veil she drew,
The watchful maid her aged mother knew.
O'erpowered with bursting joy she runs to meet
The mourning dame, and falls before her feet.
The matron with surprize her daughter rears,
Hangs on her neck, and mingles tears with tears.
Now o'er the tomb their hallow'd rites they pay,
And form with lamps an artificial day:
Erelong the damsel reach'd her native vale,
And told with joyful heart her moral tale;
Resign'd to heaven, and lost to all beside,
She liv'd contented, and contented died.

---

[7] [A typical ending of the Graveyard School—Gray, Crabbe, *et al.*]

The Seven Fountains, an Eastern Allegory[1]

Deck'd with fresh garlands, like a rural bride,
And with the crimson streamer's waving pride,
A wanton bark was floating o'er the main,
And seem'd with scorn to view the azure plain:
Smooth were the waves; and scarce a whispering gale
Fann'd with his gentle plumes the silken sail.
High on the burnish'd deck, a gilded throne
With orient pearls and beaming diamonds shone;
On which reclin'd a youth of graceful mien,
His sandals purple, and his mantle green;
His locks in ringlets o'er his shoulders roll'd,
And on his cheek appear'd the downy gold.
Around him stood a train of smiling boys,
Sporting with idle cheer and mirthful toys;
Ten comely striplings,[2] girt with spangled wings,
Blew piercing flutes, or touch'd the quivering strings;
Ten more, in cadence to the sprightly strain,
Wak'd with their golden oars the slumbering main:
The waters yielded to their guiltless blows,
And the green billows sparkled as they rose.
    Long time the barge had danc'd along the deep,
And on its glassy bosom seem'd to sleep;
But now a glittering isle[3] arose in view,
Bounded with hillocks of a verdant hue:
Fresh groves and roseate bowers appear'd above
(Fit haunts, be sure, of pleasure and of love),

And higher still a thousand blazing spires
Seem'd with gilt tops to threat the heavenly fires.
Now each fair stripling plied his labouring oar,
And straight the pinnace struck the sandy shore.
The youth arose, and, leaping on the strand,
Took his lone way along the silver sand;
While the light bark, and all the airy crew,
Sunk like a mist beneath the briny dew.
    With eager steps the young adventurer stray'd
Through many a grove, and many a winding glade:
At length he heard the chime of tuneful strings,
That sweetly floated on the Zephyr's wings;
And soon a band of damsels[4] blithe and fair,
With flowing mantles and dishevel'd hair,
Rush'd with quick pace along the solemn wood,
Where rapt in wonder and delight he stood:
In loose transparent robes they were array'd,
Which half their beauties hid, and half display'd.
    A lovely nymph approach'd him with a smile,
And said, 'O, welcome to this blissful isle!
For thou art he, whom ancient bards foretold,
Doom'd in our clime to bring an age of gold:
Hail, sacred king! and from thy subject's hand,
Accept the robes and sceptre of the land'.
    'Sweet maid', said he, 'fair learning's heavenly
        beam
O'er my young mind ne'er shed her favouring gleam;
Nor has my arm e'er hurl'd the fatal lance,
While desperate legions o'er the plain advance.
How should a simple youth, unfit to bear
The steely mail, that splendid mantle wear!'
'Ah!' said the damsel, 'from this happy shore,
We banish wisdom, and her idle lore;
No clarions here the strains of battle sing,
With notes of mirth our joyful valleys ring.
Peace to the brave! o'er us the beauteous reign,
And ever-charming pleasures form our train'.
    This said, a diadem, inlay'd with pearls,
She plac'd respectful on his golden curls;
Another o'er his graceful shoulder threw
A silken mantle of the rose's hue,
Which, clasp'd with studs of gold, behind him flow'd,
And through the folds his glowing bosom show'd.
Then in a car, by snow-white coursers drawn,
They led him o'er the dew-besprinkled lawn,
Through groves of joy and arbours of delight,
With all that could allure his ravish'd sight;
Green hillocks, meads, and rosy grots, he view'd,

[1] [The poem, written in 1767, was based on a tale in Ibn Arabshah's *Fakihatu'l-Khulafá*. Jones engrafted onto this the Prince Agib episode from the *Arabian Nights' Tales*. The poem narrates the experiences of a young prince who enjoys the pleasures of the senses—allegorically portrayed by the six gates of the mansion—until he is rescued by an old man symbolizing religion. Each gate, allegorizing a major station in sense perception, advances the organic unity of the poem into a sequence of sensory perceptions, as Jones combines the Bunyan tradition of *Pilgrim's Progress* with the Lockean mode of sensory psychology. What makes the poem fascinating as allegory is that Jones was able to demonstrate his poetic powers to blend Eastern images, prior to his going to the Orient, with Western theories of sense perception, from the Age of Enlightenment.]

[2] The follies of youth.

[3] The world.

[4] The follies and vanities of the world.

And verdurous plains with winding streams bedew'd.
On every bank, and under every shade,
A thousand youths, a thousand damsels play'd;
Some wantonly were tripping in a ring
On the soft border of a gushing spring;
While some, reclining in the shady vales,
Told to their smiling loves their amorous tales:
But when the sportful train beheld from far
The nymphs returning with the stately car,
O'er the smooth plain with hasty steps they came,
And hail'd their youthful king with loud acclaim;
With flowers of every tint the paths they strow'd,
And cast their chaplets on the hallow'd road.

    At last they reach'd the bosom of a wood,
Where on a hill a radiant palace stood;
A sumptuous dome, by hands immortal made,
Which on its walls and on its gates display'd
The gems that in the rocks of Tibet glow,
The pearls that in the shells of Ormus grow.
And now a numerous train advance to meet
The youth, descending from his regal seat;
Whom to a rich and spacious hall they led,
With silken carpets delicately spread:
There on a throne, with gems unnumber'd grac'd,
Their lovely king six blooming damsels plac'd,[5]
And, meekly kneeling, to his modest hand
They gave the glittering sceptre of command;
Then on six smaller thrones they sat reclin'd,
And watch'd the rising transports of his mind:
When thus the youth a blushing nymph address'd,
And, as he spoke, her hand with rapture press'd:
'Say, gentle damsel, may I ask unblam'd,
How this gay isle, and splendid seats are nam'd?
And you, fair queens of beauty and of grace,
Are you of earthly or celestial race?
To me the world's bright treasures were unknown,
Where late I wander'd, pensive and alone;
And, slowly winding on my native shore,
Saw the vast ocean roll, but saw no more;
Till from the waves with many a charming song,
A barge arose, and gayly mov'd along;
The jolly rowers reach'd the yielding sands,
Allur'd my steps, and wav'd their shining hands:
I went, saluted by the vocal train,
And the swift pinnace cleav'd the waves again;
When on this island struck the gilded prow,
I landed full of joy: the rest you know.
Short is the story of my tender years:

<hr/>

[5] The pleasures of the senses.

Now speak, sweet nymph, and charm my listening
    ears'.
    'These are the groves, for ever deck'd with
    flowers',
The maid replied, 'and these the fragrant bowers,
Where Love and Pleasure hold their airy court,
The seat of bliss, of sprightliness, and sport;
And we, dear youth, are nymphs of heavenly line;
Our souls immortal, as our forms divine:
For Maia, fill'd with Zephyr's warm embrace,[6]
In caves and forests cover'd her disgrace;
At last she rested on this peaceful shore,
Where in yon grot a lovely boy she bore,
Whom fresh and wild and frolick from his birth
She nurs'd in myrtle bowers, and call'd him Mirth.
He on a summer's morning chanc'd to rove
Through the green labyrinth of some shady grove,
Where, by a dimpled rivulet's verdant side,
A rising bank, with woodbine edg'd, he spied:
There, veil'd with flowerets of a thousand hues,
A nymph lay bath'd in slumber's balmy dews;
(This maid by some, for some our race defame,
Was Folly call'd, but Pleasure was her name:)
Her mantle, like the sky in April, blue,
Hung on a blossom'd branch that near her grew;
For, long disporting in the silver stream,
She shunn'd the blazing day-star's sultry beam;
And, ere she could conceal her naked charms,
Sleep caught her trembling in his downy arms:
Borne on the wings of Love, he flew, and press'd
Her breathing bosom to his eager breast.
At his wild theft the rosy morning blush'd,
The rivulet smil'd, and all the woods were hush'd,
Of these fair parents on this blissful coast
(Parents like Mirth and Pleasure who can boast?)
I with five sisters, on one happy morn,
All fair alike, behold us now, were born.
When they to brighter regions took their way,
By Love invited to the realms of day,
To us they gave this large, this gay domain,
And said, departing, Here let Beauty reign.
Then reign, fair prince, in thee all beauties shine,
And, ah! we know thee of no mortal line'.

    She said; the king with rapid ardour glow'd,
And the swift poison through his bosom flow'd:
But while she spoke he cast his eyes around
To view the dazzling roof, and spangled ground;

<hr/>

[6] [Here Jones bonds the West to his Eastern source—the
Hindu deity Maya embracing the mild, sylvan Zephyr.]

Then, turning with amaze from side to side,
Seven golden doors,[7] that richly shone, he spied,
And said, 'Fair nymph (but let me not be bold),
What mean those doors that blaze with burnish'd gold?'
'To six gay bowers', the maid replied, 'they lead,
Where Spring eternal crowns the glowing mead;
Six fountains there, that glitter as they play,
Rise to the sun with many a colour'd ray'.
'But the seventh door', said he, 'what beauties grace?'
'O, 'tis a cave, a dark and joyless place,
A scene of nameless deeds, and magick spells,
Where day ne'er shines, and pleasure never dwells:
Think not of that. But come, my royal friend,
And see what joys thy favour'd steps attend'.
She spoke, and pointed to the nearest door:
Swift he descends; the damsel flies before;
She turns the lock; it opens at command;
The maid and stripling enter hand in hand.
      The wondering youth beheld an opening glade.
Where in the midst a crystal fountain play'd;[8]
The silver sands, that on its bottom grew,
Were strown with pearls and gems of varied hue;
The diamond sparkled like the star of day,
And the soft topaz shed a golden ray;
Clear amethysts combin'd their purple gleam
With the mild emerald's sight-refreshing beam;
The sapphire smil'd like yon blue plain above,
And rubies spread the blushing tint of love.
'These are the waters of eternal light',
The damsel said, 'the stream of heavenly sight;
See, in this cup (she spoke, and stoop'd to fill
A vase of jasper with the sacred rill),
See, how the living waters bound and shine,
Which this well-polish'd gem can scarce confine!'
From her soft hand the lucid urn he took,
And quaff'd the nectar with a tender look:
Straight from his eyes a cloud of darkness flew,
And all the scene was open'd to his view;
Not all the groves, where ancient bards have told,
Of vegetable gems, and blooming gold;
Not all the bowers which oft in flowery lays
And solemn tales Arabian poets praise;
Though streams of honey flow'd through every mead,
Though balm and amber dropp'd from every reed;
Held half the sweets that Nature's ample hand
Had pour'd luxuriant o'er this wondrous land.
All flowerets here their mingled rays diffuse,

The rainbow's tints to these were vulgar hues;
All birds that in the stream their pinion dip,
Or from the brink the liquid crystal sip,
Or show their beauties to the sunny skies,
Here wav'd their plumes that shone with varying dyes;
But chiefly he, that o'er the verdant plain
Spreads the gay eyes which grace his spangled train;
And he, who, proudly sailing, loves to show
His mantling wings and neck of downy snow;
Nor absent he, who learns the human sound,
With wavy gold and moving emeralds crown'd;
Whose head and breast with polish'd sapphires glow,
And on whose wing the gems of Indus grow.
The monarch view'd their beauties o'er and o'er,
He was all eye, and look'd from every pore.
But now the damsel calls him from his trance;
And o'er the lawn delighted they advance:
They pass the hall adorn'd with royal state,
And enter now with joy the second gate.[9]
      A soothing sound he heard (but tasted first
The gushing stream that from the valley burst),
And in the shade beheld a youthful quire
That touch'd with flying hands the trembling lyre:
Melodious notes, drawn out with magick art,
Caught with sweet extasy his ravish'd heart;
An hundred nymphs their charming descants play'd,
And melting voices died along the glade;
The tuneful stream that murmur'd as it rose,
The birds that on the trees bewail'd their woes,
The boughs, made vocal by the whispering gale,
Join'd their soft strain, and warbled through the vale.
The concert ends: and now the stripling hears
A tender voice that strikes his wondering ears;
A beauteous bird, in our rude climes unknown,
That on a leafy arbour sits alone,
Strains his sweet throat, and waves his purple wings,
And thus in human accents softly sings:
      'Rise, lovely pair, a sweeter bower invites
Your eager steps, a bower of new delights;
Ah! crop the flowers of pleasure while they blow,
Ere winter hides them in a veil of snow.
Youth, like a thin anemone, displays
His silken leaf, and in a morn decays.
See, gentle youth, a lily-bosom'd bride!
See, nymph, a blooming stripling by thy side!
Then haste, and bathe your souls in soft delights,
A sweeter bow'r your wandering steps invites'.
      He ceas'd; the slender branch, from which he flew,

---

[7] [Symbol of seven senses, six sensory and one mental.]
[8] Sight.

[9] Hearing.

Bent its fair head, and sprinkled pearly dew.
The damsel smil'd; the blushing youth was pleas'd,
And by her willing hand his charmer seiz'd:
The lovely nymph, who sigh'd for sweeter joy,
To the third gate[10] conducts the amorous boy;
She turns the key; her cheeks like roses bloom,
And on the lock her fingers drop perfume.

His ravish'd sense a scene of pleasure meets,
A maze of joy, a paradise of sweets;
But first his lips had touch'd th' alluring stream,
That through the grove display'd a silver gleam.
Through jasmine bowers, and violet-scented vales,
On silken pinions flew the wanton gales,
Arabian odours on the plants they left,
And whisper'd to the woods their spicy theft;
Beneath the shrubs, that spread a trembling shade,
The musky roes, and fragrant civets, play'd.
As when at eve an Eastern merchant roves
From Hadramut to Aden's spikenard groves,
Where some rich caravan not long before
Has pass'd, with cassia fraught, and balmy store,
Charm'd with the scent that hills and vales diffuse,
His grateful journey gayly he pursues;
Thus pleas'd, the monarch fed his eager soul,
And from each breeze a cloud of fragrance stole:
Soon the fourth door[11] he pass'd with eager haste,
And the fourth stream was nectar to his taste.

Before his eyes, on agate columns rear'd,
On high a purple canopy appear'd;
And under it in stately form was plac'd
A table with a thousand vases grac'd;
Laden with all the dainties that are found
In air, in seas, or on the fruitful ground.
Here the fair youth reclin'd with decent pride,
His wanton nymph was seated by his side:
All that could please the taste the happy pair
Cull'd from the loaded board with curious care;
O'er their enchanted heads a mantling vine
His curling tendrils wove with amorous twine;
From the green stalks the glowing clusters hung
Like rubies on a thread of emeralds strung;
With these were other fruits of every hue,
The pale, the red, the golden, and the blue.
An hundred smiling pages stood around,
Their shining brows with wreaths of myrtle bound:
They, in transparent cups of agate, bore
Of sweetly-sparkling wines a precious store;

The stripling sipp'd and revel'd, till the sun
Down heaven's blue vault his daily course had run;
Then rose, and, follow'd by the gentle maid,
Op'd the fifth door:[12] a stream before them play'd.

The king, impatient for the cooling draught,
In a full cup the mystic nectar quaff'd;
Then with a smile (he knew no higher bliss)
From her sweet lip he stole a balmy kiss:
On the smooth bank of violets they reclin'd;
And, whilst a chaplet for his brow she twin'd,
With his soft cheek her softer cheek he press'd,
His pliant arms were folded round her breast.
She smil'd, soft lightning darted from her eyes,
And from his fragrant seat she bade him rise;
Then, while a brighter blush her face o'erspread,
To the sixth gate[13] her willing guest she led.

The golden lock she softly turn'd around;
The moving hinges gave a pleasing sound:
The boy delighted ran with eager haste,
And to his lips the living fountain plac'd;
The magick water pierc'd his kindled brain,
And a strange venom shot from vein to vein.
Whatever charms he saw in other bowers,
Were here combin'd, fruits, musick, odours, flowers;
A couch besides, with softest silk o'erlaid;
And, sweeter still, a lovely yielding maid,
Who now more charming seem'd, and not so coy,
And in her arms infolds the blushing boy:
They sport and wanton, till, with sleep oppress'd,
Like two fresh rose-buds on one stalk, they rest.

When morning spread around her purple flame,
To the sweet couch the five fair sisters came;
They hail'd the bridegroom with a cheerful voice,
And bade him make with speed a second choice.
Hard task to choose, when all alike were fair!
Now this, now that, engag'd his anxious care:
Then to the first who spoke his hand he lent;
The rest retir'd, and whisper'd as they went.
The prince enamour'd view'd his second bride;
They left the bower, and wander'd side by side,
With her he charm'd his ears, with her his sight;
With her he pass'd the day, with her the night.
Thus all by turns the sprightly stranger led,
And all by turns partook his nuptial bed;
Hours, days, and months, in pleasure flow'd away;
All laugh'd, all sweetly sung, and all were gay.

So had he wanton'd threescore days and seven,

---

[10] Smell.
[11] Taste.

[12] Touch.
[13] The sensual pleasures united.

More blest, he thought, than any son of heaven:
Till on a morn, with sighs and streaming tears,
The train of nymphs before his bed appears;
And thus the youngest of the sisters speaks,
Whilst a sad shower runs trickling down her cheeks:
   'A custom which we cannot, dare not fail
(Such are the laws that in our isle prevail),
Compels us, prince, to leave thee here alone,
Till thrice the sun his rising front has shown:
Our parents, whom, alas! we must obey,
Expect us at a splendid feast to-day;
What joy to us can all their splendour give?
With thee, with only thee, we wish to live.
Yet may we hope, these gardens will afford
Some pleasing solace to our absent lord?
Six golden keys, that ope you blissful gates,
Where joy, eternal joy, thy steps awaits,
Accept: the seventh (but that you heard before)
Leads to a cave, where ravening monsters roar;
A sullen, dire, inhospitable cell,
Where deathful spirits and magicians dwell.
Farewel, dear youth; how will our bosoms burn
For the sweet moment of our blest return!'
   The king, who wept, yet knew his tears were vain,
Took the seven keys, and kiss'd the parting train.
A glittering car, which bounding coursers drew,
They mounted straight, and through the forest flew.
   The youth, unknowing how to pass the day,
Review'd the bowers, and heard the fountains play;
By hands unseen whate'er he wish'd was brought;
And pleasures rose obedient to his thought.
Yet all the sweets, that ravish'd him before,
Were tedious now, and charm'd his soul no more:
Less lovely still, and still less gay they grew;
He sigh'd, he wish'd, and long'd for something new:
Back to the hall he turn'd his weary feet,
And sat repining on his royal seat.
Now on the seventh bright gate he casts his eyes
And in his bosom rose a bold surmise:
'The nymph', said he, 'was sure dispos'd to jest,
Who talk'd of dungeons in a place so blest:
What harm to open, if it be a cell
Where deathful spirits and magicians dwell?
If dark or foul, I need not pass the door;
If new or strange, my soul desires no more'.
He said, and rose; then took the golden keys,
And op'd the door: the hinges mov'd with ease.
   Before his eyes appear'd a sullen gloom,
Thick, hideous, wild; a cavern, or a tomb.
Yet as he longer gaz'd, he saw afar

A light that sparkled like a shooting star.
He paus'd: at last, by some kind angel led,
He enter'd, and advanc'd with cautious tread.
Still as he walk'd, the light appear'd more clear;
Hope sooth'd him then, and scarcely left a fear.
At length an aged sire surpriz'd he saw,
Who fill'd his bosom with a sacred awe:[14]
A book he held, which, as reclin'd he lay,
He read, assisted by a taper's ray;
His beard, more white than snow on winter's breast,
Hung to the zone that bound his sable vest;
A pleasing calmness on his brow was seen,
Mild was his look, majestick was his mien.
Soon as the youth approach'd the reverend sage,
He rais'd his head, and clos'd the serious page;
Then spoke: 'O son, what chance has turn'd thy feet
To this dull solitude, and lone retreat?'To whom the
youth: 'First, holy father, tell,
What force detains thee in this gloomy cell?
This isle, this palace, and those balmy bowers,
Where six sweet fountains fall on living flowers,
Are mine; a train of damsels chose me king,
And through my kingdom smiles perpetual spring.
For some important cause to me unknown,
This day they left me joyless and alone;
But, ere three morns with roses strow the skies,
My lovely brides will charm my longing eyes'.
   'Youth', said the sire, 'on this auspicious day
Some angel hither led thy erring way:
Hear a strange tale, and tremble at the snare,
Which for thy steps thy pleasing foes prepare.
Know, in this isle prevails a bloody law;
List, stripling, list! (the youth stood fix'd with awe:)
But seventy days the hapless monarchs reign,[15]
Then close their lives in exile and in pain;
Doom'd in a deep and frightful cave to rove,
Where darkness hovers o'er the iron grove.
Yet know, thy prudence and thy timely care
May save thee, son, from this destructive snare.
Not far from this a lovelier island lies,[16]
Too rich, too splendid, for unhallow'd eyes:
On that blest shore a sweeter fountain flows
Than this vain clime, or this gay palace knows,
Which if thou taste, whate'er was sweet before
Will bitter seem, and steal thy soul no more.
But, ere these happy waters thou canst reach,

[14] Religion.
[15] The life of man.
[16] Heaven.

Thy weary steps must pass yon rugged beach,
Where the dark sea with angry billows raves,[17]
And, fraught with monsters, curls his howling waves;
If to my words obedient thou attend,
Behold in me thy pilot and thy friend.
A bark I keep, supplied with plenteous store,
That now lies anchor'd on the rocky shore;
And, when of all thy regal toys bereft,
In the rude cave an exile thou art left,
Myself will find thee on the gloomy lea,
And waft thee safely o'er the dangerous sea'.

The boy was fill'd with wonder as he spake,
And from a dream of folly seem'd to wake:
All day the sage his tainted thoughts refin'd;
His reason brighten'd, and reform'd his mind:
Through the dim cavern hand in hand they walk'd,
And much of truth, and much of heaven, they talk'd.
At night the stripling to the hall return'd;
With other fires his alter'd bosom burn'd.
O! to his wiser soul how low, how mean,
Seem'd all he e'er had heard, had felt, had seen!
He view'd the stars, he view'd the crystal skies,
And bless'd the power all-good, all-great, all-wise;
How lowly now appear'd the purple robe,
The rubied sceptre, and the ivory globe!
How dim the rays that gild the brittle earth!
How vile the brood of Folly, and of Mirth!

When the third morning, clad in mantle gray,
Brought in her rosy car the seventieth day,
A band of slaves, who rush'd with furious sound,
In chains of steel the willing captive bound;
From his young head the diadem they tore,
And cast his pearly bracelets on the floor;
They rent his robe that bore the rose's hue,
And o'er his breast a hairy mantle threw;
Then dragg'd him to the damp and dreary cave,
Drench'd by the gloomy sea's resounding wave.
Meanwhile the voices of a numerous croud
Pierc'd the dun air, as thunder breaks a cloud:
The nymphs another hapless youth had found,
And then were leading o'er the guilty ground:
They hail'd him king (alas, how short his reign!)
And with fresh chaplets strow'd the fatal plain.

The happy exile, monarch now no more,
Was roving slowly o'er the lonely shore;
At last the sire's expected voice he knew,
And tow'rd the sound with hasty rapture flew,
The promis'd pinnace just afloat he found,

And the glad sage his fetter'd hands unbound;
But when he saw the foaming billows rave,
And dragons rolling o'er the fiery wave,
He stopp'd: his guardian caught his lingering hand,
And gently led him o'er the rocky strand;
Soon as he touch'd the bark, the ocean smil'd,
The dragons vanish'd, and the waves were mild.[18]

For many an hour with vigorous arms they row'd,
While not a star one friendly sparkle show'd;
At length a glimmering brightness they behold,
Like a thin cloud which morning dyes with gold:
To that they steer; and now, rejoic'd, they view
A shore begirt with cliffs of radiant hue.
They land: a train, in shining mantles clad,
Hail their approach, and bid the youth be glad;
They led him o'er the lea with easy pace,
And floated as they went with heavenly grace.
A golden fountain soon appear'd in sight,
That o'er the border cast a sunny light.

The sage, impatient, scoop'd the lucid wave
In a rich vase, which to the youth he gave;
He drank: and straight a bright celestial beam
Before his eyes display'd a dazzling gleam;
Myriads of airy shapes around him gaz'd;
Some prais'd his wisdom, some his courage prais'd;
Then o'er his limbs a starry robe they spread,
And plac'd a crown of diamonds on his head.

His aged guide was gone, and in his place
Stood a fair cherub flush'd with rosy grace;
Who, smiling, spake: 'Here ever wilt thou rest,
Admir'd belov'd, our brother and our guest;
So all shall end, whom vice can charm no more
With the gay follies of that perilous shore.
See yon immortal towers their gates unfold,
With rubies flaming, and no earthly gold!
There joys, before unknown, thy steps invite;
Bliss without care, and morn without a night.
But now farewel! my duty calls me hence;
Some injur'd mortal asks my just defence.
To yon pernicious island I repair,
Swift as a star'. He speaks, and melts in air.

The youth o'er walks of jasper takes his flight;
And bounds and blazes in eternal light.

---

[17] Death.

[18] [Thus the prince lost his sensory ties, preparing for the heavenly gleam to be made possible by the old sage.]

Caïssa, or, The Game at Chess

Caïssa[1]

Advertisement

The first idea of the following piece was taken from a Latin poem of Vida, entitled 'SCACCHIA LUDUS', which was translated into Italian by Marino, and inserted in the fifteenth Canto of his Adonis: the author thought it fair to make an acknowledgment in the notes for the passages which he borrowed from those two poets; but he must also do them the justice to declare, that most of the descriptions, and the whole story of Caïssa, which is written in imitation of Ovid, are his own, and their faults must be imputed to him only. The characters in the poem are no less imaginary than those in the episode; in which the invention of Chess is poetically ascribed to Mars, though it is certain that the game was originally brought from India.

Of armies on the chequer'd field array'd,
And guiltless war in pleasing form display'd;
When two bold kings contend with vain alarms,
In ivory this, and that in ebon arms;
Sing, sportive maids, that haunt the sacred hill
Of Pindus,[2] and the fam'd Pierian rill.
  Thou, joy of all below, and all above,
Mild Venus, queen of laughter, queen of love;
Leave thy bright island, where on many a rose
And many a pink thy blooming train repose:
Assist me, goddess! since a lovely pair
Command my song, like thee divinely fair.
  Near yon cool stream, whose living waters play,
And rise translucent in the solar ray;
Beneath the covert of a fragrant bower,
Where spring's soft influence purpled every flower;
Two smiling nymphs reclin'd in calm retreat,
And envying blossoms crowded round their seat:
Here Delia was enthron'd, and by her side
The sweet Sirena, both in beauty's pride:
Thus shine two roses, fresh with early bloom,
That from their native stalk dispense perfume;
Their leaves unfolding to the dawning day
Gems of the glowing mead, and eyes of May.
A band of youths and damsels sat around,
Their flowing locks with braided myrtle bound;
Agatis, in the graceful dance admir'd,
And gentle Thyrsis, by the muse inspir'd;
With Sylvia, fairest of the mirthful train;
And Daphnis, doom'd to love, yet love in vain.
Now, whilst a purer blush o'erspreads her cheeks,
With soothing accents thus Sirena speaks:
  'The meads and lawns are ting'd with beamy light,

---

[1] [Jones wrote this poem at Harrow, when he was sixteen, basing it on the chess game in Vida, in a losing competition with Pope's earlier *The Rape of the Lock*, which was also based on Vida. It appeared in *Poems* (1772) and was immediately famous, going through several editions in chess books. See Cannon, *Sir William Jones, A Bibliography* (1979): 8-9. I omit Jones's quo- tations from Vida.]

[2] [One of numerous names from Greek mythology used in this imaginative poem.]

And wakeful larks begin their vocal flight;
Whilst on each bank the dewdrops sweetly smile;
What sport, my Delia, shall the hours beguile?
Shall heavenly notes, prolong'd with various art,
Charm the fond ear, and warm the rapturous heart?
At distance shall we view the sylvan chace?
Or catch with silken lines the finny race?'
Then Delia thus: 'Or rather, since we meet
By chance assembled in this cool retreat,
In artful contest let our warlike train
Move well-directed o'er the colour'd plain;
Daphnis, who taught us first, the play shall guide;
Explain its laws, and o'er the field preside:
No prize we need, our ardour to inflame;
We fight with pleasure, if we fight for fame'.
  The nymph consents: the maids and youths prepare
To view the combat, and the sport to share;
But Daphnis most approv'd the bold design,
Whom Love instructed, and the tuneful Nine.
He rose, and on the cedar table plac'd
A polish'd board, with differing colours grac'd;
Squares eight times eight in equal order lie;
These bright as snow, those dark with sable dye;
Like the broad target by the tortoise born,
Or like the hide by spotted panthers worn.
Then from a chest, with harmless heroes stor'd,
O'er the smooth plain two well-wrought hosts he
      pour'd;
The champions burn'd their rivals to assail,
Twice eight in black, twice eight in milkwhite mail;
In shape and station different, as in name,
Their motions various, nor their power the same.
Say muse! (for Jove has nought from thee conceal'd)
Who form'd the legions on the level field?
  High in the midst the reverend kings appear,
And o'er the rest their pearly scepters rear:
One solemn step, majestically slow,
They gravely move, and shun the dangerous foe;
If e'er they call, the watchful subjects spring,
And die with rapture if they save their king;
On him the glory of the day depends,
He once imprison'd, all the conflict ends.
  The queens exulting near their consorts stand;
Each bears a deadly falchion in her hand;
Now here, now there, they bound with furious pride,
And thin the trembling ranks from side to side;
Swift as Camilla flying o'er the main,
Or lightly skimming o'er the dewy plain:
Fierce as they seem, some bold Plebeian spear
May pierce their shield, or stop their full career.

  The valiant guards, their minds on havock bent,
Fill the next squares, and watch the royal tent;
Though weak their spears, though dwarfish be their
      height,
Compact they move, the bulwark of the fight.[3]
  To right and left the martial wings display
Their shining arms, and stand in close array.
Behold, four archers, eager to advance,
Send the light reed, and rush with sidelong glance;
Through angles ever they assault the foes,
True to the colour, which at first they chose.
Then four bold knights for courage fam'd and speed,
Each knight exalted on a prancing steed:
Their arching course no vulgar limit knows,
Transverse they leap, and aim insidious blows:
Nor friends, nor foes, their rapid force restrain,
By one quick bound two changing squares they gain;
From varying hues renew the fierce attack,
And rush from black to white, from white to black.
Four solemn elephants the sides defend;
Beneath the load of ponderous towers they bend:
In one unalter'd line they tempt the fight;
Now crush the left, and now o'erwhelm the right.
Bright in the front the dauntless soldiers raise
Their polish'd spears; their steely helmets blaze:
Prepar'd they stand the daring foe to strike,
Direct their progress, but their wounds oblique.
  Now swell th' embattled troops with hostile rage,
And clang their shields, impatient to engage;
When Daphnis thus: A varied plain behold,
Where fairy kings their mimick tents unfold,
As Oberon, and Mab, his wayward queen,
Lead forth their armies on the daisied green.
No mortal hand the wonderous sport contriv'd,
By Gods invented, and from Gods deriv'd:
From them the British nymphs receiv'd the game,
And play each morn beneath the crystal Thame;
Hear then the tale, which they to Colin sung,
As idling o'er the lucid wave he hung.
  A lovely Dryad rang'd the Thracian wild,

[3] The chief art in the Tacticks of Chess consists in the nice conduct of the royal pawns; in supporting them against every attack; and, if they are taken, in supplying their places with others equally supported; a principle, on which the success of the game in great measure depends, though it seems to be omitted by the very accurate Vida. [Jones was an excellent chess-player, having studied Philidor's *Analyse du jeu des echecs* (London, 1749) and probably having played chess with Franklin.]

Her air enchanting, and her aspect mild;
To chase the bounding hart was all her joy,
Averse from Hymen, and the Cyprian boy;
O'er hills and valleys was her beauty fam'd,
And fair Caissa was the damsel nam'd.
Mars saw the maid; with deep surprize he gaz'd,
Admir'd her shape, and every gesture prais'd:
His golden bow the child of Venus bent,
And through his breast a piercing arrow sent:
The reed was Hope; the feathers, keen Desire;
The point, her eyes; the barbs, ethereal fire.
Soon to the nymph he pour'd his tender strain;
The haughty Dryad scorn'd his amorous pain:
He told his woes, where'er the maid he found,
And still he press'd, yet still Caissa frown'd;
But ev'n her frowns (ah, what might smiles have done!)
Fir'd all his soul, and all his senses won.
He left his car, by raging tigers drawn,
And lonely wander'd o'er the dusky lawn;
Then lay desponding near a murmuring stream,
And fair Caissa was his plaintive theme.
A Naiad heard him from her mossy bed,
And through the crystal rais'd her placid head;
Then mildly spake: 'O thou, whom love inspires,
Thy tears will nourish, not allay thy fires.
The smiling blossoms drink the pearly dew;
And ripening fruit the feather'd race pursue;
The scaly shoals devour the silken weeds;
Love on our sighs, and on our sorrow feeds.
Then weep no more; but, ere thou canst obtain
Balm to thy wounds, and solace to thy pain,
With gentle art thy martial look beguile;
Be mild, and teach thy rugged brow to smile.
Canst thou no play, no soothing game devise,
To make thee lovely in the damsel's eyes?
So may thy prayers assuage the scornful dame,
And ev'n Caissa own a mutual flame'.
'Kind nymph', said Mars, 'thy counsel I approve,
Art, only art, her ruthless breast can move.
But when? or how? Thy dark discourse explain:
So may thy stream ne'er swell with gushing rain;
So may thy waves in one pure current flow,
And flowers eternal on thy border blow!'
        To whom the maid replied with smiling mien:
'Above the palace of the Paphian queen
Love's brother dwells, a boy of graceful port,
By gods nam'd Euphron, and by mortals Sport:
Seek him; to faithful ears unfold thy grief,
And hope, ere morn return, a sweet relief.
His temple hangs below the azure skies;

Seest thou yon argent cloud? 'Tis there it lies'.
This said, she sunk beneath the liquid plain,
And sought the mansion of her blue-hair'd train.
        Meantime the god, elate with heart-felt joy,
Had reach'd the temple of the sportful boy;
He told Caissa's charms, his kindled fire,
The Naiad's counsel, and his warm desire.
        'Be swift', he added, 'give my passion aid;
A god requests'.—He spake, and Sport obey'd,
He fram'd a tablet of celestial mold,
Inlay'd with squares of silver and of gold;
Then of two metals form'd the warlike band,
That here compact in show of battle stand;
He taught the rules that guide the pensive game,
And call'd it *Cassa* from the Dryad's name:
(Whence Albion's sons, who most its praise confess,
Approv'd the play, and nam'd it thoughtful *Chess*.)
The god delighted thank'd indulgent Sport;
Then grasp'd the board, and left his airy court.
With radiant feet he pierc'd the clouds; nor stay'd,
Till in the woods he saw the beauteous maid:
        Tir'd with the chase the damsel sat reclin'd,
Her girdle loose, her bosom unconfin'd.
He took the figure of a wanton Faun,
And stood before her on the flowery lawn;
Then show'd his tablet: pleas'd the nymph survey'd
The lifeless troops in glittering ranks display'd;
She ask'd the wily sylvan to explain
The various motions of the splendid train;
With eager heart she caught the winning lore,
And thought ev'n Mars less hateful than before:
'What spell', said she, 'deceiv'd my careless mind?
The god was fair, and I was most unkind'.
She spoke, and saw the changing Faun assume
A milder aspect, and a fairer bloom;
His wreathing horns, that from his temples grew,
Flow'd down in curls of bright celestial hue;
The dappled hairs, that veil'd his loveless face,
Blaz'd into beams, and show'd a heavenly grace;
The shaggy hide, that mantled o'er his breast,
Was soften'd to a smooth transparent vest,
That through its folds his vigorous bosom show'd,
And nervous limbs, where youthful ardour glow'd:
(Had Venus view'd him in those blooming charms,
Not Vulcan's net had forc'd her from his arms.)
With goatlike feet no more he mark'd the ground,
But braided flowers his silken sandals bound.
The Dryad blush'd; and, as he press'd her, smil'd,
Whilst all his cares one tender glance beguil'd.
        He ends: *To arms*, the maids and striplings cry;

To arms, the groves and sounding vales reply.
Sirena led to war the swarthy crew,
And Delia those that bore the lily's hue.
Who first, O muse, began the bold attack;
The white refulgent, or the mournful black?
Fair Delia first, as favouring lots ordain,
Moves her pale legions tow'rd the sable train:
From thought to thought her lively fancy flies,
Whilst o'er the board she darts her sparkling eyes.

At length the warrior moves with haughty strides;
Who from the plain the snowy king divides:
With equal haste his swarthy rival bounds;
His quiver rattles, and his buckler sounds:
Ah! hapless youths, with fatal warmth you burn;
Laws, ever fix'd, forbid you to return.
Then from the wing a short-liv'd spearman flies,
Unsafely bold, and see! he dies, he dies:
The dark-brow'd hero, with one vengeful blow
Of life and place deprives his ivory foe.
Now rush both armies o'er burnish'd field,
Hurl the swift dart, and rend the bursting shield.
Here furious knights on fiery coursers prance,
Here archers spring, and lofty towers advance.
But see! the white-rob'd Amazon beholds
Where the dark host its opening van unfolds:
Soon as her eye discerns the hostile maid,
By ebon shield, and ebon helm betray'd;
Seven squares she passes with majestick mien,
And stands triumphant o'er the falling queen.
Perplex'd and sorrowing at his consort's fate,
The monarch burn'd with rage, despair, and hate:
Swift from his zone th' avenging blade he drew,
And, mad with ire, the proud virago slew.
Meanwhile sweet-smiling Delia's wary king
Retir'd from fight behind his circling wing.

Long time the war in equal balance hung;
Till, unforeseen, an ivory courser sprung,
And, wildly prancing in an evil hour,
Attack'd at once the monarch and the tower;
Sirena blush'd; for, as the rules requir'd,
Her injur'd sovereign to his tent retir'd;
Whilst her lost castle leaves his threatening height,
And adds new glory to th' exulting knight.

At this, pale fear oppress'd the drooping maid,
And on her cheek the rose began to fade:
A crystal tear, that stood prepar'd to fall,
She wip'd in silence, and conceal'd from all;
From all but Daphnis: He remark'd her pain,
And saw the weakness of her ebon train;
Then gently spoke: 'Let me your loss supply,

And either nobly win or nobly die:
Me oft has fortune crown'd with fair success,
And led to triumph in the fields of Chess'.
He said: the willing nymph her place resign'd,
And sat at distance on the bank reclin'd.
Thus when Minerva call'd her chief to arms,
And Troy's high turret shook with dire alarms,
The Cyprian goddess wounded left the plain,
And Mars engag'd a mightier force in vain.

Straight Daphnis leads his squadron to the field;
(To Delia's arms 'tis ev'n a joy to yield.)
Each guileful snare, and subtle art he tries,
But finds his art less powerful than her eyes:
Wisdom and strength superiour charms obey;
And beauty, beauty, wins the long-fought day.
By this a hoary chief, on slaughter bent,
Approach'd the gloomy king's unguarded tent;
Where, late, his consort spread dismay around,
Now her dark corse lies bleeding on the ground.
Hail, happy youth! thy glories not unsung
Shall live eternal on the poet's tongue;
For thou shalt soon receive a splendid change,
And o'er the plain with nobler fury range.
The swarthy leaders saw the storm impend,
And strove in vain their sovereign to defend:
Th' invader wav'd his silver lance in air,
And flew like lightning to the fatal square;
His limbs dilated in a moment grew
To stately height, and widen'd to the view:
More fierce his look, more lion-like his mien,
Sublime he mov'd, and seem'd a warriour queen.
As when the sage on some unfolding plant
Has caught a wandering fly, or frugal ant,
His hand the microscopick frame applies,
And lo! a bright-haird'd monster meets his eyes;
He sees new plumes in slender cases roll'd;
Here stain'd with azure, there bedropp'd with gold;
Thus, on the alter'd chief both armies gaze,
And both the kings are fix'd with deep amaze.
The sword, which arm'd the snow-white maid before,
He now assumes, and hurls the spear no more;
Then springs indignant on the dark-rob'd band,
And knights and archers feel his deadly hand.
Now flies the monarch of the sable shield,
His legions vanquish'd, o'er the lonely field:
So when the morn, by rosy coursers drawn,
With pearls and rubies sows the verdant lawn,
Whilst each pale star from heaven's blue vault retires,
Still Venus gleams, and last of all expires.
He hears, where'er he moves, the dreadful sound;

Check the deep vales, and *Check* the woods rebound.
No place remains: he sees the certain fate,
And yields his throne to ruin, and Checkmate.

    A brighter blush o'erspreads the damsel's cheeks,
And mildly thus the conquer'd stripling speaks:
'A double triumph, Delia, hast thou won,
By Mars protected, and by Venus' son;
The first with conquest crowns thy matchless art,
The second points those eyes at Daphnis' heart'.

    She smil'd, the nymphs and amorous youths arise,
And own, that beauty gain'd the nobler prize.

    Low in their chest the mimick troops were lay'd,
And peaceful slept the sable hero's shade.[4]

An Ode in Imitation of Alcaeus[1]

What constitutes a State?
Not high-rais'd battlement or labour'd mound,
Thick wall or moated gate;
Not cities proud with spires and turrets crown'd;
Not bays and broad-arm'd ports,
Where, laughing at the storm, rich navies ride,
Not starr'd and spangled courts,
Where low-brow'd baseness wafts perfume to pride.
No:—MEN, high-minded MEN,
With pow'rs as far above dull brutes endued
In forest, brake, or den,
As beasts excel cold rocks and brambles rude;
Men, who their *duties* know,
But know their *rights*, and, knowing, dare maintain,
Prevent the long-aim'd blow,
And crush the tyrant while they rend the chain:
These constitute a State,[2]
And sov'reign LAW, *that state's collected will*,
O'er thrones and globes elate
Sits Empress, crowning good, repressing ill;
Smit by her sacred frown

---

[1] [This ode was Jones's greatest and most popular political poem, published by the radical Society for Constitutional Information in 1782. It poetically answers the question that has concerned many countries immemorially—what constitutes a state? Drawing on the Greek poet Alcaeus, Jones develops his system of government in this oft-printed poem, which movingly defines the concept of liberty in a very modern way while questioning contemporary governmental restrictions on citizens' individual freedoms. Milton had a viable influence on Jones's poetic fancy. While reading James Thomson's Preface to the *Areopagitica; a Speech of John Milton* (London, 1738) iv, Jones was struck by a passage cited from Alcaeus: 'What makes a city? Not walls and buildings; no—but men, who know themselves to be men, and are sensible that liberty alone exalts them above brutes'. Jones began an expanded version of Alcaeus's fragment in July 1780. The final draft was composed in March 1781, after the Althorp wedding, while riding 'in my chaise between Abergavenny and Brecon' [(*Letters* 2 : 463).]

[2] [Such clear references to the American rebels and their declaration of independence antagonized the Tory Government and helped to delay Jones's long-sought appointment to the Bengal Supreme Court.]

[4] A parody of the last line in Pope's translation of the Iliad, 'And peaceful slept the mighty Hector's shade'.

The fiend *Discretion* like a vapour sinks,
And e'en th' all-dazzling *Crown*
Hides his faint rays, and at her bidding shrinks.
Such *was* this heav'n-lov'd isle,
Than *Lesbos* fairer and the *Cretan* shore!
No more shall Freedom smile?
Shall *Britons* languish, and be MEN no more?
Since all must life resign,
Those sweet rewards, which decorate the brave,
'Tis folly to decline,
And steal inglorious to the silent grave.[3]

Abergavenny, March 31, 1781.

The Muse Recalled,[1]

An Ode on the Nuptials of Lord
Viscount Althorp and Miss Lavinia Bingham,
Eldest Daughter of Charles Lord Lucan March VI,
MDCCLXXXI

Return, celestial Muse,
By whose bright fingers o'er my infant head,
Lull'd with immortal symphony, were spread
Fresh bays and flow'rets of a thousand hues;
Return! thy golden lyre,
Chorded with sunny rays of temper'd fire,
Which in Astraea's fane I fondly hung,
Bold I reclaim: but ah, sweet maid,
Bereft of thy propitious aid
My voice is tuneless, and my harp unstrung.
In vain I call . . . What charm, what potent spell
Shall kindle into life the long-unwaken'd shell?

Haste! the well-wrought[2] basket bring,
Which two sister Graces wove,
When the third, whose praise I sing,
Blushing sought the bridal grove,
Where the slow-descending sun
Gilt the bow'rs of WIMBLEDON.
In the vase mysterious fling
Pinks and roses gemm'd with dew,
Flow'rs of ev'ry varied hue,
Daughters fair of early spring,
Laughing sweet with sapphire eyes,

---

[1] [Jones's most distinguished occasional poem, written at the wedding of his former pupil, had been composed at the insistence of the Duchess of Devonshire, Countess Spencer, the bride, and others. Jones wrote it 'almost without any pre-meditation' and called it 'poetry by compulsion' in a letter to Edmund Cartwright (*Letter* 2 : 511-12). But he perceived in the occasion the kind of larger horizons which Milton had found in 'Lycidas', synthesizing his political views in the latter stanzas. The ode was printed at Horace Walpole's Strawberry Hill Press in 1781 and is still one of Jones's most popular poems.]

[2] Miss Louisa Bingham, and Miss Frances Molesworth her cousin, decked a basket with ribbands and flowers to hold the nuptial presents.

[3] [An echo of Thomas Gray's famous 'Elegy'.]

Or with Iris' mingled dyes:
Then around the basket go,
Tripping light with silent pace,
While, with solemn voice and slow
Thrice pronouncing, thrice I trace
On the silken texture bright,
Character'd in beamy light,
Names of more than mortal pow'r,
Sweetest influence to diffuse;
Names, that from her shadiest bow'r
Draw the soft reluctant muse.

First, I with living gems enchase
The name of Her, whom for this festive day
With zone and mantle elegantly gay
The Graces have adorn'd, herself a Grace,
MOLESWORTH . . . hark! a swelling note
Seems on Zephyr's wing to float,
Or has vain hope my flatter'd sense beguil'd?
Next Her, who braided many a flow'r
To deck her sister's nuptial bow'r,
BINGHAM, with gentle heart and aspect mild:
The charm prevails . . . I hear, I hear
Strains nearer yet, and yet more near.
Still, ye nymphs and youths, advance,
Sprinkle still the balmy show'r,
Mingle still the mazy dance.
Two names of unresisted pow'r,
Behold, in radiant characters I write:
O rise! O leave thy secret shrine,
For they, who all thy nymphal train outshine,
DUNCANNON,[3] heav'nly Muse, and DEVONSHIRE[4] invite.

Saw ye not yon myrtle wave?
Heard ye not a warbled strain?
Yes! the harp, which Clio gave,
Shall his ancient sound regain.
One dearer name remains. Prepare, prepare!
She comes . . . how swift th' impatient air
Drinks the rising accents sweet!
Soon the charm shall be complete.
Return, and wake the silent string;
Return, sweet Muse, for ALTHORP bids me sing.
'Tis she . . . and, as she smiles, the breathing lyre

---

[3] Lady Henrietta Spencer, second daughter of John Earl
Spencer, and wife of the lord Viscount Duncannon, eldest son
of the earl of Besborough.

[4] Lady Georgiana, eldest daughter of Earl Spencer, and wife
of William Cavendish, fifth duke of Devonshire.

Leaps from his silken bands, and darts ethereal fire.

Bright son of ev'ning, lucid star,
Auspicious rise thy soften'd beam,
Admir'd ere Cynthia's pearly car
O'er heav'n's pure azure spreads her gleam:
Thou saw'st the blooming pair,
Like thee serenely fair,
By love united and the nuptial vow,
Thou seest the mirthful train
Dance to th' unlabour'd strain,
Seest bound with myrtle ev'ry youthful brow.
Shine forth, ye silver eyes of night,
And gaze on virtues crown'd with treasures of delight.

And thou, the golden-tressed child of morn,
Whene'er thy all-inspiring heat
Bids bursting rose-buds hill and mead adorn,
See them with ev'ry gift that Jove bestows,
With ev'ry joy replete,[5]
Save, when they melt at sight of human woes.
Flow smoothly, circling hours,
And o'er their heads unblended pleasure pour;
Nor let your fleeting round
Their mortal transports bound,
But fill their cup of bliss, eternal pow'rs,
Till time himself shall cease, and suns shall blaze no
          more.

Each morn, reclin'd on many a rose,
LAVINIA's[6] pencil shall disclose

---

[5] [In a letter to Viscount Althorp on 28 Sept. 1781, Jones
wrote a critical comment on his ode: 'In the sixth stanza the
word *replete* occurs in a good sense, though our classical writers
uniformly, I believe, use it in a bad one as 'replete with guile or
with danger' [*Paradise Lost* 9 : 733]; and in the seventh (a
singular error, which never struck me, till I saw it printed) the
verb *touch* has no nominative, for *pencil*, which relate to
musick, is the only nominative that precedes, unless the *proper
name* (which I so highly respect) be understood by a figure
hardly allowable in the wildest flights of poetry. In the ninth
and tenth stanzas are two errors of Mr. Kirgate the printer; one
trifling, *chrystal* for crysal [corrected in the editions of Jones's
Works]; the other more important (as it occasions an equivoque
in our language)*reign* for *rain*, which last verb is used for
'shedding abundantly', as Milton says of his beauties that 'their
eyes rain influence' ['L'Allegro', 11. 121-22]. The last objection to the Ode, as I have before observed, is fatal; namely, that
it is very *unpoetical*, because it is very *true*, and consequently
wants the essence of poetry' (*Letters* 2 : 497-98).]

New forms of dignity and grace,
Th' expressive air, th' impassion'd face,
The curled smile, the bubbling tear,
The bloom of hope, the snow of fear,
To some poetick tale fresh beauty give,
And bid the starting tablet rise and live;
Or with swift fingers shall she touch the strings.
And in the magick loom of harmony
Notes of such wond'rous texture weave,
As lifts the soul on seraph wings,
Which, as they soar above the jasper sky,
Below them suns unknown and worlds unnumber'd
    leave.

While thou, by list'ning crowds approv'd,
Lov'd by the Muse and by the poet lov'd,
ALTHORP, shouldst emulate the fame
Of Roman patriots and th' Athenian name;
Shouldst charm with full persuasive eloquence,
With all thy mother's grace,[7]and all thy father's sense,
Th' applauding senate; whilst, above thy head,
Exulting Liberty should smile,
Then, bidding dragon-born Contention cease,
Should knit the dance with meek-ey'd Peace,
And by thy voice impell'd should spread
An universal joy around her cherish'd isle.
But ah! thy publick virtues, youth, are vain
In this voluptuous, this abandon'd age,
When Albion's sons with frantick rage,
In crimes alone and recreant baseness bold,
Freedom and Concord, with their weeping train,
Repudiate; slaves of vice, and slaves of gold!
They, on starry pinions sailing
Through the crystal fields of air,
Mourn their efforts unavailing,
Lost persuasions, fruitless care:
Truth, Justice, Reason, Valour, with them fly
To seek a purer soil, a more congenial sky.

Beyond the vast Atlantick deep
A dome by viewless genii shall be rais'd,
The walls of adamant compact and steep,
The portals with sky-tinctur'd gems emblazed:

There on a lofty throne shall Virtue stand;
To her the youth of Delaware shall kneel;
And, when her smiles rain plenty o'er the land,
Bow, tyrants, bow beneath th' avenging steel!
Commerce with fleets shall mock the waves,
And Arts, that flourish not with slaves,
Dancing with ev'ry Grace and ev'ry Muse,
Shall bid the valleys laugh and heav'nly beams
    diffuse.[8]
She ceases; and a strange delight
Still vibrates on my ravish'd ear:
What floods of glory drown my sight!
What scenes I view! What sounds I hear!
This for my friend . . . but, gentle nymphs, no more
Dare I with spells divine the Muse recall:
Then, fatal harp, thy transient rapture o'er,
Calm I replace thee on the sacred wall.
Ah, see how lifeless hangs the lyre,
Not lightning now, but glitt'ring wire!
Me to the brawling bar and wrangles high
Bright-hair'd Sabrina calls and rosy-bosom'd Wye.

---

[6] Lady Althorp has an extraordinary talent for drawing historick subjects, and expressing the passions in the most simple manner. [Jones's praises were quite conventional for the daughter of a man who was helping to finance the war against the American Colonies, which Jones strongly opposed.]

[7] Georgiana Poyntz countess Spencer.

[8] [Here Jones boldly prophesies the defeat of tyranny by virtuous America with magnificent commercial fleets and flourishing arts in her future. This prophetic idea shocked Samuel Johnson and some other Tories. Walpole praised it: 'There are many beautiful and poetic expressions in it. A wedding, to be sure, is neither a new nor a promising subject, nor will outlast the favours: still I think Mr. Jones's Ode is uncommonly good for the occasion'. In *The Letters of Horace Walpole*. Ed. Mrs. Paget Toynbee (Oxford, 1903-5), 12 : 43-44]

Damsels of Cardigan[1]

Fair Tivy, how sweet are thy waves gently flowing,
Thy wild oaken woods, and green eglantine bow'rs,
Thy banks with the blush-rose and amaranth glowing,
While friendship and mirth claim these labourless hours!
Yet weak is our vaunt, while something we want,
More sweet than the pleasure which *prospects* can give;
Come, smile, damsels of Cardigan,
Love can alone make it blissful to live.

How sweet is the odour of jasmine and roses,
That Zaphyr around us so lavishly flings!
Perhaps for Bleanpant[2] fresh perfume he composes,
Or tidings from Brownwith[3] auspiciously brings;
Yet weak is our vaunt, while something we want,
More sweet than the pleasure which *odours* can give:
Come, smile, damsels of Cardigan,
Love can alone make it blissful to live.

How sweet was the strain that enliven'd the spirit,
And cheer'd us with numbers so frolic and free!
The poet is absent: be just to his merit;
Ah! may he in love be more happy than we;
For weak is our vaunt, while something we want,
More sweet than the pleasure the *muses* can give:
Come, smile, damsels of Cardigan,
Love can along make it blissful to live.

How gay is the circle of friends round a table,
Where stately Kilgarran[4] o'erhangs the brown dale;
Where none are unwilling, and few are unable,

To sing a wild song, or repeat a wild tale!
Yet weak is our vaunt, while something we want,
More sweet than the pleasure that *friendship* can give:
Come, smile, damsels of Cardigan,
Love can alone make it blissful to live.

No longer then pore over dark gothic pages,
To cull a rude gibberish from Neatham or Brooke;[5]
Leave year-books and parchments to grey-bearded sages;
Be nature and love, and fair woman, our book;
For weak is our vaunt, while something we want,
More sweet than the pleasure that *learning* can give:
Come, smile, damsels of Cardigan,
Love can alone make it blissful to live.

Admit that out labours were crown'd with full measure,
And gold were the fruit of rhetorical flow'rs,
That India supplied us with long-hoarded treasure,
That Dinevor,[6] Slebeck,[7] and Coidsmore[8] were ours;
Yet weak is our vaunt, while something we want,
More sweet than the pleasure that *riches* can give:
Come, smile, damsels of Cardigan,
Love can alone make it blissful to live.

Or say, that, preferring fair Thames to Fair Tivy,
We gain'd the bright ermine robes, purple and red,
And peep'd thro' long perukes, like owlets thro' ivy,
Or say, that bright coronets blaz'd on our head;
Yet weak is our vaunt, while something we want,
More sweet than the pleasure that *honours* can give:
Come, smile, damsels of Cardigan,
Love can alone make it blissful to live.

---

[1] [Beside a ruined castle along the Wye, Jones composed this charming poem to be at a *fête-champêtre* of the Druids, a social group of his fellow judges, on one of his first circuits. It was set to the time of 'Carrick-fergus' and was published in *Gentleman's Magazine* of Sept. 1782 (p. 446), in Teignmouth's *Memoirs* (*Works*, 1 : 357-59), and elsewhere.]

[2] The seat of W. Brigstocke, Esq.

[3] The seat of Thomas Lloyd, Esq. [a London attorney, who owned the manor of Edgmond Shropshire, on which Jones bought a mortgage in Aug. 1789 for f5,000.]

[4] A ruin of a castle on the banks of the Tivy.

[5] [The rulings of Sir John Needham (d. 1480) and Sir Robert Brooke (d. 1558) are recorded in masses of year-books during their long tenures.]

[6] Seat of Lord Dinevor's, near Landelo, in Carmarthen.

[7] Seat of—Philips, Esq., near Haverford West.

[8] Seat of Thomas Lloyd, Esq., near Cardigan.

THE MOALLAK'AT[1]

[Extract]

Jones's Advertisement

The Discourse will comprise observations on the antiquity of the *Arabian* language and letters; on the dialects and characters of *Himyar* and *Koraish*, with accounts of some *Himyarick* poets; on the manners of the *Arabs* in the age immediately preceding that of *Mahomed*; on the temple at Mecca, and the *Moallak'at*, or pieces of poetry *suspended* on its walls or gate; lastly, on the lives of the *Seven Poets*, with a critical history of their works, and the various copies or editions of them preserved in *Europe*, *Asia*, and *Africa*.

The Notes will contain authorities and reasons for the translation of controverted passages; will elucidate all the obscure couplets, and exhibit or propose amendments of the text; will direct the reader's attention to particular beauties, or point out remarkable defects; and will throw light on the images, figures, and allusions of the *Arabian* poets, by citations either from writers of their own country, or from such of our *European* travellers as best illustrate the ideas and customs of eastern nations.

But the *Discourse* and *Notes* are ornamental only, not essential, to the work; and, by sending it abroad in its present form, the translator may reap no small advantage, if the learned here or on the Continent will favour him in the course of the summer with their strictures and annotations, and will transmit them for that purpose to the publisher. It is hoped, that the war will raise no obstacle to this intercourse with the scholars of *Leyden*, *Paris*, and *Madrid*; for men of letters, as such, ought, in all places and at all times, to carry *flags of truce*.[2]

A.D. 1783.

Jones's Argument to "The Poem of Amriolkais"

The poet, after the manner of his countrymen, supposes himself attend on a journey by a company of friends; and, as they pass near a place, where his mistress had lately dwelled, but from which her tribe was then removed, *he desires them to stop awhile*, that he might indulge the painful pleasure of weeping over the deserted remains of her tent. They comply with his request, but exhort him to show more strength of mind, and urge two topics of consolation; namely, *that he had before been equally unhappy*, and *that he had enjoyed his full share of pleasures*: thus by the recollection of his passed delight his imagination is kindled, and his grief suspended.

He then gives his friends a lively account of his juvenile frolicks, to one of which they had alluded. It seems, he had been in love with a girl named *Onaiza*, and had in vain sought an occasion to declare his passion: one day, when her tribe has struck their tents, and were changing their station, the women, as usual, came behind the rest, with the servants and baggage, in carriages fixed on the backs of camels. *Amriolkais* advanced slowly at a distance, and, when the men were out of sight, had the pleasure of seeing *Onaiza* retire with a party of damsels to a rivulet or pool, called *Daratjuljul*, where they undressed themselves, and were bathing, when the lover appeared, dismounted from his camel, and sat upon their clothes, proclaiming aloud, that *whoever could redeem her dress, must present herself naked before him.*

They adjured, entreated, expostulated; but, when it grew late, they found themselves obliged to submit, and all of them recovered their clothes except *Onaiza*, who renewed her adjurations, and continued a long time in the water: at length she also performed the condition, and

---

[1] [Jones was the first to translate this seven-poem collection from the Arabic (London, 1782), the text of which was obscure and highly difficult at the time. His capable version was a landmark in introducing the motif of kismet to the West and evoked many echoes in later poetry. Thus Tennyson took the plot and theme for 'Locksley Hall' from 'The Poem of Amriolkais', which is extracted here. In the opening scene of this haunting dramatic poem, the aging Imr-al-Qais stops at the deserted tent-site of a former mistress, where he recounts his amatory adventures to his friends until a violent rain sweeps down on them. It still offers charming pictures of the old Bedouin life and scenery and evokes the bitter-sweet universality of youthful reminiscing by the aging. See Reynold A. Nicholson, *A Literary History of the Arabs* (Cambridge, 1969), 101; Note 2 in *Letters*, 1 : 446; and A.J. Arberry's scholarly Prologue and modern English version of this complex pre-Islamic collection, in *The Seven Odes* (London, 1957).]

[2] [Jones strongly opposed war, feeling that such funds should be spent on scholarship that might advance humanity.]

dressed herself. Some hours had passed, when the girls complained of cold and hunger: *Amriokais* therefore instantly *killed the young camel on which he had ridden*, and, having called the female attendant together, made a fire and roasted him. The afternoon was spent in gay conversation, not without a cheerful cup, for he was provided with wine in a leathern bottle; but, when it was time to follow the tribe, the prince (for such was his rank) had neither camel nor horse; and *Onaiza*, after much importunity, consented to take him *on her camel before the carriage*, while the other damsels divided among themselves the less agreeable burden of his arms, and the furniture of his beast.

He next relates his courtship of *Fathima*, and his more dangerous amour with a girl of a tribe at war with his own, *whose beauties he very minutely and luxuriantly delineates*. From these love-tales he proceeds to the commendation of his own fortitude, when he was passing a desert in the darkest night; and the mention of morning, which succeeded, leads him to *a long description of his hunter, and of a chase in the forest*, followed by a feast on the game, which had been pierced by his javelins.

Here his narrative seems to be interrupted by *a storm of lightning and violent rain*: he nobly describes the shower and the torrent, which it produced down all the adjacent mountains; and, his companions retiring to avoid the storm, the drama (for the poem has the form of a dramatick pastoral) ends abruptly.

The metre is of the *first* species, called *long verse*, and consists of the *bacchius*, or *amphibrachys*, followed by the first *epitrite*; or, in the *fourth* and *eighth* places, of the distich, by the *double iambus*, the last syllable being considered as a long one: the regular form, taken from the second chapter of *Commentaries on Asiatick Poetry*, is this;

'Amator \ puellarum \ miser sae \ pe fallitur
Ocellis \ nigris,labris \ odoris, \ nigris comis'.

The Poem of Amriolkais

1   'Stay-Let us weep at the remembrance of our beloved, *at the sight of* the station *where her tent was raised*, by the edge of yon bending sands between DAHUL and HAUMEL,[3]

2   'TUDAM and MIKRA; *a station*, the marks of which are not wholly effaced, though the south wind and the north have woven the twisted sand'.

3   Thus I spoke, when my companions stopped their coursers by my side, and said, 'Perish not through despair: only be patient'.

4   'A profusion of tears', answered I, 'is my sole relief; but what avails it to shed them over the remains of a deserted mansion'?

5   'Thy condition', they replied, 'is not more painful than when thou leftest HOWAIRA, before thy present passion, and her neighbour REBABA, *on the hills of* MASEL'.

6   'Yes', I rejoined, 'when those two damsels departed, musk was diffused from their robes, as the eastern gale sheds the scent of clove-gillyflowers':

7   Then gushed the tears from my eyes, through excess of regret, and flowed down my neck, till my sword-belt was drenched in the stream.

8   'Yet hast thou passed many days in sweet converse with the fair; but none so sweet as the day, which thou spentest by *the pool of* DARATJULJUL'.

9   On that day I killed my camel to give the virgins a feast; and oh! how strange was it, that they should carry his trappings and furniture!

10   The damsels continued till evening helping one another to the roasted flesh, and to the delicate fat like the fringe of white silk finely woven.

11   On that happy day I entered the carriage, the carriage of ONAIZA, who said, 'Wo to thee! thou wilt compel me to travel on foot'.

12   She added (while the vehicle was bent aside with our weight), 'O AMRIOLKAIS, descend, or my beast also will be killed'.

13   I answered: 'Proceed, and loosen his rein; nor with-

[3] [Pre-Islamic poetry is rich in local colour, with the realistic use of place names like Ed-Dakhool and Haumal to enhance flashing images and anatomically precise descriptions.]

hold from me the fruits of thy love, which again and again may be tasted with rapture.

14 'Many a fair one like thee, though not *like thee* a virgin, have I visited by night; and many a lovely mother have I diverted from the care of her yearling infant adorned with amulets:'

15 When the suckling behind her cried, she turned round to him with half her body; but half of it, pressed beneath my embrace, was not turned from me.

16 Delightful too was the day, when FATHIMA at first rejected me on the summit of yon sand-hill, and took an oath, which she declared inviolable.

17 'O FATHIMA', said I, 'away with so much coyness; and, if thou hadst resolved to abandon me, yet at last relent.

18 'If, indeed, my disposition and manners are unpleasing to thee, rend at once the mantle of my heart, that it may be detached from thy love.

19 'Art thou so haughty, because my passion for thee destroys me; and because whatever thou commandest, my heart performs?

20 '*Thou weepest*—yet thy tears flow merely to wound my heart with the shafts of thine eyes; my heart, already broken to pieces and agonizing'.

21 *Besides these*—with many a spotless virgin, whose tent had not yet been frequented, have I holden soft dalliance at perfect leisure.

22 *To visit one of them*, I passed the guards of her bower and a hostile tribe, who would have been eager to proclaim my death.

23 It was the hour, when the Pleiads appeared in the firmament, like the folds of a silken sash variously decked with gems.

24 I approached—she stood *expecting me* by the curtain; and, *as if she was preparing* for sleep, had put off all her vesture, but her night-dress.

25 She said—'By him who created me (and gave me her lovely hand), I am unable to refuse thee: for I perceive, that the blindness of thy passion is not to be removed'.

26 Then I rose with her; and, as we walked, she drew over our footsteps the train of her pictured robe.

27 Soon as we had passed the habitations of her tribe, and come to the bosom of a vale surrounded with hillocks of spiry sand,

28 I gently drew her towards me by her curled locks, and she softly inclined to my embrace: her waist was gracefully slender; but sweetly swelled the part encircled with ornaments of gold.

29 Delicate was her shape; fair her skin; and her body well proportioned: her bosom was as smooth as a mirror,

30 Or like the pure egg of an ostrich of a yellowish tint blended with white, and nourished by a stream of wholesome water not yet disturbed.

31 She turned aside, and displayed her soft cheek: she gave a timid glance with languishing eyes, like those of a roe in *the groves of* WEGERA looking tenderly at her young.

32 Her neck was like that of a milk-white hind, but, when she raised it, exceeded not the justest symmetry; nor was the neck of my beloved so unadorned.

33 Her long coal-black hair decorated her back, thick and diffused like bunches of dates clustering on the palm-tree.

34 Her locks were elegantly turned above her head; and the riband, which bound them, was lost in her tresses, part braided, part dishevelled.

35 She discovered a waist tapered as a well-twisted cord; and a leg both as white and as smooth as the stem of a young palm, or a fresh reed, bending over the rivulet.

36 When she sleeps at noon, her bed is besprinkled with musk: she puts on her robe of undress, but leaves the apron *to her handmaids*.

37 She dispenses gifts with small delicate fingers, sweetly glowing at their tips, like the white and crimson worm of DABIA, or dentifrices made of ESEL-WOOD.

38 The brightness of her face illumines the veil of night, like the evening taper of a recluse hermit.

39 On a girl like her, a girl of a moderate height, between those who wear a frock and those who wear a gown, the most bashful man must look with an enamoured eye.

40 The blind passions of men for common objects of affection are soon dispersed; but from the love of thee my heart cannot be released.

41 O how oft have I rejected the admonitions of a morose adviser, vehement in censuring my passion for thee; nor have I been moved by his reproaches!

42 Often has the night drawn her skirts around me like the billows of the ocean, to make trial of my fortitude in a variety of cares;

43 And I said to her (when she seemed to extend her sides, to drag on her unwiedly length, and to advance slowly with her breast),

44 'Dispel thy gloom, O tedious night, that the morn may rise; although my sorrows are such, that the morning-light will not give me more comfort than thy shades.

45 'O hideous night! a night in which the stars are

prevented from rising, as if they were bound to a solid cliff with strong cables!'

46   Often too have I risen at early dawn, while the birds were yet in their nests, and mounted a hunter with smooth short hair, of a full height, and so fleet as to make captive the beasts of the forest;

47   Ready in turning, quick in pursuing, bold in advancing, firm in backing; and performing the whole with the strength and swiftness of a vast rock, which a torrent has pushed from its lofty base;

48   A bright bay steed, from whose polished back the trappings slide, as drops of rain glide hastily down the slippery marble.

49   Even in his weakest state he seems to boil while he runs; and the sound, which he makes in his rage, is like that of bubbling cauldron.

50   When other horses, that swim through the air, are languid and kick the dust, he rushes on like a flood, and strikes the hard earth with a firm hoof.

51   He makes the light youth slide from his seat, and violently shakes the skirts of a heavier and more stubborn rider;

52   Rapid as the pierced wood in the hands of a playful child, which he whirls quickly round with a well-fastened cord.

53   He has the loins of an antelope, and the thighs of an ostrich; he trots like a wolf, and gallops like a young fox.

54   Firm are his haunches; and, when his hinder parts are turned towards you, he fills the space between his legs with a long thick tail, which touches not the ground, and inclines not to either side.

55   His back, when he stands in his stall, resembles the smooth stone on which perfumes are mixed for a bride, or the seeds of coloquinteda are bruised.

56   The blood of the swift game, which remains on his neck, is like the crimson juice of *Henna* on grey flowing locks.

57   He bears us speedily to a herd of wild cattle, in which the heifers are fair as the virgins in black trailing robes, who dance round *the idol* DEWAAR:

58   They turn their backs, and appear like the variegated shells of YEMEN on the neck of a youth distinguished in his tribe for a multitude of noble kinsmen.

59   He soon brings us up to the foremost of the beasts, and leaves the rest far behind; nor has the herd time to disperse itself.

60   He runs from wild bulls to wild heifers, and overpowers them in a single heat, without being bathed, or even moistened, with sweat.

61   Then the busy cook dresses the game, roasting part,

baking part on hot stones, and quickly boiling the rest in a vessel of iron.

62   In the evening we depart; and, when the beholder's eye ascends to the head of my hunter, and then descends to his feet, it is unable at once to take in all his beauties.

63   His trappings and girths are still upon him: he stands erect before me, not yet loosed for pasture.

64   O friend, seest thou the lightning, whose flashes resemble the quick glance of two hands amid clouds raised above clouds?

65   The fire of it gleams like the lamps of a hermit, when the oil, poured on them, shakes the cord by which they are suspended.

66   I sit gazing at it, while my companions stand between DAARIDGE and ODHAIB; but far distant is the cloud on which my eyes are fixed.

67   Its right side seems to pour its rain on *the hills of* KATAN, and its left on *the mountains of* SITAAR and YADBUL.

68   It continues to discharge its waters over COTAIFA till the rushing torrent lays prostrate the groves of *Canahbel-*trees.

69   It passes over *mount* KENAAN, which it deluges in its course, and forces the wild goats to descend from every cliff.

70   On *mount* TAIMA it leaves not one trunk of a palm-tree, nor a single edifice, which is not built with well-cemented stone.

71   *Mount* TEBEIR stands in the heights of the flood like a venerable chief wrapped in a striped mantle.

72   The summit of MOGAIMIR, covered with the rubbish which the torrent has rolled down, looks in the morning like the top of a spindle encircled with wool.

73   The cloud unloads its freight on the desert of GHABEIT, like a merchant of YEMEN alighting with his bales of rich apparel.

74   The small birds of the valley warble at day-break, as if they had taken their early draught of generous wine mixed with spice.

75   The beasts of the wood, drowned in the floods of night, float, like the roots of wild onions, at the distant edge of the lake.

Lines from the Arabic[1]

(To Lady Jones)

While sad suspense and chill delay
Bereave my wounded soul of rest,
New hopes, new fears, from day to day,
By turns assail my lab'ring breast.

My heart, which ardent love consumes,
Throbs with each agonizing thought;
So flutters with entangled plumes,
The lark in wily meshes caught.

There she, with unavailing strain,
Pours thro' the night her warbled grief:
The gloom retires, but not her pain;
The dawn appears, but not relief.

Two younglings wait the parent bird,
Their thrilling sorrows to appease:
She comes—ah! no: the sound they heard
Was but a whisper of the breeze.

On Parent Knees[1]

On parent knees, a naked, a new-born child,
Weeping thou sat'st, while all around thee smil'd:
So live, that, sinking in thy last long sleep,
Calm thou may'st smile, when all around thee weep.

---

[1] [In 1783 Jones translated a small poem, which he polished into a minor ode expressing a lover's agonizing over his beloved, like a mother lark caught in meshes that prevent her return to her nestlings. Comparing his anguish over Lady Jones's recent illness with the trapped lark was in perfect harmony with bird imagery in Sanskrit and Arabic poetry. As three quatrains are in the Warren Hastings Papers, Jones may have presented them to the Governor-General (British Library Add. MS. 39898, f. 30). Teignmouth published Jones's poem in *Memoirs* (*Works*, 1 : 407-8). The haunting last stanza may have partly inspired Shelley and Keats toward their great odes to a skylark and a nightingale, respectively.]

[1] [This didactic quatrain, translated from the Persian about 1784, soon became a standard poem, brilliant for its compactness and universal theme, which accorded with Jones's moral perspective on life. Published in *The Asiatick Miscellany* 2 (Calcutta, 1786), it joined 'A Persian Song of Hafiz' and 'An Ode in Imitation of Alcaeus' as one of his three best poems. Jones's old friends like Mrs. Hester Thrale copied it into their diaries. His quatrain was later translated into German, French, Latin, and Arabic.]

The Concluding Sentence of Berkeley's Siris, Imitated[1]

Before thy mystic altar, heav'nly Truth;[2]
I kneel in manhood, as I knelt in youth:
Thus let me kneel, till this dull form decay,
And life's last shade be brighten'd by thy ray:
Then shall my soul, now lost in clouds below,
Soar without bound, without consuming glow.

Written After a Perusal of The Eighth Sermon of Barrow, 1786[1]

As meadows parch'd, brown groves, and withering flow'rs,
Imbibe the sparkling dew and genial show'rs;
As chill dark air inhales the morning beam,
As thirsty harts enjoy the gelid stream;
Thus to man's grateful soul from heav'n descend,
The mercies of his FATHER, LORD, and FRIEND.

---

[1] [Jones wrote a 'long poetical epistle' in 1786, apparently eulogizing Earl Spencer, his former pupil, in his six-line addendum (in *Letters*, 2: 709-10). The 'poetical epistle' is untraced, but the sestet was revised and written into Jones's copy of *Siris*, appearing in Alexander Chalmers's *The Works of the English Poets* (London, 1810), 18: 462, and elsewhere. The philosopher George Berkeley (1685-1753) is remembered for *Siris* (London, 1744) and *A Treatise Concerning the Principles of Human Knowledge* (Dublin, 1710).]

[2] [Jones may be associating the Hindu concept of 'heav'nly Truth' with Berkeley's Biblical phrase 'the altar of Truth'.]

[1] [Jones was a devout Christian, but his great intellectual powers and wide scholarship led him to appreciate Islam and Hinduism. Often turning to Christian divines to intensify his intuitive faith, he owned the works of Isaac Barrow (1630-77) and in 1786 composed this pious sestet on Barrow's idea of pantheism (see *Letters*, 2: 902). This aroused his deep interest in Hindu pantheism as expressed in the Vedas. The sestet appeared in Chalmers, ed., *The Works of the English Poets*, (London, 1810), 18: 462.]

The Enchanted Fruit; or, The Hindu Wife:[1]
An Antediluvian Tale Written in The Province of Bahar

'O lovely age,[2] by *Brahmens* fam'd
Pure *Setye Yug*[3] in *Sanscrit* nam'd!
Delightful! Not for cups of *gold*,
Or wives *a thousand centuries* old;
Or men, degenerate now and small,
Then *one and twenty cubits* tall:
Not that plump *cows* full udders bore,
And bowls with *holy curd*[4] ran o'er;
Not that, by Deities defended
Fish, Boar, Snake, Lion,[5] heav'n-descended,
Learn'd *Pendits*, now grown sticks and clods,
Redde fast the *Nagry of the Gods*[6]
And laymen, faithful to *Narayn*[7]
Believ'd in *Brahmā*'s mystick strain;[8]

---

[1] [This famous, sometimes too erudite verse-tale was composed in 1784, after pundits had told Jones the story of the *Sabhāparva* from the *Mahābhārata*, where all the five Pāṇḍava brothers and their wife-in-common Draupadi confess their worst sins of revenge, rage, intemperance, avarice, pride, and lust. In 287 couplets, the poem appeared in Francis Gladwin's *Asiatick Miscellany* 1 (Calcutta, 1785): 188-211. With this poem he inaugurated a tradition of verse-tales that was later followed by Byron and other Romantics. Jones's scholarly footnotes explaining allusions to Hinduism, Indian geography, and foods, together with the transliteration of many Sanskrit words, still sometimes present an obstacle to Western readers. The poem reveals a rich appreciation of the *Mahābhārata*, despite his brief residence in India to that point.]

[2] A parody on the Ode in *Tasso's Aminta*, beginning, *O bella età dell' oro!*

[3] The *Golden Age* of the *Hindus*.

[4] Called *Joghrāt*, the food of CRISHNA in his infancy and youth.

[5] The four first *Avatārs*, or *Incarnations* of the *Divine Spirit*.

[6] The *Sanscrit*, or *Sengscrit*, is written in letters so named.

[7] *Narayn* or *Nārāyan*, the *spirit* of GOD.

[8] The *Vayds*, or *Sacred Writings* of *Brahma*, called *Rig*, *Sām*, and *Yejar*: doubts have been raised concerning the authority of the *fourth*, or *At'herven, Vayd*.

Not that all Subjects spoke plain truth,
While *Rajas* cherish'd eld and youth,
No—yet delightful times! because
Nature then reign'd, and *Nature*'s *Laws;*
When females of the softest kind
Were unaffected, unconfin'd;
And this grand rule from none was hidden;[9]
WHAT PLEASETH, HATH NO LAW FORBIDDEN'.

Thus with a lyre in *India* strung,
Aminta's poet would have sung;
And thus too, in a modest way,
All virtuous males will sing or say:
But swarthy nymphs of *Hindustan*
Look deeper than short-sighted man,
And thus, in some poetick chime,
Would speak with reason, as with rhyme:
'O lovelier age, by *Brahmens* fam'd,
Gay *Dwāpar Yug*[10] in *Sanscrit* nam'd!
Delightful! though impure with *brass*
In many a green ill-scented mass;
Though husbands, but *sev'n* cubits high,
Must in *a thousand summers* die;
Though, in the lives of dwindled men,
Ten parts were Sin; Religion, *ten*;
Though *cows* would rarely fill the pail,
But made th' expected creambowl fail;
Though lazy *Pendits* ill could read
(No care of ours) their *Yejar Veid*;
Though *Rajas* look'd a little proud,
And *Ranies* rather spoke too loud;
Though *Gods*, display'd to mortal view
In mortal forms, were only *two*;
(Yet CRISHNA,[11] sweetest youth, was one,
Crishna, whose cheeks outblaz'd the sun)
Delightful, ne'ertheless! because
Not bound by vile unnatural laws,
Which curse this age from *Cāley*[12] nam'd,
By some base woman-hater fram'd.
Prepost'rous! that one biped vain
Should drag ten house-wives in his train,
And stuff them in a gaudy cage,
Slaves to weak lust or potent rage!

---

[9] 'Se piace, ei l'ice'. *Tasso.*

[10] The *Brazen Age*, or that in which Vice and Virtue were in *equal* proportion.

[11] The *Apollo* of *India*.

[12] The *Earthen Age*, or that of *Caly* or *Impurity*: this verse alludes to *Cāley*, The *Hecate* of the *Indians*.

Not such the *Dwāpar Yug*! oh then
ONE BUXOM DAME MIGHT WED FIVE MEN'.

True History, in solemn terms,
This Philosophick lore confirms;
For *India* once, as now cold *Tibet*,[13]
A groupe unusual might exhibit,
Of sev'ral husbands, free from strife,
Link'd fairly to a single wife!
Thus Botanists, with eyes acute
To see prolifick dust minute,
Taught by their learned northern *Brahmen*[14]
To class by *pistil* and by *stamen*,
Produce from nature's rich dominion
Flow'rs *Polyandrian Monogynian*,
Where embryon blossoms, fruits, and leaves
Twenty prepare, and ONE receives.

But, lest my word should nought avail,
Ye Fair, to no unholy tale
Attend.[15]*Five thousand* years[16] ago,
As annals in *Benares* show,
When *Pāndu* chiefs with Curus fought,[17]
And each the throne imperial sought,
Five brothers of the regal line
Blaz'd high with qualities divine.
The first a prince without his peer,
Just, pious, lib'ral *Yudhishteir*;[18]
Then *Erjun*, to the base a rod,
An Hero favour'd by a *God*;[19]

Bheima, like mountain-leopard strong,
Unrival'd in th' embattled throng,
Bold *Nacul*, fir'd by noble shame
To emulate fraternal fame;
And *Sehdeo*, flush'd with manly grace,
Bright virtue dawning in his face:
To these a dame devoid of care,
Blythe *Draupady*, the debonair,
Renown'd for beauty, and for wit,
In wedlock's pleasing chain was knit.[20]

It fortun'd, at an idle hour,
This five-mal'd single-femal'd flow'r
One balmy morn of fruitful May
Through vales and meadows took its way.
A low thatch'd mansion met their eye
In trees umbrageous bosom'd high;
Near it (no sight, young maids, for you)
A temple rose to *Mahadew*.[21]
A thorny hedge and reedy gate
Enclos'd the garden's homely state;
Plain in its neatness: thither wend
The princes and their lovely friend.
Light-pinion'd gales, to charm the sense,
Their odorif'rous breath dispense;
From *Bēla's*[22] pearl'd, or pointed, bloom,
And *Mālty* rich, they steal perfume:
There honey-scented *Singarhār*,
And *Jūhy*, like a rising star,
Strong *Chempā*, darted by *Cāmdew*,
And Mulsery of paler hue,
Cayora,[23] which the *Ranies* wear
In tangles of their silken hair,
Round[24] *Bābul*-flow'rs, and *Gulachein*
Dyed like the shell of Beauty's Queen,
Sweet *Mindy*[25] press'd for crimson stains,

[13] See the accounts published in the *Philosophical Transactions* from the papers of Mr. *Bogle*.

[14] *Linnaeus*.

[15] The story is told by the *Jesuit* BOUCHET, in his Letter to HUET, Bishop of *Avranches*.

[16] A round number is chosen; but the *Caly Yug*, a little before which *Crishna* disappeared from this world, began *four thousand, eight hundred*, and *eighty-four* years ago, that is, according to our Chronologists, *seven hudred* and *forty-seven* before the flood; and by the calculation of M. Bailly, but *four hundred* and *fifty-four* after the foundation of the *Indian* empire.

[17] This war, which *Crishna* fomented in favour of the *Pandu Prince*, *Yudhishtir*, supplied *Vyās* with the subject of his noble Epick Poem, *Mahābhārat*.

[18] This word is commonly pronounced with a strong accent on the last letter, but the preceding vowel is short in *Sengscrit*. The prince is called on the coast *Dherme Rāj*, or Chief Magistrate.

[19] The Geita, containing Instructions to Erjun, was composed by Crishna, who peculiarly distinguished him.

[20] *Yudhishtir* and *Draupady*, called *Drobada* by *M. Sonnerat*, are deified on the Coast; and their feast, of which that writer exhibits an engraving, is named the *Procession of Fire*, because she passed *every year* from one of her *five* husbands to another, after a solemn purification by that element. In the *Bhāshā* language, her name is written, DRŌPTY.

[21] The *Indian* JUPITER.

[22] The varieties of *Bela*, and the *three* flowers next mentioned, are beautiful species of *Jasmin*.

[23] The *Indian* Spikenard.

[24] The *Mimosa*, or true *Acacia*, that produces the *Arabian* Gum.

[25] Called *Alhhinná* by the *Arabs*.

And sacred *Tulsy*[26] pride of plains,
With *Sēwty*, small unblushing rose,
Their odours mix, their tints disclose,
And, as a gemm'd tiara, bright,
Paint the fresh branches with delight.

One tree above all others tower'd
With shrubs and saplings close imbower'd,
For every blooming child of Spring
Paid homage to the verdant King:
Aloft a solitary fruit,
Full sixty cubits from the root,
Kiss'd by the breeze, luxuriant hung,
Soft chrysolite with em'ralds strung.
'Try we', said *Erjun* indiscreet,
'If yon proud fruit be sharp or sweet;
My shaft its parent stalk shall wound:
Receive it, ere it reach the ground'.

Swift as his word, an arrow flew:
The dropping prize besprent with dew
The brothers, in contention gay,
Catch, and on gather'd herbage lay.

That instant scarlet lightnings flash,
And *Jemna*'s waves her borders lash,
Crishna from *Swerga*'s[27] height descends,
Observant of his mortal friends:
Not such, as in his earliest years,
Among his wanton cowherd peers,
In *Gocul* or *Brindāben*'s[28] glades,
He sported with the dairy-maids;
Or, having pip'd and danc'd enough,
Clos'd the brisk night with *blindman's-buff*;[29]
(List, antiquaries, and record
This pastime of the *Gopia*'s Lord)[30]
But radiant with ethereal fire:
Nared[31] alone could bards inspire
In lofty *Slokes*[32] his mien to trace,
And unimaginable grace.

With human voice, in human form,
He mildy spake, and hush'd the storm:
'O mortals, ever prone to ill!
Too rashly *Erjun* prov'd his skill.
Yon fruit a pious *Muny*[33] owns,
Assistant of our heav'nly thrones.
The golden pulp, each month renew'd,
Supplies him with ambrosial food.
Should he the daring archer curse,
Not *Mentra*[34] deep, nor magick verse,
Your gorgeous palaces could save
From flames, your embers, from the wave'.[35]

The princes, whom th' immod'rate blaze
Forbids their sightless eyes to raise,
With doubled hands his aid implore,
And vow submission to his lore.
'One remedy, and simply one,
Or take', said he, 'or be undone:
Let each his crimes or faults confess,
The greatest name, omit the less;
Your actions, words, e'en thoughts reveal:
No part must *Draupady* conceal:
So shall the fruit, as each applies
The faithful charm, *ten cubits* rise;
Till, if the dame be frank and true,
It join the branch, where late it grew'.
He smil'd, and shed a transient gleam;
Then vanish'd, like a morning dream.

Now, long entranc'd, each waking brother
Star'd with amazement on another,
Their consort's cheek forgot its glow,
And pearly tears began to flow;
When *Yudhishteir*, high-gifted man,
His plain confession thus began.

'Inconstant fortune's wreathed smiles,
Duryōdhen's rage, *Duryōdhen*'s wiles,[36]

---

[26] Of the kind called *Ocymum*.

[27] The heaven of *Indra*, or the Empyreum.

[28] In the district of *Mat' hura*, not far from *Agra*.

[29] This is told in the *Bhāgawat*.

[30] GOPY NAT'H, a title of *Crishna*, corresponding with *Nymphagetes*, an epithet of *Neptune*.

[31] [See Jones's Argument to 'A Hymn to Sereswaty' relative to Narad's place in Hindu mythology.]

[32] Tetrasticks without rhyme.

[33] An inspired Writer: *twenty* are so called.

[34] Incantation.

[35] This will receive illustration from a passage in the *Ramayen*: 'Even he, who cannot be slain by the ponderous arms of *Indra*, nor by those of *Cāly*, nor by the terrible *Checra* (or *Discus*), of *Vishnu*, shall be destroyed, if a *Brahmen* execrate him, *as if he were consumed by fire*'.

[36] [In the *Mahābhārata*, Duryōdhan was the chief Kaurava prince, who was hostile to the five Pāṇḍava brothers, his cousins.]

Fires rais'd for this devoted head,
E'en poison for my brethren spread,
My wand'rings through wild scenes of wo,
And persecuted life, you know.
Rude wassailers defil'd my halls,
And riot shook my palace-walls,
My treasures wasted. This and more
With resignation calm I bore;
But, when the late-descending god
Gave all I wish'd with soothing nod,
When, by his counsel and his aid,
Our banners danc'd, our clarions bray'd
(Be this my greatest crime confess'd),
Revenge sate ruler in my breast:
I panted for the tug of arms,
For skirmish hot, for fierce alarms;
Then had my shaft *Duryōdhen* rent,
This heart had glow'd with sweet content'.

He ceas'd: the living gold upsprung,
And from the bank *ten* cubits hung.

Embolden'd by this fair success,
Next *Erjun* hasten'd to confess:
'When I with *Aswatthāma* fought;[37]
My noose the fell assassin caught;
My spear transfix'd him to the ground:
His giant limbs firm cordage bound:
His holy thread extorted awe
Spar'd by religion and by law;
But, when his murd'rous I view'd
In blameless kindred gore imbued,
Fury my boiling bosom sway'd,
And *Rage* unsheath'd my willing blade:
Then, had not *Crishna*'s arm divine
With gentle touch suspended mine,
This hand a *Brahmen* had destroy'd,
And vultures with his blood been cloy'd'.

The fruit, forgiving *Erjun*'s dart,
Ten cubits rose with eager start.

Flush'd with some tints of honest shame,
Bheima to his confession came:
'Twas at a feast for battles won
From *Dhriterāshtra*'s guileful son,

High on the board in vases pil'd
All vegetable nature smil'd:
Proud *Anaras*[38] his beauties told,
His verdant crown and studs of gold,
To *Dallim*,[39] whose soft rubies laugh'd
Bursting with juice, that gods have quaff'd;
Ripe *Kellas*[40] here in heaps were seen,
Kellas, the golden and the green,
With *Ambas*[41] priz'd on distant coasts,
Whose birth the fertile *Ganga* boasts:
(Some gleam like silver, some outshine
Wrought ingots from *Besoara*'s mine):
Corindas there, too sharp alone,
With honey mix'd, impurpled shone;
Talsans[42] his liquid crystal spread
Pluck'd from high *Tara*'s tufted head;
Round *Jamas*[43] delicate as fair,
Like rose-water perfum'd the air;
Bright salvers high-rais'd *Comlas*[44] held
Like topazes, which *Amrit*[45] swell'd;
While some delicious *Attas*[46] bore,
And *Catels*[47] warm, a sugar'd store;
Others with *Bēla*'s grains were heap'd,
And mild *Papayas* honey-steep'd;
Or sweet *Ajeūrs*[48] the red and pale,
Sweet to the taste and in the gale.
Here mark'd we purest basons fraught
With sacred cream and fam'd *Joghrāt*;
Nor saw we not rich bowls contain
The *Chawls*'s[49] light nutritious grain,
Some virgin-like in native pride,
And some with strong *Haldea*[50] dyed,
Some tasteful to dull palates made
If *Merich*[51] lend his fervent aid,
Or *Langa*[52] shap'd like od'rous nails,

38 Ananas.
39 Pomegranate.
40 Plantains.
41 Mangos.
42 Palmyra-fruit.
43 Rose-apples.
44 Oranges.
45 The Hindu Nectar.
46 Custard-apples.
47 Jaik-fruit.
48 Guayavas.
49 Rice.
50 Turmeric.
51 Indian Pepper.
52 Cloves.

37 [The son of Droṇāchārya, who was the teacher of both the
Pāṇḍava and the Kaurava princes. Jones is dramatizing a famous battle in the *Mahābhārata*.]

Whose scent o'er groves of spice prevails,
Or *Adda*[53] breathing gentle heat,
Or *Joutery*[54] both warm and sweet.
Supiary[55] next (in *Pāna*[56] chew'd,
And *Catha*[57] with strong pow'rs endued,
Mix'd with *Elachy*'s[58] glowing seeds,
Which some remoter climate breeds),
Near *Jeïfel*[59] sate, like *Jeifel* fram'd
Though not for equal fragrance nam'd:
Last, *Nāryal*,[60] who all ranks esteem,
Pour'd in full cups his dulcet stream:
Long I survey'd the doubtful board
With each high delicacy stor'd;
Then freely gratified my soul,
From many a dish, and many a bowl,
Till health was lavish'd, as my time:
Intemp'rance was my fatal crime'.

Uprose the fruit; and now *mid-way*
Suspended shone like blazing day.

Nacal then spoke: (a blush o'erspread
His cheeks, and conscious droop'd his head):
'Before *Duryōdhen*, ruthless king,
Taught his fierce darts in air to sing,
With bright-arm'd ranks, by *Crishna* sent,
Elate from *Indraprest*[61] I went
Through *Eastern* realms; and vanquish'd all
From rough *Almōra* to *Nipāl*.
Where ev'ry mansion, new or old,
Flam'd with Barbarick gems and gold.
Here shone with pride the regal stores
On iv'ry roofs, and cedrine floors;
There diadems of price unknown
Blaz'd with each all-attracting stone;
Firm diamonds, like fix'd honour true,
Some pink, and some of yellow hue,
Some black, yet not the less esteem'd;
The rest like tranquil *Jemna* gleam'd,
When in her bed the *Gopia* lave

Betray'd by the pellucid wave.
Like raging fire the ruby glow'd,
Or soft, but radiant, water show'd;
Pure amethysts, in richest ore
Oft found, a purple vesture wore;
Sapphirs, like yon ethereal plain;
Em'ralds, like *Peipel*[62] fresh with rain;
Gay topazes, translucent gold;
Pale chrysolites of softer mould;
Fam'd beryls, like the surge marine,
Light-azure mix'd with modest green;
Refracted ev'ry varying dye,
Bright as yon bow, that girds the sky.
Here opals, which all hues unite,
Display'd their many-tinctur'd light,
With turcoises divinely blue
(Though doubts arise, where first they grew,
Whether chaste elephantine bone
By min'rals ting'd, or native stone),
And pearls unblemish'd, such as deck
Bhavāny's[63] wrist or *Lecshmy*'s[64] neck.
Each castle ras'd, each city storm'd,
Vast loads of pillag'd wealth I form'd,
Not for my coffers; though they bore,
As you decreed, my lot and more.
Too pleas'd the brilliant heap I stor'd,
Too charming seem'd the guarded hoard:
An odious vice this heart assail'd;
Base *Av'rice* for a time prevail'd'.

Th' enchanted orb *ten* cubits flew,
Strait as the shaft, which *Erjun* drew.

Sehdio, with youthful ardour bold,
Thus, penitent, his failings told;
'From clouds, by folly rais'd, these eyes
Experience clear'd, and made me wise;
For, when the crash of battle roar'd,
When death rain'd blood from spear and sword,
When, in the tempest of alarms,
Horse roll'd on horse, arms clash'd with arms,
Such acts I saw by others done,
Such perils brav'd, such trophies won,
That, while my patriot bosom glow'd,
Though some faint skill, some strength I show'd,
And, no dull gazer on the field,

[53] Ginger.
[54] Mace.
[55] Areca-nut.
[56] Betel-leaf.
[57] What we call Japan-earth.
[58] Cardamums.
[59] Nutmeg.
[60] Coconut.
[61] DEHLY [New Delhi].

[62] A sacred tree like an *Aspin*.
[63] The *Indian* VENUS.
[64] The *Indian* CERES.

This hero slew, that forc'd to yield,
Yet, meek humility, to thee,
When *Erjun* fought, low sank my knee:
But, ere the din of war began,
When black'ning cheeks just mark'd the man,
Myself invincible I deem'd,
And great, without a rival, seem'd.
Whene'er I sought the sportful plain,
No youth of all the martial train
With arm so strong or eye so true
The *Checra's*[65] pointed circle threw;
None, when the polish'd cane we bent,
So far the light-wing'd arrow sent;
None from the broad elastick reed,
Like me, gave *Agnyastra*[66] speed,
Or spread its flames with nicer art
In many an unextinguish'd dart;
Or, when in imitated fight
We sported till departing light,
None saw me to the ring advance
With falchion keen or quiv'ring lance,
Whose force my rooted seat could shake,
Or on my steed impression make:
No charioteer, no racer fleet
O'ertook my wheels or rapid feet.
Next, when the woody heights we sought,
With madd'ning elephants I fought:
In vain their high-priz'd tusks they gnash'd;
Their trunked heads my *Geda*[67] mash'd.
No buffalo, with phrensy strong,
Could bear my clatt'ring thunder long:
No pard or tiger, from the wood
Reluctant brought, this arm withstood.
Pride in my heart his mansion fix'd,
And with pure drops black poison mix'd'.

Swift rose the fruit, exalted now
Ten cubits from his natal bough.

Fair *Draupady*, with soft delay,
Then spake: 'Heav'n's mandate I obey;
Though nought, essential to be known,
Has heav'n to learn, or I to own.
When scarce a damsel, scarce a child,
In early bloom your handmaid smil'd,
Love of the World her fancy mov'd,

Vain pageantry her heart approv'd:
Her form, she thought, and lovely mien,
All must admire, when all had seen:
A thirst of pleasure and of praise
(With shame I speak) engross'd my days;
Nor were my night-thoughts, I confess,
Free from solicitude for dress;
How best to bind my flowing hair
With art, yet with an artless air
(My hair, like musk in scent and hue;
Oh! blacker far and sweeter too);
In what nice braid or glossy curl
To fix a diamond or a pearl,
And where to smooth the love-spread toils
With nard or jasmin's fragrant oils;
How to adjust the golden *Teic*,[68]
And most adorn my forehead sleek;
What *Condals*[69] should emblaze my ears,
Like *Seita's* waves[70] or *Seita's* tears:[71]
How elegantly to dispose
Bright circlets for my well-form'd nose;
With strings of rubies how to deck,
Or em'rald rows, my stately neck,
While some that ebon tow'r embrac'd,
Some pendent sought my slender waist;
How next my purfled veil to chuse
From silken stores of varied hues;
Which would attract the roving view,
Pink, violet, purple, orange, blue;
The loveliest mantle to select,
Or unembellish'd or bedeck'd;
And how my twisted scarf to place
With most inimitable grace;
(Too thin its warp, too fine its woof,
For eyes of males not beauty-proof);
What skirts the mantle best would suit,
Ornate with stars or tissued fruit,
The flow'r-embroider'd or the plain
With silver or with golden vein;
The *Chury*[72] bright, which gayly shows
Fair objects, aptly to compose;

---

[68] Properly *Teica*, an ornament of gold, placed above the nose.

[69] Pendents.

[70] SEITĀ CUND, or the *Pool* of *Seitā*, the wife of RAM, is the name given to the wonderful spring at *Mengeir*, with boiling water of exquisite clearness and purity.

[71] Her tears, when she was made captive by the giant *Rawān*.

[72] A small mirror worn in a ring.

---

[65] A radiated metalline ring, used as a missile weapon.

[66] Fire-arms, or rockets, early known in *India*.

[67] A mace, or club.

How each smooth arm and each soft wrist
By richest *Cosecs*[73] might be kiss'd;
While some, my taper ankles round,
With sunny radiance ting'd the ground.
O waste of many a precious hour!
O *Vanity*, how vast thy pow'r!'

Cubits twice four th' ambrosial flew,
Still from its branch disjoin'd by *two*.

Each husband now, with wild surprise,
His compeers and his consort eyes;
When *Yudhishteir*: 'Thy female breast
Some faults, perfidious, hath suppress'd.
Oh! give the close-lock'd secret room,
Unfold its bud, expand its bloom;
Lest, sinking with our crumbled halls,
We see red flames devour their walls'.
Abash'd, yet with a decent pride,
Firm *Draupady* the fact denied;
Till, through an arched alley green,
The limit of that sacred scene,
She saw the dreaded *Muny* go
With steps majestically slow;
Then said: (a stifled sigh she stole,
And show'd the conflict of her soul
By broken speech and flutt'ring heart)
'One trifle more I must impart:
A *Brahmen* learn'd, of pure intent
And look demure, one morn you sent,
With me, from *Sanscrit* old, to read
Each high *Purān*[74] each holy *Veid*.
His thread, which *Brehmā*'s lineage show'd,
O'er his left shoulder graceful flow'd;
Of *Crishna* and his nymphs he redde,
How with nine maids the dance he led;
How they ador'd, and he repaid
Their homage in the sylvan shade.
While this gay tale my spirits cheer'd,
So keen the *Pendit*'s eyes appear'd,
So sweet his voice—a blameless fire
This bosom could not but inspire.
Bright as a God he seem'd to stand:
The rev'rend volume left his hand,
With mine he press'd'—With deep despair
Brothers on brothers wildly stare:
From *Erjun* flew a wrathful glance;

[73] Bracelets.
[74] A Mythological and Historical Poem.

Tow'rd them they saw their dread advance;
Then, trembling, breathless, pale with fear,
'Hear', said the matron, 'calmly hear!
'By *Tulsy*'s leaf the truth I speak—
The *Brahmen* ONLY KISS'D MY CHEEK'.

Strait its full height the wonder rose,
Glad with its native branch to close.

Now to the walk approach'd the Sage
Exulting in his verdant age:
His hands, that touch'd his front, express'd
Due rev'rence to each princely guest,
Whom to his rural board he led
In simple delicacy spread,
With curds their palates to regale,
And cream-cups from the *Gopia*'s pail.

Could you, ye Fair, like this black wife,
Restore us to primeval life,
And bid that apple, pluck'd for *Eve*
By him, who might all wives deceive,
Hang from its parent bough once more
Divine and perfect, as before,
Would you confess your little faults?
(Great ones were never in your thoughts);
Would you the secret wish unfold,
Or in your heart's full casket hold?
Would you disclose your inmost mind,
And speak plain truth, to bless mankind?

'What!' said the Guardian of our realm,
With waving crest and fiery helm,
'What! are the fair, whose heav'nly smiles
Rain glory through my cherish'd isles,
Are they less virtuous or less true
Then *Indian* dames of sooty hue?
No, by these arms.[75] The cold surmise
And doubt injurious vainly rise.
Yet dares a bard, who better knows,
This point distrustfully propose;

[75] [Rejecting the colonial doctrine that Asians were inferior people who should be ruled by civilized Europeans so as perhaps eventually to rescue the supposedly degraded Hindus and other non-Christians from their uncivilized ways, Jones was already attacking the doctrine. His transfer of Sanskrit knowledge to Europe completely overturned that ethnocentric image and initiated a new Renaissance both in Europe and India.]

Vain fabler now! though oft before
His harp has cheer'd my sounding shore'.

With brow austere the martial maid
Spoke, and majestick trod the glade:
To that fell cave her course she held,
Where *Scandal*, bane of mortals, dwell'd.
Outstretch'd on filth the pest she found,
Black fetid venom streaming round:
A gloomy light just serv'd to show
The darkness of the den below.
Britannia with resistless might
Soon dragg'd him from his darling night:
The snakes, that o'er his body curl'd,
And flung his poison through the world,
Confounded with the flash of day,
Hiss'd horribly a hellish lay.
His eyes with flames and blood suffus'd,
Long to th' ethereal beam unus'd,
Fierce in their gory sockets roll'd;
And desperation made him bold:
Pleas'd with the thought of human woes,
On scaly dragon feet he rose.
Thus, when *Asūrs* with impious rage,
Durst horrid war with *Dēvta*'s wage,
And darted many a burning mass
E'en on the brow of gemm'd *Cailās*,
High o'er the rest, on serpents rear'd,
The grisly king of *Deits* appear'd.

The nymph beheld the fiend advance,
And couch'd her far-extending lance:
Dire drops he threw; th' infernal tide
Her helm and silver hauberk dyed:
Her moonlike shield before her hung;
The monster struck, the monster stung:
Her spear with many a griding wound
Fast nail'd him to the groaning ground.
The wretch, from juster vengeance free,
Immortal born by heav'n's decree,
With chains of adamant secur'd,
Deep in cold gloom she left immur'd.

Now reign at will, victorious Fair,
In *British*, or in *Indian*, air!
Still with each envying flow'r adorn
Your tresses radiant as the morn;
Still let each *Asiatick* dye
Rich tints for your gay robes supply;
Still through the dance's laby'rinth float,

And swell the sweetly-lengthen'd note;
Still, on proud steeds or glitt'ring cars,
Rise on the course like beamy stars;
And, when charm'd circles round you close
Of rhyming bards and smiling beaux,
Whilst all with eager looks contend
Their wit or worth to recommend,
Still let your mild, yet piercing, eyes
Impartially adjudge the prize.

A Hymn to Camdeo[1]

The Argument

The *Hindū* God, to whom the following poem is addressed, appears evidently the same with the *Grecian* EROS and the *Roman* CUPIDO; but the *Indian* description of his person and arms, his family, attendants, and attributes, has new and peculiar beauties.

According to the mythology of *Hindustān*, he was the son of MAYA, or the general *attracting* power, and married to RETTY or *Affection*; and his bosom friend is BESSENT or *Spring*: he is represented as a beautiful youth, sometimes conversing with his mother and consort in the midst of his gardens and temples; sometimes riding by moonlight on a parrot or lory, and attended by dancing-girls or nymphs, the foremost of whom bears his colours, which are a *fish* on a red ground. His favourite place of resort is a large tract of country round AGRA, and principally the plains of *Matra*, where KRISHEN also and the nine GOPIA, who are clearly the *Apollo* and *Muses* of the *Greeks*, usually spend the night with musick and dance. His bow of sugar-cane or flowers, with a string of bees, and his *five* arrows, each pointed with an *Indian* blossom of a heating quality, are allegories equally new and beautiful. He has at least twenty-three names, most of which are introduced in the hymn: that of *Cām* or *Cāma* signifies *desire*, a sense which it also bears in ancient and modern *Persian*; and it is possible, that the words *Dipuc* and *Cupid*, which have the same signification, may have the same origin; since we know, that the old *Hetruscans*, from whom great part of the *Roman* language and religion was derived, and whose system had a near affinity with that of the *Persians* and *Indians*, used to write their lines alternately forwards and backwards, as furrows are made by the plough; and, though the two last letters of *Cupido* may be only the grammatical termination, as in *libido* and *capedo*, yet the primary root of *cupio* is contained in the three first letters. The seventh stanza alludes to the bold attempt of this deity to wound the great God *Mahadeo*, for which he was punished by a flame consuming his corporeal nature and reducing him to a mental essence; and hence his chief dominion is over the *minds* of mortals, or such deities as he is permitted to subdue.

[1] [Jones correctly considered this—his first hymn on a Hindu divinity—to be the first 'correct specimen of Hindu mythology that has appeared; it is certainly new and quite original, except the form of the stanza, which is Milton's' (*Letters* 2 : 641). He wanted to introduce the West to Hinduism. Besides its accuracy in portraying the Indian love-god in English verse, the hymn dramatizes Kama's bow of sugar cane or flowers, with its string of bees and its arrows tipped with blossoms possessing the magical power of provoking amorous feelings. Its high quality was recognized by the periodicals that quickly reprinted it—thus the *Annual Register* for 1784-85, pp. 137-38.]

The Hymn

What potent God from *Agra*'s orient bow'rs
Floats thro' the lucid air, whilst living flow'rs
With sunny twine the vocal arbours wreathe,
And gales enamour'd heav'nly fragrance breathe?
Hail, pow'r unknown! for at thy beck
Vales and groves their bosoms deck,
And ev'ry laughing blossom dresses
With gems of dew his musky tresses.
I feel, I feel thy genial flame divine,
And hallow thee and kiss thy shrine.

'Knowst thou not me?' Celestial sounds I hear!
'Knowst thou not me?' Ah, spare a mortal ear!
'Behold'—My swimming eyes entranc'd I raise,
But oh! they shrink before th' excessive blaze.
Yes, son of *Maya*, yes, I know
Thy bloomy shafts and cany bow,
Cheeks with youthful glory beaming,
Locks in braids ethereal streaming,
Thy scaly standard, thy mysterious arms,
And all thy pains and all thy charms.

God of each lovely sight, each lovely sound,
Soul-kindling, world-inflaming, star-ycrown'd,
Eternal *Cāma*! Or doth *Smara* bright,
Or proud *Ananga* give thee more delight?
Whate'er thy seat, whate'er thy name,
Seas, earth, and air, thy reign proclaim;
Wreathy smiles and roseate pleasures
Are thy richest, sweetest treasures.
All animals to thee their tribute bring,
And hail thee universal king.

Thy consort mild, *Affection* ever true,
Graces thy side, her vest of glowing hue,
And in her train twelve blooming girls advance,
Touch golden strings and knit the mirthful dance.
Thy dreaded implements they bear,
And wave them in the scented air,
Each with pearls her neck adorning,
Brighter than the tears of morning.
Thy crimson ensign, which before them flies,
Decks with new stars the sapphire skies.

God of the flow'ry shafts and flow'ry bow,
Delight of all above and all below!
Thy lov'd companion, constant from his birth,
In heav'n clep'd *Bessent*, and gay *Spring* on earth,
Weaves thy green robe and flaunting bow'rs,
And from thy clouds draws balmy show'rs,
He with fresh arrows fills thy quiver,
(Sweet the gift and sweet the giver!)
And bids the many-plumed warbling throng
Burst the pent blossoms with their song.

He bends the luscious cane, and twists the string
With bees, how sweet! but ah, how keen their sting!
He with five flow'rets tips thy ruthless darts,
Which thro' five senses pierce enraptur'd hearts:
Strong *Chumpa*, rich in od'rous gold,
Warm *Amer*, nurs'd in heav'nly mould,
Dry *Nagkeser* in silver smiling,
Hot *Kiticum* our sense beguiling,
And last, to kindle fierce the scorching flame,
Loveshaft, which Gods bright *Bela* name.[2]

Can men resist thy pow'r,[3] when *Krishen* yields,
Krishen, who still in *Matra*'s holy fields
Tunes harps immortal, and to strains divine
Dances by moonlight with the *Gopia* nine?
But, when thy daring arm untam'd
At *Mahadeo* a loveshaft aim'd,
Heav'n shook, and, smit with stony wonder,
Told his deep dread in bursts of thunder,
Whilst on thy beauteous limbs an azure fire
Blaz'd forth, which never must expire.

O thou for ages born, yet ever young,
For ages may thy *Bramin*'s lay be sung!
And, when thy lory spreads his em'rald wings,
To waft thee high above the tow'rs of kings,
Whilst o'er thy throne the moon's pale light
Pours her soft radiance thro' the night,

---

[2] [The names of five Indian flowers.]
[3] Jones's predominant idea in the hymn is the irresistible power of love, something which Shelley recognized while referring to Jones's hymn in his letter to T.J. Hogg. See *The Letters of Percy Bysshe Shelley*, ed. F.L. Jones (Oxford, 1964), 1:112, See also John Holloway, *Widening Horizons in English Verse* (London, 1966), 6, 63-67; Pachori, 'Tennyson's Early Poems and Their Hindu Imagery', *Literature East & West* 9 (1978-79): 132-38; and Jones's 'Damsels of Cardigan', for its refrain: 'Love can alone make it blissful to live'.]

And to each floating cloud discovers
The haunts of blest or joyless lovers,
Thy mildest influence to thy bard impart,
To warm, but not consume, his heart.

A Hymn to Nārāyena[1]

The Argument

A complete introduction to the following Ode would be
no less than a full comment on the VAYDS and PURĀNS of
the HINDUS, the remains of *Egyptian* and *Persian* Theology,
and the tenets of the *Ionick* and *Italick* Schools; but
this is not the place for so vast a disquisition. It will be
sufficient here to premise, that the inextricable difficulties
attending the *vulgar notion of material substances*,
concerning which

'We know this only, that we nothing know',

induced many of the wisest among the Ancients, and
some of the most enlightened among the Moderns, to
believe, that the whole Creation was rather an *energy* than
a *work*, by which the Infinite Being, who is present at all
times in all places, exhibits to the minds of his creatures
a set of perceptions, like a wonderful picture or piece of
musick, always varied, yet always uniform; so that all
bodies and their qualities exist, indeed, to every wise and
useful purpose, but exist only as far as they are *perceived*;
a theory no less pious than sublime, and as different from
any principle of Atheism, as the brightest sunshine differs
from the blackest midnight. This *illusive operation* of the
Deity the *Hindu* philosophers call MĀYĀ, or *Deception*;
and the word occurs in this sense more than once in the
commentary on the *Rig Vayd*, by the great VASISHTHA, of
which Mr. HALHED has given us an admirable specimen.
The *first* stanza of the Hymn represents the sublimest

---

[1] [The hymn was composed in 1785, probably while Jones
was waiting to begin a courtroom hearing in Calcutta. It first
appeared in *The Asiatick Miscellany* (I) and is his most poetical
and profound explanation of Hinduism, as is shown in his
Argument. Emerson possibly borrowed from this hymn the
Vedantic concept of Brahman, the Supreme Reality, for his
cryptic poem 'Brahma', as Shelley may also have used the
hymn for his 'Hymn to Intellectual Beauty'. See Pachori,
'Shelley's "Hymn to Intellectual Beauty" and Sir William
Jones', *Comparatist* 10 (1987): 54-63. Jones probably derived
his moving description of Brahma, the Creator, from the
Mānava-Dharmaśāstra. The hymn contains Miltonic echoes.]

attributes of the Supreme Being, and the three forms, in which they most clearly appear to us, *Power*, *Wisdom*, and *Goodness*, or, in the language of ORPHEUS and his disciples, *Love*: the *second* comprises the *Indian* and *Egyptian* doctrine of the Divine Essence and Archetypal *Ideas*; for a distinct account of which the reader must be referred to a noble description in the sixth book of PLATO's *Republick*; and the fine explanation of that passage in an elegant discourse by the author of CYRUS, from whose learned work a hint has been borrowed for the conclusion of this piece. The *third* and *fourth* are taken from the Institutes of MENU, and the eighteenth *Puran* of VYĀSĀ, entitled *Srey Bhagawat*, part of which has been translated into *Persian*, not without elegance, but rather too paraphrastically.[2] From BREHME, or the *Great Being*, in the *neuter* gender, is formed BREHMĀ, in the *masculine*; and the second word is appropriated to the *creative power* of the Divinity.

The spirit of GOD, called NĀRĀYENA, or *moving on the water*, has a multiplicity of other epithets in *Sanscrit*, the principal of which are introduced, expressly or by allusion, in the *fifth* stanza; and two of them contain the names of the *evil beings*, who are feigned to have sprung from the ears of VISHNU: for thus the divine spirit is entitled, when considered as the *preserving power*: the *sixth* ascribes the perception of *secondary* qualities by our *senses* to the immediate influence of MĀYĀ; and the *seventh* imputes to her operation the *primary* qualities of *extension* and *solidity*.

The Hymn

SPIRIT of Spirits, who, through ev'ry part
Of space expanded and of endless time,
Beyond the stretch of lab'ring thought sublime,
Badst uproar into beauteous order start,
Before Heav'n was, Thou art:
Ere spheres beneath us roll'd or spheres above,
Ere earth in firmamental ether hung,
Thou satst alone; till, through thy mystick Love,
Things unexisting to existence sprung,
And grateful descant sung.
What first impell'd thee to exert thy might?
Goodness unlimited. What glorious light
Thy pow'r directed? Wisdom without bound.
What prov'd it first? Oh! guide my fancy right;
Oh! raise from cumbrous ground
My soul in rapture drown'd,
That fearless it may soar on wings of fire;
For Thou, who only knowst, Thou only canst inspire.

Wrapt in eternal solitary shade,
Th' impenetrable gloom of light intense,
Impervious, inaccessible, immense,
Ere spirits were infus'd or forms display'd,
BREHM his own Mind survey'd,
As mortal eyes (thus finite we compare
With infinite) in smoothest mirrors gaze:
Swift, at his look, a shape supremely fair
Leap'd into being with a boundless blaze,
That fifty suns might daze.
Primeval MAYA[3] was the Goddess nam'd,
Who to her sire, with Love divine inflam'd,
A casket gave with rich *Ideas* fill'd,
From which this gorgeous Universe he fram'd;
For, when th' Almighty will'd
Unnumber'd worlds to build,
From Unity diversified he sprang,
While gay Creation laugh'd, and procreant Nature rang.

First an all-potent all-pervading sound
Bade flow the waters—and the waters flow'd,[4]

---

2 [See Jones's letter to Charles Wilkins of 14 April 1785, concerning his additional sources (*Letters* 2: 669-70).]

3 [Goddess of worldly illusions in the Hindu pantheon.]
4 [See the chapter 'Creation' in the *Institutes of Manu*.]

Exulting in their measureless abode,
Diffusive, multitudinous, profound,
Above, beneath, around;
Then o'er the vast expanse primordial wind
Breath'd gently, till a lucid bubble rose,
Which grew in perfect shape an Egg refin'd:[5]
Created substance no such lustre shows,
Earth no such beauty knows.
Above the warring waves it danc'd elate,
Till from its bursting shell with lovely state
A form cerulean flutter'd o'er the deep,
Brightest of beings, greatest of the great:
Who, not as mortals steep,
Their eyes in dewy sleep,
But heav'nly-pensive on the Lotos lay,
That blossom'd at his touch and shed a golden ray.

Hail, primal blossom! hail empyreal gem!
KEMEL, or PEDMA, or whate'er high name
Delight thee, say, what four-form'd Godhead came,
With graceful stole and beamy diadem,
Forth from thy verdant stem?
Full-gifted BREHMA! Rapt in solemn thought
He stood, and round his eyes fire-darting threw;
But, whilst his viewless origin he sought,
One plain he saw of living waters blue,
Their spring nor saw nor knew.
Then, in his parent stalk again retir'd,
With restless pain for ages he inquir'd
What were his pow'rs, by whom, and why conferr'd:
With doubts perplex'd, with keen impatience fir'd
He rose, and rising heard
Th' unknown all-knowing Word,
'BREHMA! no more in vain research persist:
My veil[6] thou canst not move—Go; bid all worlds
    exist'.

Hail, self-existent, in celestial speech
NARAYEN,[7] from thy watry cradle, nam'd;
Or VENAMALY may I sing unblam'd,
With flow'ry braids, that to thy sandals reach,
Whose beauties, who can teach?
Or high PEITAMBER clad in yellow robes
Than sunbeams brighter in meridian glow,

That weave their heav'n-spun light o'er circling
    globes?
Unwearied, lotos-eyed, with dreadful bow,
Dire Evil's constant foe!
Great PEDMANABHA, o'er thy cherish'd world
The pointed *Checra*, by thy fingers whirl'd,
Fierce KYTABH shall destroy and MEDHU grim
To black despair and deep destruction hurl'd.
Such views my senses dim.
My eyes in darkness swim:
What eye can bear thy blaze, what utt'rance tell
Thy deeds with silver trump or many-wreathed shell?

Omniscient Spirit, whose all-ruling pow'r
Bids from each sense bright emanations beam;
Glows in the rainbow, sparkles in the stream,
Smiles in the bud, and glistens in the flow'r
That crowns each vernal bow'r;
Sighs in the gale, and warbles in the throat
Of ev'ry bird, that hails the bloomy spring,
Or tells his love in many a liquid note,
Whilst envious artists touch the rival string,
Till rocks and forests ring;
Breathes in rich fragrance from the sandal grove,
Or where the precious musk-deer playful rove;
In dulcet juice from clust'ring fruit distills,
And burns salubrious in the tasteful clove:
Soft banks and verd'rous hills
Thy present influence fills;
In air, in floods, in caverns, woods, and plains;
Thy will inspirits all, thy sov'reign MAYA reigns.

Blue crystal vault, and elemental fires,
That in th' ethereal fluid blaze and breathe;
Thou, tossing main, whose snaky branches wreathe
This pensile orb with intertwisted gyres;
Mountains, whose radiant spires
Presumptuous rear their summits to the skies,
And blend their em'rald hue with sapphire light;
Smooth meads and lawns, that glow with varying dyes
Of dew-bespangled leaves and blossoms bright,
Hence! vanish from my sight:
Delusive Pictures! unsubstantial shows!
My soul absorb'd One only Being knows,
Of all perceptions One abundant source,
Whence ev'ry object ev'ry moment flows:
Suns hence derive their force,
Hence planets learn their course;
But suns and fading worlds I view no more:
GOD only I perceive; GOD only I adore.

---

[5] [See Note 2, *Letters* 2: 669.]

[6] [The veil of Māyā spread over the world.]

[7] [Narayana is portrayed here as Vishnu relaxing on his serpent-bed in the ocean.]

A Hymn to Sereswaty[1]

The Argument

The *Hindu* goddesses are uniformly represented as the subordinate powers of their respective lords: thus LACSHMY, the consort of VISHNU the *Preserver*, is the goddess of *abundance* and *prosperity*; BHAVĀNY, the wife of MAHĀDĒV, is the genial power of *fecundity*; and SERESWATY, whose husband was the Creator BREHMĀ, possesses the powers of Imagination and Invention, which may justly be termed *creative*. She is, therefore, adored as the patroness of the Fine Arts, especially of Musick and Rhetorick, as the inventress of the SANSCRIT Language, of the *Dēvanāgry* Letter, and of the Sciences, which writing perpetuates; so that her attributes correspond with those of MINERVA MUSICA, in *Greece* and *Italy*, who invented the flute, and presided over literature. In this character she is addressed in the following Ode, and particularly as the *Goddess of Harmony*: since the *Indians* usually paint her with a musical instrument in her hand: the seven notes, an artful combination of which constitutes *Musick* and variously affects the passions, are feigned to be her earliest production; and the greatest part of the Hymn exhibits a correct delineation of the RĀGMĀLĀ, or *Necklace of Musical Modes*, which may be

---

[1] [This hymn, written in 1785, uses Jones's deepening knowledge of Indian music to extol Saraswati, Brahma's wife and inventress of the Sanskrit language. It emerges as a tour de force in which the goddess is visualized mainly as the originator of music, speech, and learning. Much of the poem reflects a pictorial outline of the Raga system of classical Indian music, of which Jones demonstrates a pioneering understanding and wishes to communicate to the West. The pictures of Indian ragas strung together in the poem might have been partly inspired by the Raga-Rangini paintings, which, if so, reflect an early Western artistic inspiration for poetry. See Jones's 'On the Musical Modes of the Hindus', *Works* 4 : 166-210. The hymn, which again employs numerous obscure names from Hindu mythology and musicology that invariably disturbed contemporary European critics, first appeared in *The Asiatick Miscellany* (I). Cannon owns the holograph of Jones's previously unpublished translation of this Sanskrit hymn to Sarasvati that was his general source.]

considered as the most pleasing invention of the ancient *Hindus*, and the most beautiful union Painting with poetical Mythology and the genuine theory of Musick.

The different position of the *two* semitones in the scale of seven notes given birth to seven *primary* modes: and, as the whole series consists of *twelve* semitones, every one of which may be made a *modal* note or *tonick*, there are in nature, (though not universally in practice) *seventy-seven* other modes, which may be called *derivative*: all the *eighty-four* are distributed by the PERSIANS, under the notion of *locality*, into three classes consisting of *twelve* rooms, *twenty-four* angles, and *forty-eight* recesses; but the HINDU arrangement is elegantly formed on the variations of the *Indian* year, and the association of ideas; a powerful auxiliary to the ordinary effect of modulation. The Modes, in this system, are deified; and, as there are *six* seasons in *India*, namely, two Springs, Summer, Autumn, and two Winters, an original RĀG, or *God of the Mode*, is conceived to preside over a particular season; each principal mode is attended by five RĀGNYS or *Nymphs of Harmony*; each has *eight* Sons or *Genii* of the same divine Art; and each RĀG, with his family, is appropriated to a distinct season, in which alone his melody can be sung or played at prescribed hours of the day and night: the mode of DEIPEC, or CUPID the *Inflamer*, is supposed to be lost; and a tradition is current in *Hindustan*, that a musician, who attempted to restore it, was consumed by fire from heaven. The natural distribution of modes would have been, *seven, thirty-three,* and *forty-four*, according to the number of the minor and major secondary tones; but this order was varied for the sake of the charming fiction above-mentioned. NĀRED, who is described in the *third* stanza, was one of the first created beings, corresponding with the MERCURY of the *Italians*, inventor of the VENE, a fretted instrument, supported by two large *gourds*, and confessedly the finest used in *Asia*.

A full discussion of so copious a subject would require a separate dissertation; but here it will be sufficient to say, that almost every allusion and every epithet in the Poem, as well as the names, are selected from approved treatises, either originally *Persian* or translated from the *Sanscrit*, which contain as lively a display of genius as human imagination ever exhibited.

The last couplet alludes to the celebrated place of pilgrimage, at the confluence of the *Gangā* and *Yamnā*, which the *Sereswaty*, another sacred river, is supposed to join under ground.

The Hymn

Sweet grace of BREHMĀ's bed!
Thou when thy glorious lord
Bade airy nothing breathe and bless his pow'r,
Set'st with illumin'd head,
And, in sublime accord,
Sev'n sprightly notes, to hail th' auspicious hour,
Led'st from their secret bow'r:
They drank the air; they came
With many a sparkling glance,
And knit the mazy dance,
Like yon bright orbs, that gird the solar flame,
Now parted, now combin'd,
Clear as thy speech, and various as thy mind.

Young Passions at the sound
In shadowy forms arose,
O'er hearts, yet uncreated, sure to reign;
Joy, that o'erleaps all bound,
Grief, that in silence grows,
Hope, that with honey blends the cup of pain,
Pale Fear, and stern Disdain,
Grim Wrath's avenging band,
Love, nurs'd in dimple smooth,
That ev'ry pang can soothe;
But, when soft Pity her meek trembling hand
Stretch'd, like a new-born girl,
Each sigh was musick, and each tear a pearl.

Thee her great parent owns,
All-ruling Eloquence,
That, like full GANGA, pours her stream divine,
Alarming states and thrones:
To fix the flying sense
Of words, thy daughters, by the varied line
(Stupendous art!) was Thine;
Thine, with pointed reed
To give primeval Truth
Th' unfading bloom of youth,
And paint on deathless leaves high Virtue's meed:
Fair Science, heav'n-born child,
And playful Fancy on thy bosom smil'd.

Who bids the fretted *Vene*

Start from his deep repose,
And wakes to melody the quiv'ring frame?
What youth with godlike mien
O'er his bright shoulder throws
The verdant gourd, that swells with struggling flame?
NĀRED, immortal name!
He, like his potent Sire,
Creative spreads around
The mighty world of sound,
And calls from speaking wood ethereal fire;
While to th' accordant strings
Of boundless heav'ns and heav'nly deeds he sings.

But look! the jocund hours
A lovelier scene display,
Young HINDOL sportive in his golden swing
High canopied with flow'rs;
While *Rāgnys* ever gay
Toss the light cordage, and in cadence sing
The sweet return of Spring:
Here dark *Virāwer* stands;
There *Rāmcary* divine
And fawn-ey'd *Lelit* shine;
But stern *Daysāsha* leads her warring bands,
And slow in ebon clouds
Petmenjary her fading beauty shrouds.

Ah! where has DEIPEC veil'd
His flame-encircled head?
Where flow his lays too sweet for mortal ears?
O loss, how long bewail'd!
Is yellow *Cāmōd* fled?
And blythe *Cārnāty* vaunting o'er her peers?
Where stream *Caydāra*'s tears
Intent on scenes above,
A beauteous anchorite?
No more shall *Daysa* bright
With gentle numbers call her tardy love?
Has *Netta*, martial maid,
Lock'd in sad slumbers her sky-temper'd blade?

Once, when the vernal noon
Blaz'd with resistless glare,
The Sun's eye sparkled, and a God was born:
He smil'd; but vanish'd soon—
Then groan'd the northern air;
The clouds, in thunder mutt'ring sullen scorn,
Delug'd the thirsty corn.
But, earth-born artist, hold!
If e'er thy soaring lyre

To *Deipec*'s notes aspire,
Thy strings, thy bow'r, thy breast with rapture bold,
Red lightning shall consume;
Nor can thy sweetest song avert the doom.

See sky-form'd MAYGH descend
In fertilising rain,
Whilst in his hand a falchion gleams unsheath'd!
Soft nymphs his car attend,
And raise the golden grain,
Their tresses dank with dusky spikenard wreath'd:
(A sweeter gale ne'er breath'd)
Tenca with laughing eyes,
And *Gujry*'s bloomy cheek,
Melār with dimple sleek,
On whose fair front two musky crescents rise:
While *Dayscār* his rich neck
And mild *Bhopāly* with fresh jasmin deck.

Is that the King of Dread
With ashy musing face,
From whose moon-silver'd locks fam'd GANGA
        springs?
Tis BHAIRAN, whose gay bed
Five blushing damsels grace,
And rouse old Autumn with immortal strings,
Till ev'ry forest rings;
Bengāly lotos-crown'd,
Vairāty like the morn,
Sindvy with looks of scorn,
And *Bhairavy*, her brow with *Champa*'s bound;
But *Medhumādha*'s eyes
Speak love, and from her breast pomegranates rise.

Sing loud, ye lucid spheres;
Ye gales, more briskly play,
And wake with harmony the drooping meads:
The cooler season cheers
Each bird, that panting lay,
And SIRY bland his dancing bevy leads
Hymning celestial deeds:
Marvā with robes like fire,
Vasant whose hair perfumes
With musk its rich-eyed plumes,
Āsāvery, who list'ning asps admire,
Dhenāsry, flow'r of glades,
And *Mālsry*, whom the branching *Amra* shades.

MALCAUS apart reclines
Bedeck'd with heav'n-strung pearls,

Blue-mantled, wanton, drunk with youthful pride;
Nor with vain love repines,
While softly-smiling girls
Melt on his cheek or frolick by his side,
And wintry winds deride;
Shambhāwty leads along
Cocabh with kerchief rent,
And *Gaūry* wine-besprent,
Warm *Guncary*, and *Toda* sweet in song,
Whom antelopes surround
With smooth tall necks, and quaff the streaming sound.

Nor deem these nuptial joys
With lovely fruit unblest:
No; from each God an equal race proceeds,
From each eight blooming boys;
Who, their high birth confess'd,
With infant lips gave breath to living reeds
In valleys, groves, and meads:
Mark how they bound and glance!
Some climb the vocal trees,
Some catch the sighing breeze,
Some, like new stars, with twinkling sandals dance;
Some the young *Shamma* snare,
Some warble wild, and some the burden bear.

These are thy wond'rous arts;
Queen of the flowing speech,
Thence SERESWATY nam'd and VĀNY bright!
Oh, joy of mortal hearts,
Thy mystick wisdom teach;
Expand thy leaves, and, with ethereal light,
Spangle the veil of night.
If LEPIT please thee more,
Or BRĀHMY, awful name,
Dread BRĀHMY's aid we claim,
And thirst, VĀCDĒVY, for thy balmy lore
Drawn from that rubied cave,
Where meek-ey'd pilgrims hail the triple wave.[2]

---

[2] [The sacred Hindu pilgrimage to Triveni, where the Ganges, Yamuna, and the invisible Sarasvati merge. By bringing the goddess of music and arts from the heavens to the earth in the form of a river, Jones mingles the flowing imagery of music with the river.]

[Jones's Original Version of a Sanskrit Hymn to Sarasvati]

1. She gave to the holy man *Yadhyrāswa*, who had presented oblations to her, a son named *Divōdasa* (Laskushe), rapid in his course, who discharged *all* debts *divine and human*;

2. She, who destroyed the opulent, selfish merchant: These O *Saraswati*, are thy mighty gifts!

3. She, by her own powers, divides the crags of mountains with vast waves, as a man, who digs for the stalks of the lotos-plant, *divides the mud*: that *Saraswati*, who destroys both her banks: let us adore, for the sake of her protection, with pious hymns *and* with acts of devotion!

4. Thou, *Saraswati*; slewest the contemness of the Gods, and *Vritra* the son of that universal artist, the deceiver (Twastri) *Vrisaya*, meaning *Viswacarma*, or *Vulian*: thou who providest sustenance also (gavest to men the lands, of which the giants had been *deprived*, and causedst pure water to flow for them, *or*) thou tookest lands from the giants, and causedst poison to flow for their destruction.

5. May *Saraswati*, the goddess who giveth nourishment, who preserveth such as devoutly meditate on her, support us abundantly with means of subsistence!

6. *Preserve* Him, O goddess *Saraswati*; who celebrates thee, no *Indra*, for their *mighty deeds* in the battle with *Vritra*, a battle occasioned by *a contest* for wealth.

7. O *Saraswati*, thou goddess liberal and powerful, preserve us in battles! give us wealth to be enjoyed, as the God, who supports us, *gives us* well-being.

8. Oh! may that *Saraswati*, who, tremendous in her golden car, destroys her foes, accept our well-composed hymn!

9. Oh may that *Saraswati* whose watery flood rushes with awful roar, unbounded, unbent, brilliant, unobstructed in its course, and *her sisters*, who preside over other holy rivers, put an end to all our foes, as the sun, who incessantly moves, brings days to an end!

10. Let that *Saraswati*, with seven sisters, the dearest among the dear, who was duly praised *by old sages*, be celebrated by us!

11. May *Saraswati*, who fills with her glory all things

on earth, inhabited worlds of great extent, *and* the firmament *itself*, defend us from *every* foe!

12. Let her be adored in every battle, who pervades the three worlds at once, who has seven companions closely united, who gives increase to five races, four classes of *men*, and one of celestial *quiristers* [choristers].

13. That *Saraswati*, who is eminently distinguished among those sisters by her vast magnificence, by her bright glories, the most rapid of rapid streams, and who was created eminent in great qualities, as a splendid car *is formed* for glory, must be celebrated by every wise man.

[Previously unpublished, from a manuscript owned by Cannon.]

A Hymn to Ganga[1]

The Argument

This poem would be rather obscure without geographical notes; but a short introductory explanation will supply the place of them, and give less interruption to the reader.

We are obliged to a late illustrious *Chinese* monarch named CAN-HÍ, who directed an accurate survey to be made of *Pŏtyid* or (as it is called by the *Arabs*) *Tebbut*, for our knowledge, that a chain of mountains nearly parallel with *Imaus*, and called *Cantésè* by the *Tartars*, forms a line of separation between the sources of two vast rivers; which, as we have abundant reason to believe, run at first in opposite directions, and, having finished a winding circuit of two thousand miles, meet a little below *Dhacà*, so as to inclose the richest and most beautiful peninsula on earth, in which the BRITISH nation, after a prosperous course of brilliant actions in peace and war, have now the principal sway. These rivers are *deified* in INDIA; that, which rises on the *western* edge of the mountain, being considered as the daughter of MAHĀDĒVA or SĪVA, and the other as the son of BRAHMĀ: their loves, wanderings, and nuptials are the chief subject of the following Ode, which is feigned to have been the work of a BRĀHMEN, in an early age of HINDU antiquity, who, by a prophetical spirit, discerns the toleration and equity of the BRITISH government, and concludes with a prayer *for its peaceful duration under good laws well administered.*

After a general description of the *Ganges*, an account is given of her fabulous birth, like that of *Pallas*, from the forehead of *Siva*, the *Jupiter Tonans* and *Genitor* of the *Latins*; and the creation of her lover by an act of

Brahmā's will is the subject of another stanza, in which his course is delineated through the country of *Pŏtyid*, by the name of *Sanpò*, or *Supreme Bliss*, where he passes near the fortress of *Rimbù*, the island of *Paltē* or *Yambrò* (known to be the seat of a high priestess almost equally venerated with the Goddess *Bhawāni*) and *Trashilhumbo* (as a *Pŏtya* or *Tebbutian* would pronounce it), or the sacred mansion of the *Lama* next in dignity to that of *Pŏtala*, who resides in a city, to the south of the *Sanpò*, which the *Italian* travellers write *Sgigatzhè*, but which, according to the letters, ought rather to be written in a manner, that would appear still more barbarous in our orthography. The *Brahmaputra* is not mentioned again till the *twelfth* stanza, where his progress is traced, by very probable conjecture, through *Rangamāti*, the ancient *Rangamriticà* or *Rangamar*, celebrated for the finest spikenard, and *Srīhàt* or *Siret*, the *Serratœ* of *Elian*, whence the fragrant essence extracted from the *Malobathrum*, called *Sadāh* by the *Persians*, and *Tējapātra* by the *Indians*, was carried by the *Persian* gulf to *Syria*, and from that coast into *Greece* and *Italy*. It is not, however, positively certain, that the *Brahmaputra* rises as it is here described: two great geographers are decidedly of opposite opinions on this very point; nor is it impossible that the *Indian* river may be one arm of the *Sanpò*, and the *Nau-cyan*, another; diverging from the mountains of *Ashām*, after they have been enriched by many rivers from the rocks of *China*.

The *fourth* and *fifth* stanzas represent the Goddess obstructed in her passage to the west by the hills of *Emodi*, so called from a *Sanscrit* word signifying *snow*, from which also are derived both *Imaus* and *Himālaya* or *Himola*. The *sixth* describes her, after her entrance into *Hindūstan* through the straits of *Cūpala*, flowing near *Sambal*, the *Sambalaca* of *Ptolemy*, famed for a beautiful plant of the like name, and thence to the once opulent city and royal place of residence, *Cānyacuvja*, erroneously named *Calinipaxa* by the *Greeks*, and *Canauj*, not very accurately, by the modern *Asiaticks*: here she is joined by the *Calinadi*, and pursues her course to Prayaga, whence the people of Bahar were named *Prasii*, and where the *Yamunà*, having received the *Sereswatì* below *Indraprest'ha* or *Dehlì*, and watered the poetical ground of *Mat'hurà* and *Agarà*, mingles her noble stream with the *Gangā* close to the modern fort of *Ilahābàd*. This place is considered as the confluence of *three* sacred rivers, and known by the name of *Trivēni*, or the *three plaited locks*; from which a number of pilgrims, who there begin the ceremonies to be completed at *Gayà*, are continually bringing vases of water, which they preserve with super-

---

[1] [This hymn, probably composed in 1785 while Jones was sailing up the very river he was celebrating, was published in *The Asiatick Miscellany* (I). It graphically portrays Ganga's birth, her loves and wanderings, and ultimate nuptials with Brahmaputra. In eulogizing the greatest and holiest of Indian rivers, Jones followed the poetic tradition of Michael Drayton, who celebrated English rivers in his *Polyalbion*. It is likely that the hymn may have influenced Coleridge—see Cannon, 'A New Probable Source for "Kubla Khan"', *College English* 17 (December 1955): 136-41.]

stitious veneration, and are greeted by all the *Hindus*, who meet them on their return.

Six of the principal rivers, which bring their tribute to the *Ganges*, are next enumerated, and are succinctly described from real properties: thus the *Gandac*, which the *Greeks* knew by a similar name, abounds, according to *Giorgi*, with *crocodiles* of enormous magnitude; and the *Mahanadi* runs by the plain of *Gaura*, once a populous district with a magnificent capital, from which the *Bengalese* were probably called *Gangaridæ*, but now the seat of desolation, and the haunt of wild beasts. From *Prayāga* she hastens to *Cāsì*, or as the *Muslimans* name it, *Benāres*; and here occasion is taken to condemn the cruel and intolerant spirit of the crafty tyrant AURANGZÍB, whom the *Hindus* of *Cashmir* call *Aurangāsùr, or the Demon*, not the *Ornament*, of the *Throne*. She next bathes the skirts of *Pātaliputra*, changed into *Patna*, which, both in situation and name, agrees better on the whole with the ancient *Palibothra*, than either *Prayāga*, or *Cānyacuvja*: if *Megasthenes* and the ambassadors of *Seleucus* visited the last-named city, and called it *Palibothra*, they were palpably mistaken. After this are introduced the beautiful hill of *Muctigiri*, or *Mengìr*, and the wonderful pool of *Sītā*, which takes its name from the wife of *Rāma*, whose conquest of *Sinhaldwīp*, or *Sīlàn*, and victory over the giant *Rāwan*, are celebrated by the immortal *Vālmīci*, and by other epick poets of *India*.

The pleasant hills of *Cāligràm* and *Gangā-presàd* are then introduced, and give occasion to deplore and extol the late excellent AUGUSTUS CLEVLAND, Esq. who nearly completed by lenity the glorious work, which severity could not have accomplished, of civilizing a ferocious race of *Indians*,[2] whose mountains were formerly, perhaps, a rocky island, or washed at least by that sea, from which the fertile champaign of *Bengal* has been gained in a course of ages. The western arm of the *Ganges* is called *Bhāgira'thì*, from a poetical fable of a demigod or holy man, named *Bhāgīrat'ha*, whose devotion had obtained from *Siva* the privilege of leading after him a great part of the heavenly water, and who drew it accordingly in two branches; which embrace the fine island, now denominated from *Kāsimbāzàr*, and famed for the defeat of the monster *Sirājuddaulah*, and, having met near the venerable *Hindu* seminary of *Nawadwìp* or *Nedīyā*, flow in a copious stream by the several *European* settlements, and reach the Bay at an island which assumes the name

of *Sāgar*, either from the *Sea* or from an ancient Raja of distinguished piety. The *Sundarabans* or *Beautiful Woods*, an appellation to which they are justly entitled, are incidentally mentioned, as lying between the *Bhāgirat'hì* and the *Great River*, or *Eastern* arm, which, by its junction with the *Brahmāputra*, forms many considerable islands; one of which, as well as a town near the conflux, derives its name from *Lacshmī*, the Goddess of Abundance.

It will soon be perceived, that the *form* of the stanza, which is partly borrowed from GRAY, and to which he was probably partial, as he uses it *six* times in *nine*, is enlarged in the following Hymn by a line of *fourteen* syllables, expressing the long and solemn march of the great *Asiatick* rivers.

---

[2] [Cleveland (1755-84), a British civilian in Bengal, died from these exertions.]

The Hymn

How sweetly GANGĀ smiles, and glides
Luxuriant o'er her broad autumnal bed!
Her waves perpetual verdure spread,
Whilst health and plenty deck her golden sides:
As when an eagle,[3] child of light,
On *Cambala*'s unmeasur'd height,
By *Pōtala*, the pontiff's throne rever'd,
O'er her eyry proudly rear'd
Sits brooding, and her plumage vast expands,
Thus GANGĀ o'er cherish'd lands,
To *Brahmà*'s grateful race endear'd,
Throws wide her fost'ring arms, and on her banks
    divine
Sees temples, groves, and glitt'ring tow'rs, that in her
    crystal shine.

Above the stretch of mortal ken,
On bless'd *Cailāsa*'s top, where ev'ry stem
Glow'd with a vegetable gem,
MAHĒSA stood, the dread and joy of men;
While *Pārvatì*, to gain a boon,
Fix'd on his locks a beamy moon,
And hid his frontal eye, in jocund play,
With reluctant sweet delay:
All nature straight was lock'd in dim eclipse
Till *Brāhmans* pure, with hallow'd lips
And warbled pray'rs restor'd the day;
When GANGĀ from his brow by heav'nly fingers
    press'd
Sprang radiant, and descending grac'd the caverns
    of the west.

The sun's car blaz'd, and laugh'd the morn;
What time near proud *Cantēsa*'s eastern bow'rs,
(While *Dēvatà*'s rain'd living flow'rs)
A river-god, so *Brahmà* will'd, was born,
And roll'd mature his vivid stream
Impetuous with celestial gleam:
The charms of GANGĀ, through all worlds proçlaim'd,

Soon his youthful breast inflam'd,
But destiny the bridal hour delay'd;
Then, distant from the west'ring maid,
He flow'd, now blissful *Sanpò* nam'd,
By *Paltè* crown'd with hills, bold *Rimbu*'s tow'ring
    state,
And where sage *Trashilhumbo* hails her *Lama*'s form
    renate.

But she, whose mind, at *Siva*'s nod,
The picture of that sov'reign youth had seen,
With graceful port and warlike mien,
In arms and vesture like his parent God,
Smit with the bright idea rush'd,
And from her sacred mansion gush'd,
Yet ah! with erring step—The western hills
Pride, not pious ardour, fills:
In fierce confed'racy the giant bands
Advance with venom-darting hands,
Fed by their own malignant rills;
Nor could her placid grace their savage fury quell:
The madding rifts and should'ring crags her foamy
    flood repell.[4]

'Confusion wild and anxious wo
Haunt your waste brow', she said, 'unholy rocks,
Far from these nectar-dropping locks!
But thou, lov'd Father, teach my waves to flow'.
Loud thunder her high birth confess'd;
Then from th' inhospitable west
She turn'd, and, gliding o'er a lovelier plain,
Cheer'd the pearled East again:
Through groves of nard she roll'd, o'er spicy reeds,
Through golden vales and em'rald meads;
Till, pleas'd with INDRA's fair domain,
She won through yielding marl her heav'n-directed
    way:
With lengthen'd notes her eddies curl'd, and pour'd
    a blaze of day.

Smoothly by *Sambal*'s flaunting bow'rs,
Smoothly she flows, where *Calinadi* brings
To *Cānyacuvja*, seat of kings,

---

[3] [An almost Homeric simile of an eagle brooding over its
aerie, projecting a fostering image of the river over the vast
territory.]

[4] [In his 21 September 1786 letter to Samuel Davis, Jones
expressed a wish to 'insert a stanza on your water-fall in my ode
to the Ganges: at least if you draw the romantick scene, as I trust
you have already drawn it, I will translate your picture into my
feebler colouring' (*Letter* 2: 705).]

On prostrate waves her tributary flow'rs;
Whilst *Yamunà*, whose waters clear
Fam'd *Indraprestha*'s vallies cheer,
With *Sereswatī* knit in mystick chain,
Gurgles o'er the vocal plain
Of *Mathurà*, by sweet *Brindāvan*'s grove,
Where *Gōpa*'s love-lorn daughters rove,
And hurls her azure stream amain,
Till blest *Prayāga*'s point beholds three mingling tides,
Where pilgrims on the far-sought bank drink nectar, as
    it glides.

From *Himola*'s perennial snow,
And southern *Palamau*'s less daring steep,
Sonorous rivers, bright though deep,
O'er thirsty deserts youth and freshness throw.
'A goddess comes', cried *Gumti* chaste,
And roll'd her flood with zealous haste:
Her follow'd *Sona* with pellucid wave
Dancing from her diamond cave,
Broad *Gogra*, rushing swift from northern hills,
Red *Gandac*, drawn by crocodiles,
(Herds, drink not there, nor, herdsmen, lave!)
Cosa, whose bounteous hand *Nēpālian* odour flings,
And *Mahanadi* laughing wild at cities, thrones, and
    kings.

Thy temples, CĀSĪ, next she sought,
And verd'rous plains by tepid breezes fann'd,
Where health extends her pinions bland,
Thy groves, where pious *Vālmic* sat and thought,
Where *Vyāsa* pour'd the strain sublime,
That laughs at all-consuming time,
And *Brāhmans* rapt the lofty *Vēda* sing.
Cease, oh! cease—a ruffian king,
The demon of his empire, not the grace,
His ruthless bandits bids deface
The shrines, whence gifts ethereal spring:
So shall his frantick sons with discord rend his throne,
And his fair-smiling realms be sway'd by nations yet
    unknown.

Less hallow'd scenes her course prolong;
But *Cāma*, restless pow'r, forbids delay:
To Love all virtues homage pay,
E'en stern religion yields. How full, how strong
Her trembling panting surges run,
Where *Pātali*'s immortal son
To domes and turrets gives his awful name
Fragrant in the gales of fame!

Nor stop, where RĀMA, bright from dire alarms,
Sinks in chaste *Sītà*'s constant arms,
While bards his wars and truth proclaim:
There from a fiery cave the bubbling crystal flows,
And *Muctigir*, delightful hill, with mirth and beauty
    glows.

Oh! rising bow'rs, great *Cālī*'s boast,
And thou, from *Gangà* nam'd, enchanting mount,
What voice your wailings can recount
Borne by shrill echoes o'er each howling coast,
When He, who bade your forests bloom,
Shall seal his eyes in iron gloom?
Exalted youth! The godless mountaineer,
Roaming round his thickets drear,
Whom rigour fir'd, nor legions could appall,
I see before thy mildness fall,
Thy wisdom love, thy justice fear:
A race, whom rapine nurs'd, whom gory murder stains,
Thy fair example wins to peace, to gentle virtue trains.

But mark, where old *Bhāgīrath* leads
(This boon his pray'rs of *Mahādèv* obtain:
Grace more distinguish'd who could gain?)
Her calmer current o'er his western meads,
Which trips the fertile plains along,
Where vengeance waits th' oppressor's wrong;
Then girds, fair *Nawadwìp*, thy shaded cells,
Where the *Pendit* musing dwells;
Thence by th' abode of arts and commerce glides,
Till *Sāgar* breasts the bitter tides:
While She, whom struggling passion swells,
Beyond the labyrinth green, where pards by moonlight
    prowl,
With rapture seeks her destin'd lord, and pours her
    mighty soul.

Meanwhile o'er *Pōtyid*'s musky dales,
Gay *Rangamar*, where sweetest spikenard blooms,
And *Siret*, fam'd for strong perfumes,
That, flung from shining tresses, lull the gales,
Wild *Brahmaputra* winding flows,
And murmurs hoarse his am'rous woes;
Then, charming GANGĀ seen, the heav'nly boy
Rushes with tumultuous joy:
(Can aught but Love to men or Gods be sweet?)
When she, the long-lost youth to greet,
Darts, not as earth-born lovers toy,
But blending her fierce waves, and teeming
    verdant isles;

While buxom *Lacshmì* crowns their bed, and sounding
   ocean smiles.

What name, sweet bride, will best allure
Thy sacred ear, and give thee honour due?
Vishnupedì? Mild *Bhìshmasù*?
Smooth *Suranimnagà*? *Trisrōtà* pure?
By that I call? Its pow'r confess;
With growing gifts thy suppliants bless,
Who with full sails in many a light-oar'd boat
On thy jasper bosom float;
Nor frown, dread Goddess, on a peerless race
With lib'ral heart and martial grace,
Wafted from colder isles remote:
As they preserve our laws, and bid our terror cease,[5]
So be their darling laws preserv'd in wealth, in joy,
   in peace!

A Hymn to Indra[1]

The Argument

So many allusions to *Hindu* Mythology occur in the
following Ode, that it would be scarce intelligible without
an explanatory introduction, which, on every account and
on all occasions, appears preferable to notes in the margin.

A distinct idea of the God, whom the poem celebrates,
may be collected from a passage in the ninth section of
the *Gītà*, where the sudden change of measure has an
effect similar to that of the finest modulation:

> *tè punyamāsādya surēndra lōcam*
> *asnanti divyān dividēvabhōgān,*
> *tè tam bhuctwà swergalōcamvisālam*
> *cshīnè punyè mertyalōcam visant*

'These, having through virtue reached the mansion of the
king of *Sura*'s, feast on the exquisite heavenly food of the
Gods: they, who have enjoyed this lofty region of SWER-
GA, *but* whose virtue is exhausted, revisit the habitation
of mortals'.

INDRA, therefore, or the *King* of Immortals, corresponds with one of the ancient *Jupiters* (for several of that
name were worshipped in *Europe*), and particularly with
Jupiter the *Conductor*, whose attributes are so nobly
described by the *Platonick* Philosophers: one of his numerous titles is *Dyupeti*, or, in the nominative case before
certain letters. *Dyupetir*; which means the *Lord of
Heaven*, and seems a more probable origin of the *Hetruscan* word than *Juvans Pater*; as *Diespiter* was, probably,
not the *Father*, but the *Lord*, of *Day*. He may be considered as the JOVE of ENNIUS in his memorable line:

'Aspice hoc sublime candens, quem invocant omnes
Jovem',

where the poet clearly means the firmament, of which

---

[5] [Jones's view of the British in India.]

[1] [Composed in 1785, with inspiration from the *Bhagavad-
Gīta* and published in *The Asiatick Miscellany* (2), this hymn
portrays a vision by the sage Vyāṣsa of Indra, the 'King of
Immortals', in all the pomp and glory of his court.]

INDRA is the personification. He is the God of thunder and the five elements, with inferior Genii under his command; and is conceived to govern the Eastern quarter of the world, but to preside, like the *Genius* or *Agathodoemon* of the Ancients, over the celestial bands, which are stationed on the summit of MĒRU, or the North-pole, where he solaces the Gods with nectar and heavenly musick: hence, perhaps, the *Hindus*, who give evidence, and the magistrates, who hear it, are directed to stand fronting the East or the North.

This imaginary mount is here feigned to have been in a vision at *Vārānasì* very improperly called *Banāris*, which takes its name from two rivulets, that embrace the city; and the bard, who was favoured with the sight, is supposed to have been VYĀSA, surnamed *Dwaipāyana*, or *Dwelling in an Island*; who, if he really composed the *Gìtà*, makes very flattering mention of himself in the tenth chapter. The plant *Latà*, which he describes weaving a net round the mountain *Mandara*, is transported by a poetical liberty to *Sumēru*, which the great author of the *Mahabhārat* has richly painted in four beautiful couplets: it is the generick name for a *creeper*, though represented here as a species, of which many elegant varieties are found in *Asia*.

The Genii named *Cinnara*'s are the male dancers in *Swerga*, or the Heaven of INDRA; and the *Apsarà*'s are his dancing-girls, answering to the *fairies* of the PERSIANS, and to the damsels called in the KORAN *hhúru'lûyùn*, or *with antelopes' eyes*. For the story of *Chitrarat'ha*, the chief musician of the *Indian* paradise, whose painted car was burned by ARJUN, and for that of the *Chaturdesaretna*, or *fourteen gems*, as they are called, which were produced by churning the ocean, the reader must be referred to Mr. WILKINS's learned annotations on his accurate version of the *Bhagavadgìtà*. The fable of the pomegranate-flower is borrowed from the popular mythology of *Nēpàl* and *Tibet*.

In this poem the same form of stanza is repeated with *variations*, on a principle entirely new in modern lyric poetry, which on some future occasion may be fully explained.

The Hymn

But ah! what glories yon blue vault emblaze?[2]
What living meteors from the zenith stream?
Or hath a rapt'rous dream
Perplex'd the isle-born bard in fiction's maze?
He wakes; he hears; he views no fancied rays.
'Tis INDRA mounted on the sun's bright beam;
And round him revels his empyreal train:
How rich their tints! how sweet their strain!

Like shooting stars around his regal seat
A veil of many-colour'd light they weave,
That eyes unholy would of sense bereave:
Their sparkling hands and lightly-tripping feet
Tir'd gales and panting clouds behind them leave.
With love of song and sacred beauty smit
The mystick dance they knit;
Pursuing, circling, whirling, twining, leading,
Now chasing, now receding;
Till the gay pageant from the sky descends
On charm'd *Sumēru*, who with homage bends.

Hail, mountain of delight,
Palace of glory, bless'd by glory's king!
With prosp'ring shade embow'r me, whilst I sing
Thy wonders yet unreach'd by mortal flight.
Sky-piercing mountain! In thy bow'rs of love
No tears are seen, save where medicinal stalks

2 [Jones had sent the following stanza to Charles Wilkins and had drastically revised it perhaps as a result of his suggestions:

But ah! what glories from the zenith break?
What lucid forms yon jasper vault emblaze?
Like living suns their airy course they take:
Fall back, ye nations, and enraptured gaze!
Mazy dances briskly knitting,
Now they meet, and now retire,
Round their Prince, in splendour sitting,
Weaving veils of heav'nly fire:
High on a milk-white Elephant he rides,
Whose agate hoof the buxom air divides.

[See Jones's letter of 11 May 1785 to Wilkins, *Letters* 2: 671.]

Weep drops balsamick o'er the silver'd walks;
No plaints are heard, save where the restless dove
Of coy repulse and mild reluctance talks;
Mantled in woven gold, with gems enchas'd,
With em'rald hillocks grac'd,
From whose fresh laps in young fantastick mazes
Soft crystal bounds and blazes
Bathing the lithe convolvulus, that winds
Obsequious, and each flaunting arbour binds.

When sapient BRAHMĀ this new world approv'd,
On woody wings eight primal mountains mov'd;
But INDRA mark'd *Sumēru* for his own,
And motionless was ev'ry stone.

Dazzling the moon he rears his golden head:
Nor bards inspir'd, nor heav'n's all-perfect speech
Less may unhallow'd rhyme his beauties teach,
Or paint the pavement which th' immortals tread;
Nor thought of man his awful height can reach:
Who sees it, maddens; who approaches, dies;
For, with flame-darting eyes,
Around it roll a thousand sleepless dragons;
While from their diamond flagons
The feasting Gods exhaustless nectar sip,
Which glows and sparkles on each fragrant lip.

This feast, in mem'ry of the churned wave
Great INDRA gave, when *Amrit* first was won
From impious demons, who to *Māyā*'s eyes
Resign'd the prize, and rued the fight begun.

Now, while each ardent *Cinnara* persuades
The soft-ey'd *Apsarà* to break the dance,
And leads her loth, yet with love-beaming glance,
To banks of marjoram and *Champac* shades,
Celestial *Genii* tow'rd their king advance
(So call'd by men, in heav'n *Gandharva*'s nam'd)
For matchless musick fam'd.
Soon, where the bands in lucid rows assemble,
Flutes breathe, and citherns tremble;
Till CHITRARATHA sings—His painted car,
Yet unconsum'd, gleams like an orient star.

Hush'd was ev'ry breezy pinion,
Ev'ry stream his fall suspended:
Silence reign'd; whose sole dominion
Soon was rais'd, but soon was ended.

He sings, how 'whilom from the troubled main

The sov'reign elephant *Airāvan* sprang;
The breathing shell, that peals of conquest rang;
The parent cow, whom none implores in vain;
The milkwhite steed, the bow with deaf'ning clang;
The Goddesses of beauty, wealth, and wine;
Flow'rs, that unfading shine,
NĀRĀYAN's gem, the moonlight's tender languish;
Blue venom, source of anguish;
The solemn leech, slow-moving o'er the strand,
A vase of long-sought *Amrit* in his hand.

'To soften human ills dread SIVA drank
The pois'nous flood, that stain'd his azure neck;
The rest thy mansions deck,
High *Swerga*, stor'd in many a blazing rank.

'Thou, God of thunder, satst on *Mēru* thron'd,
Cloud-riding, mountain-piercing, thousand-ey'd,
With young PULŌMAJÀ, thy blooming bride,
Whilst air and skies thy boundless empire own'd;
Hail, DYUPETIR, dismay to BALA's pride!
Or speaks PURANDER best thy martial fame,
Or SACRA, mystick name?[3]
With various praise in odes and hallow'd story
Sweet bards shall hymn thy glory.
Thou, VĀSAVA, from this unmeasur'd height
Shedst pearl, shedst odours o'er the sons of light!'
The Genius rested; for his pow'rful art
Had swell'd the monarch's heart with ardour vain,
That threaten'd rash disdain, and seem'd to low'r
On Gods of loftier pow'r and ampler reign.

He smil'd; and, warbling in a softer mode,
Sang 'the red light 'ning, hail, and whelming rain
O'er *Gōcul* green and *Vraja*'s nymph-lov'd plain
By INDRA hurl'd, whose altars ne'er had glow'd,
Since infant CRISHNA rul'd the rustick train
Now thrill'd with terrour—Them the heav'nly child
Call'd, and with looks ambrosial smil'd,
Then with one finger rear'd the vast *Govērdhen*,
Beneath whose rocky burden
On pastures dry the maids and herdsmen trod:
The Lord of thunder felt a mightier God!'

---

[3] [Jones thanked Wilkins for furnishing various epithets of
Indra, which was 'quite in the manner of the very ancient
Orphick Hymns to the same Deities' (*Letters* 2:678). The *Rig
Veda* Hymns comprise an originally orally transmitted collec-
tion as large as the *Odyssey* and *Iliad* combined.]

What furies potent modulation soothes!
E'en the dilated heart of INDRA shrinks:
His ruffled brow he smoothes,
His lance half-rais'd with listless languor sinks.

A sweeter strain the sage musician chose:
He told, how 'SACHI, soft as morning light,
Blythe SACHI, from her Lord INDRĀNĪ hight,
When through clear skies their car ethereal rose,
Fix'd on a garden trim her wand'ring sight,
Where gay pomegranates, fresh with early dew,
Vaunted their blossoms new:
'Oh! pluck', she said, 'yon gems, which nature dresses
To grace my darker tresses'.
'In form a shepherd's boy, a God in soul,
He hasten'd, and the bloomy treasure stole.

'The reckless peasant, who those glowing flow'rs,
Hopeful of rubied fruit, had foster'd long,
Seiz'd and with cordage strong
Shackled the God, who gave him show'rs.

'Straight from sev'n winds immortal Genii flew,
Green *Varuna*, whom foamy waves obey,
Bright *Vahni* flaming like the lamp of day,
Cuvēra sought by all, enjoyed by few,
Marut, who bids the winged breezes play,
Stern *Yama*, ruthless judge, and *Isa* cold
With *Nairrit* mildly bold:
They with the ruddy flash, that points his thunder,
Rend his vain bands asunder.
Th' exulting God resumes his thousand eyes,
Four arms divine, and robes of changing dyes'.

Soft memory retrac'd the youthful scene:
The thund'rer yielded to resistless charms,
Then smil'd enamour'd on his blushing queen,
And melted in her arms.

Such was the vision, which, on *Varan*'s breast
Or *Asì* pure with offer'd blossoms fill'd,
DWAIPĀYAN slumb'ring saw; (thus NĀRED will'd)
For waking eye such glory never bless'd,
Nor waking ear such musick ever thrill'd.
It vanish'd with light sleep: he, rising, prais'd
The guarded mount high-raised,
And pray'd the thund'ring pow'r, that sheafy treasures,
Mild show'rs and vernal pleasures,
The lab'ring youth in mead and vale might cheer,
And cherish'd herdsmen bless th' abundant year.

Thee, darter of the swift blue bolt, he sang;
Sprinkler of genial dews and fruitful rains
O'er hills and thirsty plains!
'When through the waves of war thy charger sprang,
Each rock rebellow'd and each forest rang,
Till vanquish'd *Asurs* felt avenging pains.
Send o'er their seats the snake, that never dies,
But waft the virtuous to thy skies!'

A Hymn to Sūrya[1]

The Argument

A plausible opinion has been entertained by learned men, that the principal source of idolatry among the ancients was their enthusiastick admiration of the Sun; and that, when the primitive religion of mankind was lost amid the distractions of establishing regal government, or neglected amid the allurements of vice, they ascribed to the great visible luminary, or to the wonderful fluid, of which it is the general reservoir, those powers of pervading all space and animating all nature, which their wiser ancestors had attributed to one eternal MIND, by whom the substance of fire had been created as an inanimate and secondary cause of natural phenomena. The Mythology of the East confirms this opinion; and it is probable, that the *triple Divinity* of the *Hindus* was originally no more than a personification of the Sun, whom they call *Treyitenu,* or *Three-bodied,* in his triple capacity of producing forms by his genial *heat,* preserving them by his *light,* or destroying them by the concentrated force of his *igneous* matter: this, with the wilder conceit of a *female power* united with the Godhead, and ruling nature by his authority, will account for nearly the whole system of *Egyptian, Indian,* and *Grecian* polytheism, distinguished from the sublime Theology of the Philosophers, whose understandings were too strong to admit the popular belief, but whose influence was too weak to reform it. SŪRYA, the PHŒBUS of *European* heathens, has near fifty names or epithets in the *Sanscrit* language; most of which, or at least the meanings of them, are introduced in the following Ode; and every image, that seemed capable of poetical ornament, has been selected from books of the highest authority among the *Hindus:* the title *Arca* is very singular; and it is remarkable, that the *Tibetians* represent the Sun's car in the form of a *boat.*

It will be necessary to explain a few other particulars of the *Hindu* Mythology, to which allusions are made in the poem. SOMA, or the Moon, is a *male* Deity in the *Indian* system, as *Mona* was, I believe, among the *Saxons,* and *Lunus* among some of the nations, who settled in *Italy:* his titles also, with one or two of the ancient fables, to which they refer, are exhibited in the second stanza. Most of the *Lunar mansions* are believed to be the daughters of *Casyapa,* the first production of *Brahmà's* head, and from their names are derived those of the twelve months, who are here feigned to have married as many constellations: this primeval *Brāhman* and *Vinatà* are also supposed to have been the parents of *Arun,* the charioteer of the Sun, and of the bird *Garuda,* the eagle of the great *Indian* JOVE, one of whose epithets is *Mādhava.*

After this explanation the Hymn will have few or no difficulties, especially if the reader has perused and studied the *Bhagavadgītà,* with which our literature has been lately enriched, and the fine episode from the *Mahābhārat,* on the production of the *Amrita,* which seems to be almost wholly astronomical, but abounds with poetical beauties. Let the following description of the demon *Rāhu,* decapitated by *Nārāyan,* be compared with similar passages in *Hesiod* and *Milton:*

> *tach ch'hailasringapratiman dānavasya sirò*
> *   mahat*
> *chacrach'hinnam c'hamutpatya nenādīti*
> *   bhayancaram,*
> *tat cabandham pepātāsya visp'hurad dharanītalè*
> *sapervatavanadwīpān daityasyācampa-*
> *   yanmahīm.*

---

[1] [The hymn, written in 1786 and published in *The Asiatick Miscellany* (2), is a tribute to the Hindu sun god, the primal source of energy and principle of solar power. Jones blends the scientific dimensions with the mythological view of the sun in a prayerful manner, similar to what Shelley later did in 'Ode to the West Wind' and 'The Cloud'. In his 1 March 1785 letter to Wilkins, Jones somewhat facetiously refers to the sun god: 'The powerful *Surye,* whom I worship only that he may do me no harm, confines me to my house, as long as he appears in the heavens' (*Letters* 2: 665). He may also be alluding to the hot sun in Calcutta.]

The Hymn

Fountain of living light,
That o'er all nature streams,
Of this vast microcosm both nerve and soul;
Whose swift and subtil beams,
Eluding mortal sight,
Pervade, attract, sustain th' effulgent whole,
Unite, impel, dilate, calcine,
Give to gold its weight and blaze,
Dart from the diamond many-tinted rays,
Condense, protrude, transform, concoct, refine
The sparkling daughters of the mine;
Lord of the lotos, father, friend, and king,
O Sun, thy pow'rs I sing:
Thy substance *Indra* with his heav'nly bands
Nor sings nor understands;
Nor e'en the *Vēdas* three to man explain
Thy mystick orb triform, though *Brahmā* tun'd the
    strain.

Thou, nectar-beaming Moon,
Regent of dewy night,
From yon black roe, that in thy bosom sleeps,
Fawn-spotted *Sasin* hight;
Wilt thou desert so soon
Thy night-flow'rs pale, whom liquid odour steeps,
And *Oshadhi's* transcendent beam
Burning in the darkest glade?
Will no lov'd name thy gentle mind persuade
Yet one short hour to shed thy cooling stream?
But ah! we court a passing dream:
Our pray'r nor *Indu* nor *Himānsu* hears;
He fades; he disappears—
E'en *Casyapa's* gay daughters twinkling die,
And silence lulls the sky,
Till *Chātacs* twitter from the moving brake,
And sandal-breathing gales on beds of ether wake.

Burst into song, ye spheres;
A greater light proclaim,
And hymn, concentrick orbs, with sev'nfold chime
The God with many a name;
Nor let unhallow'd ears
Drink life and rapture from your charm sublime:

'Our bosoms, *Aryama*, inspire,
Gem of heav'n, and flow'r of day,
Vivaswat, lancer of the golden ray,
Divācara, pure source of holy fire,
Victorious *Rāma's* fervid fire,
Dread child of *Aditi*, *Martunda* bless'd,
Or *Sūra* be address'd,
Ravi, or *Mihira*, or *Bhānu* bold,
Or *Arca*, title old,
Or *Heridaswa* drawn by green-hair'd steeds,
Or *Carmasacshi* keen, attesting secret deeds.

'What fiend, what monster fierce
E'er durst thy throne invade?
Malignant *Rāhu*.[2] Him thy wakeful sight,
That could the deepest shade
Of snaky *Narac* pierce,
Mark'd quaffing nectar; when by magick sleight
A *Sūra's* lovely form he wore,
Rob'd in light, with lotos crown'd,
What time th' immortals peerless treasures found
On the churn'd Ocean's gem-bespangled shore,
And *Mandar's* load the tortoise bore:
Thy voice reveal'd the daring sacrilege;
Then, by the deathful edge
Of bright *Sudersan* cleft, his dragon head
Dismay and horror spread
Kicking the skies, and struggling to impair
The radiance of thy robes, and stain thy golden hair.

'With smiles of stern disdain
Thou, sov'reign victor, seest
His impious rage: soon from the mad assault
Thy coursers fly releas'd;
Then toss each verdant mane,
And gallop o'er the smooth aerial vault;
Whilst in charm'd *Gōcul's* od'rous vale
Blue-ey'd *Yamunà* descends
Exulting, and her tripping tide suspends,
The triumph of her mighty sire to hail:
So must they fall, who Gods assail!
For now the demon rues his rash emprise,
Yet, bellowing blasphemies
With pois'nous throat, for horrid vengeance thirsts,
And oft with tempest bursts,
As oft repell'd he groans in fiery chains,

[2] [Jones graphically portrays a scene from the myth relating
to the sun's eclipse, in the fight between Surya and Rāhu, the
planetary demon.]

And o'er the realms of day unvanquish'd *Sūrya* reigns'.

Ye clouds, in wavy wreathes
Your dusky van unfold;
O'er dimpled sands, ye surges, gently flow,
With sapphires edg'd and gold!
Loose-tressed morning breathes,
And spreads her blushes with expansive glow;
But chiefly where heav'n's op'ning eye
Sparkles at her saffron gate,
How rich, how regal in his orient state!
Erelong he shall imblaze th' unbounded sky:
The fiends of darkness yelling fly;
While birds of liveliest note and lightest wing
The rising daystar sing,
Who skirts th' horizon with a blazing line
Of topazes divine;
E'en, in their prelude, brighter and more bright,
Flames the red east, and pours insufferable light.[3]

First o'er blue hills appear,
With many an agate hoof
And pasterns fring'd with pearl, sev'n coursers green;
Nor boasts yon arched woof,
That girds the show'ry sphere,
Such heav'n-spun threads of colour'd light serene,
As tinge the reins, which *Arun* guides,
Glowing with immortal grace,
Young *Arun*, loveliest of *Vinatian* race,
Though younger He, whom *Mādhava* bestrides,
When high on eagle-plumes he rides:
But oh! what pencil of a living star
Could paint that gorgeous car,
In which, as in an ark supremely bright,
The lord of boundless light
Ascending calm o'er th' empyrean sails,
And with ten thousand beams his awful beauty veils.

Behind the glowing wheels
Six jocund seasons dance,
A radiant month in each quick-shifting hand;
Alternate they advance,
While buxom nature feels
The grateful changes of the frolick band:
Each month a constellation fair
Knit in youthful wedlock holds,
And o'er each bed a varied sun unfolds,
Lest one vast blaze our visual force impair,

A canopy of woven air.
Vasanta blythe with many a laughing flow'r
Decks his *Candarpa*'s bow'r;
The drooping pastures thirsty *Grīshma* dries,
Till *Vershà* bids them rise;
Then *Sarat* with full sheaves the champaign fills,
Which *Sisira* bedews, and stern *Hēmanta* chills.[4]

Mark, how th' all-kindling orb
Meridian glory gains!
Round *Mēru*'s breathing zone he winds oblique
O'er pure cerulean plains:
His jealous flames absorb
All meaner lights, and unresisted strike
The world with rapt'rous joy and dread.
Ocean, smit with melting pain,
Shrinks, and the fiercest monster of the main
Mantles in caves profound his tusky head
With sea-weeds dank and coral spread:
Less can mild earth and her green daughters bear
The noon's wide-wasting glare;
To rocks the panther creeps; to woody night
The vulture steals his flight;
E'en cold cameleons pant in thickets dun,
And o'er the burning grit th' unwinged locusts run!

But when thy foaming steeds
Descend with rapid pace
Thy fervent axle hast'ning to allay,
What majesty, what grace
Dart o'er the western meads
From thy relenting eye their blended ray!
Soon may th' undazzled sense behold
Rich as *Vishnu*'s diadem,
Or *Amrit* sparkling in an azure gem.
Thy horizontal globe of molten gold,
Which pearl'd and rubied clouds infold.
It sinks; and myriads of diffusive dyes
Stream o'er the tissued skies,
Till *Soma* smiles, attracted by the song
Of many a plumed throng
In groves, meads, vales; and, whilst he glides above,
Each bush and dancing bough quaffs harmony and
    love.

3 See GRAY's Letters, p. 382, 4to. and the note.

4 [Jones accurately transliterates the names for the six Indian
seasons, probably from the *Ṛitusaṃhāra*, thus: 'Vasanta' for the
spring, 'Grishma' for the summer, 'Versha' for the monsoon,
'Sarat' for the autumn, 'Sisira' for the winter, and 'Hemanta'
for the cold weather.]

Then roves thy poet free,
Who with no borrow'd art
Dares hymn thy pow'r, and dust provoke thy blaze,
But felt thy thrilling dart;
And now, on lowly knee,[5]
From him, who gave the wound, the balsam prays.
Herbs, that assuage the fever's pain,
Scatter from thy rolling car,
Cull'd by sage *Aswin* and divine *Cumàr*;
And, if they ask, 'What mortal pours the strain?'
Say (for thou seest earth, air, and main)
Say: 'From the bosom of yon silver isle,
Where skies more softly smile,
He came; and, lisping our celestial tongue,
Though not from *Brahmā* sprung,
Draws orient knowledge from its fountains pure,
Through caves obstructed long, and paths too long
        obscure'.[6]

Yes; though the *Sanscrit* song
Be strown with fancy's wreathes,
And emblems rich, beyond low thoughts refin'd,
Yet heav'nly truth it breathes
With attestation strong,
That, loftier than thy sphere, th' Eternal Mind,
Unmov'd, unrival'd, undefil'd,
Reigns with providence benign:
He still'd the rude abyss, and bade it shine
(While Sapience with approving aspect mild
Saw the stupendous work, and smil'd);
Next thee, his flaming minister, bade rise
O'er young and wondering skies.
Since thou, great orb, with all-enlight'ning ray
Rulest the golden day,
How far more glorious He, who said serene,
BE, and *thou wast*—Himself unform'd, unchang'd,
        unseen![7]

---

[5] [Jones's attitude toward the sun is ever devotional.]

[6] [Here Jones directly alludes to himself and his Sanskrit study.]

[7] [Jones descends from myth to truth, possibly Vedantic, and moralizes his hymn with a planetary religious fact that the sun is governed by the first cause, 'the Eternal Mind'.]

A Hymn to Lacshmī[1]

The Argument

Most of the allusions to *Indian* Geography and Mythology, which occur in the following Ode to the Goddess of Abundance, have been explained on former occasions; and the rest are sufficiently clear. LACSHMĪ, or SRĪ, the CERES of *India*, is the *preserving power* of nature, or, in the language of allegory, the consort of VISHNU or HERI, a personification of the divine goodness; and her origin is variously deduced in the several *Purānās*, as we might expect from a system wholly figurative and emblematical. Some represent her as the daughter of BHRIGU, a son of BRAHMĀ; but, in the *Mārcandēya Purān*, the *Indian* ISIS, or *Nature*, is said to have assumed three transcendent forms, according to her three *guna's* or *qualities*, and, in each of them, to have produced a pair of divinities, BRAHMĀ and LACSHMĪ, MAHĒSA and SERESWATĪ, VISHNU and CĀLĪ: after whose intermarriage, BRAHMĀ and SERESWAĪ formed the mundane Egg, which MAHĒSA and CĀLĪ divided into halves; and VISHNU together with LACSHMĪ preserved it from destruction: a third story supposes her to have sprung from the *Sea of milk*, when it was churned on the second incarnation of HERI, who is often painted reclining on the serpent ANANTA, the emblem of eternity; and this fable, whatever may be the meaning of it, has been chosen as the most poetical. The other names of SRĪ, or *Prosperity*, are HERIPRIYĀ, PEDMĀLAYĀ, or PEDMĀ, and CAMALĀ; the first implying the wife of VISHNU, and the rest derived from the names of the Lotos. As to the tale of SUDĀMAN, whose wealth is proverbial among the *Hindus*, it is related at considerable length in the *Bhāgavat*, or great *Purān* on the Acheivements of CRISHNA: the *Brāhmen*, who read it with me, was frequently stopped by his tears. We may be inclined

---

[1] [This hymn, written in 1788 and published in the *New Asiatic Miscellany* (Calcutta, 1789), celebrates the goddess of prosperity and abudance, one of the most popular deities in Hinduism. Deriving partly from the *Bhagavad-Gita*, it allegorizes Lakshmi's qualities as the world's great mother and preserving power of nature, foreshadowing Shelley's similar concept. Jones's 'Argument' suggests that Europeans should study and benefit from Hinduism.]

perhaps to think, that the wild fables of idolaters are not
worth knowing, and that we may be satisfied with mis-
spending our time in learning the Pagan Theology of old
Greece and *Rome*; but we must consider, that the allegories contained in the Hymn to LACSHMĪ constitute at
this moment the prevailing religion of a most extensive
and celebrated Empire, and are devoutly believed by
many millions, whose industry adds to the revenue of
Britain, and whose manners, which are interwoven with
their religious opinions, nearly affect all *Europeans*, who
reside among them.

The Hymn

Daughter of Ocean and primeval Night,
Who, fed with moonbeams dropping silver dew,
And cradled in a wild wave dancing light,
Saw'st with a smile new shores and creatures new,
Thee, Goddess, I salute; thy gifts I sing,
And, not with idle wing,
Soar from this fragrant bow'r through tepid skies,
Ere yet the steeds of noon's effulgent king
Shake their green manes and blaze with rubied eyes:
Hence, floating o'er the smooth expanse of day,
Thy bounties I survey,
See through man's oval realm thy charms display'd,
See clouds, air, earth, performing thy behest,
Plains by soft show'rs, thy tripping handmaids, dress'd,
And fruitful woods, in gold and gems array'd,
Spangling the mingled shade;
While autumn boon his yellow ensign rears,
And stores the world's true wealth in rip'ning ears.

But most that central tract thy smile adorns,
Which old *Himāla* clips with fost'ring arms,
As with a waxing moon's half-circling horns,
And shields from bandits fell, or worse alarms
Of *Tatar* horse from *Yunan* late subdued,
Or *Bactrian* bowmen rude;
Snow-crown'd *Himāla*, whence, with wavy wings
Far spread, as falcons o'er their nestlings brood,
Fam'd *Brahmaputra* joy and verdure brings,
And *Sindhu*'s five-arm'd flood from *Cashghar* hastes,
To cheer the rocky wastes,
Through western this and that through orient plains;
While bluish *Yamunà* between them streams,
And *Gangà* pure with sunny radiance gleams,
Till *Vānì*, whom a russet ochre stains,
Their destin'd confluence gains:
Then flows in mazy knot the triple pow'r.
O'er laughing *Magadh* and the vales of *Gour*.

Not long inswath'd the sacred infant lay
(Celestial forms full soon their prime attain):
Her eyes, oft darted o'er the liquid way,
With golden light emblaz'd the darkling main;
And those firm breasts, whence all our comforts well,

Rose with enchanting swell;
Her loose hair with the bounding billows play'd,
And caught in charming toils each pearly shell,
That idling through the surgy forest stray'd;
When ocean suffer'd a portentous change,
Toss'd with convulsion strange;
For lofty *Mandar* from his base was torn,
With streams, rocks, woods, by God and Demons
    whirl'd,
While round his craggy sides the mad spray curl'd,
Huge mountain, by the passive Tortoise borne:
Then sole, but not forlorn,
Shipp'd in a flow'r, that balmy sweets exhal'd,
O'er waves of dulcet cream PEDMĀLĀ sail'd.

So name the Goddess from her Lotos blue,
Or CAMALĀ, if more auspicious deem'd:
With many-petal'd wings the blossom flew,
And from the mount a flutt'ring sea-bird seem'd,
Till on the shore it stopp'd, the heav'n-lov'd shore,
Bright with unvalued store
Of gems marine by mirthful INDRA won;
But she, (What brighter gem had shone before?)
No bride for old MĀRĪCHA's frolick son,
On azure HERI fix'd her prosp'ring eyes:
Love bade the bridegroom rise;
Straight o'er the deep, then dimpling smooth,
    he rush'd;
And tow'rd th' unmeasur'd snake, stupendous bed,
The world's great mother, not reluctant, led:
All nature glow'd, whene'er she smil'd or blush'd;
The king of serpents hush'd
His thousand heads, where diamond mirrors blaz'd,
That multiplied her image, as he gaz'd.

Thus multiplied, thus wedded, they pervade,
In varying myriads of ethereal forms,
This pendent Egg by dovelike MĀYĀ laid,
And quell MAHĒSA's ire, when most it storms;
Ride on keen lightning and disarm its flash,
Or bid loud surges lash
Th' impassive rock, and leave the rolling barque
With oars unshatter'd milder seas to dash;
And oft, as man's unnumber'd woes they mark,
They spring to birth in some high-favour'd line,
Half human, half divine,
And tread life's maze transfigur'd, unimpair'd:
As when, through blest *Vrindāvan*'s od'rous grove,
They deign'd with hinds and village girls to rove,
And myrth or toil in field or dairy shar'd,

As lowly rusticks far'd:
Blythe RĀDHĀ she, with speaking eyes, was nam'd,
He CRISHNA, lov'd in youth, in manhood fam'd.

Though long in *Mathurā* with milkmaids bred,
Each bush attuning with his past'ral flute,
ANANDA's holy steers the Herdsman fed,
His nobler mind aspir'd to nobler fruit:
The fiercest monsters of each brake or wood
His youthful arm withstood,
And from the rank mire of the stagnant lake
Drew the crush'd serpent with ensanguin'd hood;
Then, worse than rav'ning beast or fenny snake,
A ruthless king his pond'rous mace laid low,
And heav'n approv'd the blow:
No more in bow'r or wattled cabin pent,
By rills he scorn'd and flow'ry banks to dwell;
His pipe lay tuneless, and his wreathy shell
With martial clangor hills and forests rent;
On crimson wars intent
He sway'd high *Dwāracā*, that fronts the mouth
Of gulfy *Sindhu* from the burning south.

A Brāhmen young, who, when the heav'nly boy
In *Vraja* green and scented *Gōcul* play'd,
Partook each transient care, each flitting joy,
And hand in hand through dale or thicket stray'd,
By fortune sever'd from the blissful seat,
Had sought a lone retreat;
Where in a costless hut sad hours he pass'd,
Its mean thatch pervious to the daystar's heat,
And fenceless from night's dew or pinching blast:
Firm virtue he possess'd and vig'rous health,
But they were all his wealth.
SUDĀMAN was he nam'd; and many a year
(If glowing song can life and honour give)
From sun to sun his honour'd name shall live:
Oft strove his consort wife their gloom to cheer,
And hide the stealing tear;
But all her thrift could scarce each eve afford
The needful sprinkling of their scanty board.

Now Fame, who rides on sunbeams, and conveys
To woods and antres deep her spreading gleam,
Illumin'd earth and heav'n with CRISHNA's praise:
Each forest echoed loud the joyous theme,
But keener joy SUDĀMAN's bosom thrill'd,
And tears ecstatick rill'd:
'My friend', he cried, 'is monarch of the skies!'
Then counsell'd she, who nought unseemly will'd:

'Oh! haste; oh! seek the God with lotos eyes;
The pow'r, that stoops to soften human pain,
None e'er implor'd in vain'.
To *Dwāracà*'s rich tow'rs the pilgrim sped,
Though bashful penury his hope depress'd;
A tatter'd cincture was his only vest,
And o'er his weaker shoulder loosely spread
Floated the mystick thread:
Secure from scorn the crowded paths he trode
Through yielding ranks, and hail'd the Shepherd God.

'Friend of my childhood, lov'd in riper age,
A dearer guest these mansions never grac'd:
O meek in social hours, in council sage!'
So spake the Warriour, and his neck embrac'd;
And e'en the Goddess left her golden seat
Her lord's compeer to greet:
He charm'd, but prostrate on the hallow'd floor,
Their purfled vestment kiss'd and radiant feet;
Then from a small fresh leaf, a borrow'd store
(Such off'rings e'en to mortal kings are due)
Of modest rice he drew.
Some proffer'd grains the soft-ey'd Hero ate,
And more had eaten, but, with placid mien,
Bright RUCMINĪ (thus name th' all-bounteous Queen)
Exclaim'd: 'Ah, hold! enough for mortal state!'
Then grave on themes elate
Discoursing, or on past adventures gay,
They clos'd with converse mild the rapt'rous day.

At smile of dawn dismiss'd, ungifted, home
The hermit plodded, till sublimely rais'd
On granite columns many a sumptuous dome
He view'd, and many a spire, that richly blaz'd,
And seem'd, impurpled by the blush of morn,
The lowlier plains to scorn
Imperious: they, with conscious worth serene,
Laugh'd at vain pride, and bade new gems adorn
Each rising shrub, that clad them. Lovely scene
And more than human! His astonish'd sight
Drank deep the strange delight:
He saw brisk fountains dance, crisp riv'lets wind
O'er borders trim, and round inwoven bow'rs,
Where sportive creepers, threading ruby flow'rs
On em'rald stalks, each vernal arch intwin'd,
Luxuriant though confin'd;
And heard sweet-breathing gales in whispers tell
From what young bloom they sipp'd their spicy smell.

Soon from the palace-gate in broad array

A maiden legion, touching tuneful strings,
Descending strow'd with flow'rs the brighten'd way,
And straight, their jocund van in equal wings
Unfolding, in their vacant centre show'd
Their chief, whose vesture glow'd
With carbuncles and smiling pearls atween;
And o'er her head a veil translucent flow'd,
Which, dropping light, disclos'd a beauteous queen,
Who, breathing love, and swift with timid grace,
Sprang to her lord's embrace
With ardent greeting and sweet blandishment;
His were the marble tow'rs, th' officious train,
The gems unequal'd and the large domain:
When bursting joy its rapid stream had spent,
The stores, which heav'n had lent,
He spread unsparing, unattach'd employ'd
With meekness view'd, with temp'rate bliss enjoy'd.

Such were thy gifts, PEDMĀLĀ, such thy pow'r!
For, when thy smile irradiates yon blue fields,
Observant INDRA sheds the genial show'r,
And pregnant earth her springing tribute yields
Of spiry blades, that clothe the champaign dank,
Or skirt the verd'rous bank,
That in th' o'erflowing rill allays his thirst:
Then, rising gay in many a waving rank,
The stalks redundant into laughter burst;
The rivers broad, like busy should'ring bands,
Clap their applauding hands;
The marish dances and the forest sings;
The vaunting trees their bloomy banners rear;
And shouting hills proclaim th' abundant year,
That food to herds, to herdsmen plenty brings,
And wealth to guardian kings.
Shall man unthankful riot on thy stores?
Ah, no! he bends, he blesses, he adores.

But, when his vices rank thy frown excite,
Excessive show'rs the plains and valleys drench,
Or warping insects heath and coppice blight,
Or drought unceasing, which no streams can quench,
The germin shrivels or contracts the shoot,
Or burns the wasted root:
Then fade the groves with gather'd crust imbrown'd,
The hills lie gasping, and the woods are mute,
Low sink the riv'lets from the yawning ground;
Till Famine gaunt her screaming pack lets slip,
And shakes her scorpion whip;
Dire forms of death spread havock, as she flies,
Pain at her skirts and Mis'ry by her side,

And jabb'ring spectres o'er her traces glide;
The mother clasps her babe, with livid eyes,
Then, faintly shrieking, dies:
He drops expiring, or but lives to feel
The vultures bick'ring for their horrid meal.[2]

From ills, that, painted, harrow up the breast,
(What agonies, if real, must they give!)
Preserve thy vot'ries:[3] be their labours blest!
Oh! bid the patient *Hindu* rise and live.
His erring mind, that wizard lore beguiles
Clouded by priestly wiles,
To senseless nature bows for nature's GOD.
Now, stretch'd o'er ocean's vast from happier isles,
He sees the wand of empire, not the rod:
Ah, may those beams, that western skies illume,
Disperse th' unholy gloom!
Meanwhile may laws, by myriads long rever'd,
Their strife appease, their gentler claims decide;
So shall their victors, mild with virtuous pride,
To many a cherish'd grateful race endear'd,
With temper'd love be fear'd:
Though mists profane obscure their narrow ken,
They err, yet feel; though pagans, they are men.

Two Hymns to Pracriti[1]

The Argument

In all our conversations with learned *Hindus* we find them enthusiastick admirers of Poetry, which they consider as a divine art, that had been practised for numberless ages in heaven, before it was revealed on earth by VĀLMĪC, whose great Heroick Poem is fortunately preserved: the *Brāhmans* of course prefer that poetry, which they believe to have been *actually inspired*; while the *Vaidyas*, who are in general perfect grammarians and good poets, but are not suffered to read any of the *sacred* writings except the *Ayurvēda*, or *Body of Medical Tracts*, speak with rapture of their innumerable *popular* poems, *Epick*, *Lyrick*, and *Dramatick*, which were composed by men not literally inspired, but called, metaphorically, the sons of SERESWATI, or MINERVA; among whom the *Pandits* of all sects, nations, and degrees are unanimous in giving the prize of glory to CĀLĪDĀSA, who flourished in the court of VICRAMĀDITYA, fifty-seven years before Christ. He wrote several *Dramas*, one of which, entitled SACONTALĀ, is in my possession; and the subject of it appears to be as interesting as the composition is beautiful: besides these he published the *Mēghadūta*, or cloud messenger, and the *Nalōdaya*, or rise of NALA, both elegant love-tales; the *Raghuvansa*, an Heroick Poem; and the *Cumāra Sambhava*, or birth of CUMĀRA, which supplied me with materials for the first of the following Odes. I have not indeed yet read it; since it could not be correctly

[2] [A vivid portrayal of famine and floods which are likely to result if the goddess in angered.]

[3] [Jones makes a plea to Lakshmi for protection and preservation of the devotees: thus a prayerful exhortation to the goddess of wealth and happiness.]

[1] [The two hymns, composed about 1788 in an especially Oriental richness of style and published posthumously in *Works* (13:241-65), address the dual aspects of the goddess who represents the prime principle of nature: destroyer of evil through Durgā, and the gracious disposer of the plenty in nature through Bhavanī. 'The Hymn to Durgā' tells a story from Kālidāsa's *Kumārasambhava*, made up of a series of units, each consisting of strophe, antistrophe, and epode, and each devoted to an episode of Kālī. The other hymn describes Bhavāni's power of fecundity and productivity in nature, another character of the Indian Isis. Her mild, mysterious powers cause even the iron breasts of river dragons to melt with passion. The predominant image in the poem is the image of dance, a symbol of activity in nature.]

copied for me during the short interval, in which it is in my power to amuse myself with literature; but I have heard the story told both in *Sanscrit* and *Persian*, by many *Pandits*, who had no communication with each other; and their outline of it coincided so perfectly, that I am convinced of its correctness: that outline is here filled up, and exhibited in a lyric form, partly in the *Indian*, partly in the *Grecian*, taste; and great will be my pleasure, when I can again find time for such amusements, in reading the whole poem of CĀLĪDĀSA, and in comparing my descriptions with the original composition. To anticipate the story in a preface would be to destroy the interest, that may be taken in the poem; a disadvantage attending all prefatory arguments, of which those prefixed to the several books of TASSO, and to the Dramas of METASTASIO, are obvious instances; but, that any interest may be taken in the two hymns addressed to PRACRITI, under different names, it is necessary to render them intelligible by a previous explanation of the mythological allusions, which could not but occur in them.

ISWARA, or ĪSA, and ĪSĀNI or ĪSĪ, are unquestionably the OSIRIS and ISIS of Egypt; for, though neither a resemblance of names, nor a similarity of character, would separately prove the identity of *Indian* and *Egyptian* Deities, yet, when they both concur, with the addition of numberless corroborating circumstances, they form a proof little short of demonstration. The *female* divinity, in the mythological systems of the East, represents the active *power* of the *male*; and that ĪSĪ means *active nature*, appears evidently from the word *sácta*, which is derived from *sacti*, or *power*, and applied to those *Hindus*, who direct their adoration principally to that goddess: this feminine character of PRACRITI, or *created nature*, is so familiar in most languages, and even in our own, that the gravest *English* writers, on the most serious subjects of religion and philosophy, speak of *her* operations, as if *she* were actually an animated being; but such personifications are easily misconceived by the multitude, and have a strong tendency to polytheism. The principal operations of nature are, not the absolute annihilation and new creation of what we call *material substances*, but the temporary extinction and reproduction, or, rather in one word, the *transmutation*, of *forms*; whence the epithet *Polymorphos* is aptly given to nature by *European* philosophers: hence ĪSWARA, SIVA, HARA (for those are his names and near a thousand more), united with ĪSĪ, represent the secondary causes, whatever thay may be, of natural phenomena, and principally those of temporary *destruction* and *regeneration*; but the *Indian* ISIS appears in a variety of characters, especially in those of

PĀRVATĪ, CĀLĪ, DURGĀ, and BHĀVANĪ, which bear a strong resemblance to the JUNO of HOMER, to HECATE, to the armed PALLAS, and to the *Lucretian* VENUS.

The name PĀRVATĪ took its rise from a wild poetical fiction. HIMĀLAYA, or the *Mansion of Snow*, is the title given by the *Hindus* to that vast chain of mountains, which limits *India* to the north, and embraces it with its eastern and western arms, both extending to the ocean; the former of those arms is called *Chandrasēc'hara*, or the *Moon's Rock*; and the second, which reaches as far west as the mouths of the *Indus*, was named by the ancients *Montes Parveti*. These hills are held sacred by the *Indians*, who suppose them to be the terrestrial haunt of the God ĪSWARA. The mountain *Himālaya*, being personified, is represented as a powerful monarch, whose wife was MĒNĀ: their daughter is named PĀRVATĪ, or *Mountain-born*, and DURGĀ, or of *difficult access*; but the *Hindus* believe her to have been married to SIVA in a pre-existent state, when she bore the name of SATĪ. The daughter of HIMĀLAYA had two sons; GANĒSA, or the *Lord of Spirits*, adored as the wisest of Deities, and always invoked at the beginning of every literary work, and CUMĀRA, SCANDA, or CĀRTICĒYA, commander of the celestial armies.

The pleasing fiction of CĀMA, the *Indian* CUPID, and his friend VASANTA, or the Spring, has been the subject of another poem;[2] and here it must be remembered, that the God of Love is named also SMARA, CANDARPA, and ANANGA. One of his arrows is called *Mellicà*, the *Nyctanthes* of our *Botanists*, who very unadvisedly reject the vernacular names of most *Asiatick* plants: it is beautifully introduced by CĀLĪDĀSA into this lively couplet:

> *mellicāmuculè bhāti gunjanmattamadhuvratah,*
> *Prayānè panchaōānasya sanc'hamāpūrayanniva.*

'The intoxicated bee shines and murmurs in the fresh-blown *Mellicà*, like him who gives breath to a white conch in the procession of the God with five arrows'.

A critick, to whom CĀLĪDĀSA repeated this verse, observed, that the comparison was not exact: since the bee sits on the blossom itself, and does not murmur at the end of the tube, like him who blows a conch: 'I was aware of that', said the poet, 'and, therefore, described the bee as *intoxicated*: a drunken musician would blow the shell at the wrong end:' There was more than wit in this answer: it was a just rebuke to a dull critick; for poetry delights in *general* images, and is so far from being a

---

[2] ['A Hymn to Camdeo']

perfect imitation, that a scrupulous exactness of descriptions and similes, by leaving nothing for the imagination to supply, never fails to diminish or destroy the pleasure of every reader, who has an imagination to be gratified.

It may here be observed, that *Nymphæa*, not *Lotos*, is the *generick* name in *Europe* of the flower consecrated to ISIS: the *Persians* know by the name of *Nīlūfer* that species of it, which the Botanists ridiculously call *Nelumbo*, and which is remarkable for its curious *pericarpium*, where each of the seeds contains in miniature the leaves of a perfect vegetable. The *lotos* of HOMER was probably the *sugar-cane*, and that of LINNÆUS is papilionaceous plant; but he gives the same name to another species of the *Nymphæa*; and the word is so constantly applied among us in *India* to the *Nīlūfer*, that any other would be hardly intelligible: the *blue* lotos grows in *Cashmīr* and in *Persia*, but not in *Bengal*, where we see only the *red* and the *white*; and hence occasion is taken to feign, that the lotos of *Hindustan* was dyed crimson by the blood of SIVA.

CUVERA, mentioned in the fourteenth stanza, is the God of Wealth, supposed to reside in a magnificent city, called *Alacà*; and VRIHASPATI, or the Genius of the planet *Jupiter*, is the preceptor of the Gods in *Swerga* or the firmament: he is usually represented as their orator, when any message is carried from them to one of the three superior Deities.

The lamentations of RETI, the wife of CAMA, fill a whole book in the *Sanscrit* poem, as I am informed by my teacher, a learned *Vaidya*; who is restrained only from reading the book, which contains a description of the nuptials; for the ceremonies of a marriage where BRAHMA himself officiated as the father of the bridegroom, are too holy to be known by any but *Brāhmans*.

The achievements of DURGA in her martial character as the patroness of *Virtue*, and her battle with a demon in the shape of a buffalo, are the subject of many episodes in the *Purānas* and *Cāvyas*, or *sacred* and *popular* poems; but a full account of them would have destroyed the unity of the Ode, and they are barely alluded to in the last stanza.

It seemed proper to change the measure, when the goddess was to be addressed as BHAVANI, or the *power of fecundity*; but such a change, though very common in *Sanscrit*, has its inconveniences in *European* poetry: a distinct Hymn is therefore appropriated to her in that capacity; for the explanation of which we need only premise, that LACSHMI is the Goddess of *Abundance*; that the *Cētaca* is a fragrant and beautiful plant of the *Diœcian*

kind, known to Botanists by the name of *Pandanus*; and that the *Dūrgōtsava*, or great festival of BHAVANI at the close of the rains, ends in throwing the image of the goddess into the *Ganges* or other sacred water.

I am not conscious of having left unexplained any difficult allusion in the two poems; and have only to add (lest *European* criticks should consider a few of the images as inapplicable to *Indian* manners), that the ideas of *snow* and *ice* are familiar to the *Hindus*; that the mountains of Himalaya may be clearly discerned from a part of *Bengal*; that the *Grecian* HÆMUS is the *Sanscrit* word *haimas*, meaning *snowy*; and that funeral *urns* may be seen perpetually on the banks of the river.

The two Hymns are neither translations from any other poems, nor imitations of any; and have nothing of PINDAR in them except the measures, which are nearly the same, syllable for syllable, with those of the first and second *Nemean* Odes: more musical stanzas might perhaps have been formed; but, in every art, variety and novelty are considerable sources of pleasure. The style and manner of PINDAR have been greatly mistaken; and, that a distinct idea of them may be conceived by such, as have not access to that inimitable poet in his own language, I cannot refrain from subjoining the first *Nemean* Ode, not only in the same measure as nearly as possible, but almost word for word, with the original;[3] those epithets and phrases only being necessarily added, which are printed in *Italick* letters.

---

[3] [Not reprinted here. In *Works* 13:337-41]

The Hymn to Durgā

I.1

From thee begins the solemn air,
Ador'd GANEŚA; next, thy sire we praise
(Him, from whose red clust'ring hair
A new-born crescent sheds propitious rays,
Fair as GANGĀ's curling foam),
Dread ISWARA; who lov'd o'er awful mountains,
Rapt in prescience deep, to roam,
But chiefly those, whence holy rivers gush,
Bright from their secret fountains,
And o'er the realms of BRAHMĀ rush.

I. 2.

Rock above rock they ride sublime,
And lose their summits in blue fields of day,
Fashion'd first, when rolling time,
Vast infant, in his golden cradle lay,
Bidding endless ages run
And wreathe their giant heads in snows eternal
Gilt by each revolving sun;
Though neither morning beam, nor noontide glare,
In wintry sign or vernal,
Their adamantine strength impair;

I. 3.

Nor e'en the fiercest summer heat
Could thrill the palace, where their Monarch reign'd
On his frost-impearled seat,
(Such height had unremitted virtue gain'd!)
HIMĀLAYA, to whom a lovely child,
Sweet PARVATĪ, sage MĒNĀ bore,
Who now, in earliest bloom, saw heav'n adore
Her charms; earth languish, till she smil'd.

II. 1.

But she to love no tribute paid;
Great ISWARA her pious cares engag'd:
Him, who Gods and fiends dismay'd,
She sooth'd with off'rings meek, when most he rag'd.
On a morn, when, edg'd with light,
The lake-born flow'rs their sapphire cups expanded
Laughing at the scatter'd night,
A vale remote and silent pool she sought,
Smooth-footed, lotos-handed,
And braids of sacred blossoms wrought;

II. 2.

Not for her neck, which, unadorn'd,
Bade envying antelopes their beauties hide:
Art she knew not, or she scorn'd;
Nor had her language e'en a name for pride.
To the God, who, fix'd in thought,
Sat in a crystal cave new worlds designing,
Softly sweet her gift she brought,
And spread the garland o'er his shoulders broad,
Where serpents huge lay twining,
Whose hiss the round creation aw'd.

II. 3.

He view'd, half-smiling, half-severe,
The prostrate maid—That moment through the rocks
He, who decks the purple year,
VASANTA, vain of odorif'rous locks,
With CĀMA, hors'd on infant breezes flew:
(Who knows not CĀMA, nature's king?)
VASANTA barb'd the shaft and fix'd the string;
The living bow CANDARPA drew.

III. 1.

Dire sacrilege! The chosen reed,
That SMARA pointed with transcendent art,
Glanc'd with unimagin'd speed,
And ting'd its blooming barb in .SIVA's heart:
Glorious flow'r, in heav'n proclaim'd
Rich *Mellicà*, with balmy breath delicious,
And on earth *Nyctanthes* nam'd!
Some drops divine, that o'er the lotos blue
Trickled in rills auspicious,
Still mark it with a crimson hue.

III. 2.

Soon clos'd the wound its hallow'd lips;
But nature felt the pain: heav'n's blazing eye
Sank absorb'd in sad eclipse,
And meteors rare betray'd the trembling sky;
When a flame, to which compar'd
The keenest lightnings were but idle flashes,
From that orb all-piercing glar'd,
Which in the front of wrathful HARA rolls,
And soon to silver ashes
Reduc'd th' inflamer of our souls.

III. 3.

VASANT, for thee a milder doom,
Accomplice rash, a thund'ring voice decreed:
'With'ring live in joyless gloom,
While ten gay signs the dancing seasons lead.
Thy flow'rs, perennial once, now annual made,
The Fish and Ram shall still adorn;
But, when the Bull has rear'd his golden horn,

Shall, like yon idling rainbow, fade'.
IV. 1.
The thunder ceas'd; the day return'd;
But SIVA from terrestrial haunts had fled:
Smit with rapt'rous love he burn'd,
And sigh'd on gemm'd *Cailāsa*'s viewless head.
Lonely down the mountain steep,
With flutt'ring heart, soft PARVATI descended;
Nor in drops of nectar'd sleep
Drank solace through the night, but lay alarm'd,
Lest her mean gifts offended
The God her pow'rful beauty charm'd.
IV. 2.
All arts her sorr'wing damsels tried,
Her brow, where wrinkled anguish low'r'd, to smoothe,
And, her troubled soul to soothe,
Sagacious MĒNĀ mild reproof applied;
But nor art nor counsel sage,
Nor e'en her sacred parent's tender chiding,
Could her only pain assuage:
The mountain drear she sought, in mantling shade
Her tears and transports hiding,
And oft to her adorer pray'd.
IV. 3.
There on a crag, whose icy rift
Hurl'd night and horror o'er the pool profound,
That with madding eddy swift
Revengeful bark'd his rugged base around,
The beauteous hermit sat; but soon perceiv'd
A *Brāhmen* old before her stand,
His rude staff quiv'ring in his wither'd hand,
Who, falt'ring, ask'd for whom she griev'd.
V. 1.
'What graceful youth with accents mild,
Eyes like twin stars, and lips like early morn,
Has thy pensive heart beguil'd?'
'No mortal youth', she said with modest scorn,
'E'er beguil'd my guiltless heart:
Him have I lost, who to these mountains hoary
Bloom celestial could impart.
Thee I salute, thee ven'rate, thee deplore,
Dread SIVA, source of glory,
Which on these rocks must gleam no more!'
V. 2.
'Rare object of a damsel's love',
The wizard bold replied, 'who, rude and wild,
Leaves eternal bliss above,
And roves o'er wastes where nature never smil'd,
Mounted on his milkwhite bull!
Seek INDRA with aerial bow victorious,

Who from vases ever full
Quaffs love and nectar; seek the festive hall,
Rich caves, and mansion glorious
Of young CUVĒRA, lov'd by all;
V. 3.
'But spurn that sullen wayward God,
That three-ey'd monster, hideous, fierce, untam'd,
Unattir'd, ill-girt, unshod . . . '
'Such fell impiety, the nymph exclaim'd,
Who speaks, must agonize; who hears, must die;
Nor can this vital frame sustain
The pois'nous taint, that runs from vein to vein;
Death may atone the blasphemy'.
VI. 1.
She spoke, and o'er the rifted rocks
Her lovely form with pious phrensy threw;
But beneath her floating locks
And waving robes a thousand breezes flew,
Knitting close their silky plumes,
And in mid-air a downy pillow spreading;
Till, in clouds of rich perfumes
Embalm'd, they bore her to a mystick wood;
Where streams of glory shedding,
The well-feign'd *Brāhmen*, SIVA stood.
VI. 2.
The rest, my song conceal:
Unhallow'd ears the sacrilege might rue.
Gods alone to Gods reveal
In what stupendous notes th' immortals woo.
Straight the sons of light prepar'd
The nuptial feast, heav'n's opal gates unfolding,
Which th' empyreal army shar'd;
And sage HIMĀLAYA shed blissful tears
With aged eyes beholding
His daughter empress of the spheres.
VI. 3.
Whilst ev'ry lip with nectar glow'd,
The bridegroom blithe his transformation told:
Round the mirthful goblets flow'd,
And laughter free o'er plains of ether roll'd:
'Thee too, like VISHNU', said the blushing queen,
'Soft MĀYĀ, guileful maid, attends;
But in delight supreme the phantasm ends;
Love crowns the visionary scene'.
VII. 1.
Then rose VRIHASPATI, who reigns
Beyond red MANGALA's terrifick sphere,
Wand'ring o'er cerulean plains:
His periods eloquent heav'n loves to hear
Soft as dew on waking flow'rs.

He told, how TĀRACA with snaky legions,
Envious of supernal pow'rs,
Had menac'd long old MĒRU's golden head,
And INDRA's beaming regions
With desolation wild had spread:
VII. 2.
How, when the Gods to BRAHMĀ flew
In routed squadrons, and his help implor'd;
'Sons', he said, 'from vengeance due
The fiend must wield secure his fiery sword
(Thus th' unerring Will ordains),
Till from the Great Destroyer's pure embraces,
Knit in love's mysterious chains
With her, who, daughter to the mountain-king,
Yon snowy mansion graces,
CUMĀRA, warrior-child, shall spring;
VII. 3.
'Who, bright in arms of heav'nly proof,
His crest a blazing star, his diamond mail
Colour'd in the rainbow's woof,
The rash invaders fiercely shall assail,
And, on a stately peacock borne, shall rush
Against the dragons of the deep;
Nor shall his thund'ring mace insatiate sleep,
Till their infernal chief it crush'.
VIII. 1.
'The splendid host with solemn state
(Still spoke th' ethereal orator unblam'd)
Reason'd high in long debate;
Till, through my counsel provident, they claim'd
Hapless CĀMA's potent aid:
At INDRA's wish appear'd the soul's inflamer,
And, in vernal arms array'd,
Engag'd (ah, thoughtless!) in the bold emprise
To tame wide nature's tamer,
And soften Him, who shakes the skies.
VIII. 2.
'See now the God, whom all ador'd,
An ashy heap, the jest of ev'ry gale!
Loss by heav'n and earth deplor'd!
For, love extinguish'd, earth and heav'n must fail.
Mark, how RETĪ bears his urn,
And tow'rd her widow'd pile with piercing ditty
Points the flames—ah, see it burn!
How ill the fun'ral with the feast agrees!
Come, love's pale sister, pity;
Come, and the lover's wrath appease'.
VIII. 3.
Tumultuous passions, whilst he spoke,
In heav'nly bosoms mix'd their bursting fire,

Scorning frigid wisdom's yoke,
Disdain, revenge, devotion, hope, desire:
Then grief prevail'd; but pity won the prize.
Not SIVA could the charm resist:
'Rise, holy love!' he said; and kiss'd
The pearls, that gush'd from DURGĀ's eyes.
IX. 1.
That instant through the blest abode,
His youthful charms renew'd, ANANGA came:
High on em'rald plumes he rode
With RETĪ brighten'd by th' eluded flame;
Nor could young VASANTA mourn
(Officious friend!) his darling lord attending,
Though of annual beauty shorn:
'Love-shafts enow one season shall supply,
He menac'd unoffending,
To rule the rulers of the sky'.
IX. 2.
With shouts the boundless mansion rang;
And, in sublime accord, the radiant quire
Strains of bridal rapture sang
With glowing conquest join'd and martial ire:
'Spring to life, triumphant son,
Hell's future dread, and heav'n's eternal wonder!
Helm and flaming habergeon
For thee, behold, immortal artists weave,
And edge with keen blue thunder
The blade, that shall th' oppressor cleave'.
IX. 3.
O DURGĀ, thou hast deign'd to shield
Man's feeble virtue with celestial might,
Gliding from yon jasper field,
And on a lion borne, hast brav'd the fight;
For, when the demon Vice thy realms defied,
And arm'd with death each arched horn,
Thy golden lance, O goddess mountain-born,
Touch but the pest—He roar'd and died.

The Hymn to Bhavānī

When time was drown'd in sacred sleep,
And raven darkness brooded o'er the deep,
Reposing on primeval pillows
Of tossing billows,
The forms of animated nature lay;
Till o'er the wild abyss, where love
Sat like a nestling dove,
From heav'n's dun concave shot a golden ray.

Still brighter and more bright it stream'd,
Then, like a thousand suns, resistless gleam'd;
Whilst on the placid waters blooming,
The sky perfuming,
An op'ning Lotos rose, and smiling spread
His azure skirts and vase of gold,
While o'er his foliage roll'd
Drops, that impearl BHAVĀNĪ's orient bed.

Mother of Gods, rich nature's queen,
Thy genial fire emblaz'd the bursting scene;
For, on th' expanded blossom sitting,
With sun-beams knitting
That mystick veil for ever unremov'd,
Thou badst the softly kindling flame
Pervade this peopled frame
And smiles, with blushes ting'd, the work approv'd.

Goddess, around thy radiant throne
The scaly shoals in spangled vesture shone,
Some slowly through green waves advancing,
Some swiftly glancing,
As each thy mild mysterious pow'r impell'd:
E'en ores and river-dragons felt
Their iron bosoms melt
With scorching heat; for love the mightiest quell'd.

But straight ascending vapours rare
O'ercanopied thy seat with lucid air,
While, through young INDRA's new dominions
Unnumber'd pinions
Mix'd with thy beams a thousand varying dyes,
Of birds or insects, who pursued
Their flying loves, or woo'd

Them yielding, and with musick fill'd the skies.

And now bedeck'd with sparkling isles
Like rising stars, the watry desert smiles;
Smooth plains by waving forests bounded,
With hillocks rounded,
Send forth a shaggy brood, who, frisking light
In mingled flocks or faithful pairs,
Impart their tender cares:
All animals to love their kind invite.

Nor they alone: those vivid gems,
That dance and glitter on their leafy stems,
Thy voice inspires, thy bounty dresses,
Thy rapture blesses,
From yon tall palm, who, like a sunborn king,
His proud tiara spreads elate,
To those, who throng his gate,
Where purple chieftains vernal tribute bring.

A gale so sweet o'er GANGĀ breathes,
That in soft smiles her graceful cheek she wreathes.
Mark, where her argent brow she raises,
And blushing gazes
On yon fresh *Cētaca*, whose am'rous flow'r
Throws fragrance from his flaunting hair,
While with his blooming fair
He blends perfume, and multiplies the bow'r.

Thus, in one vast eternal gyre,
Compact of fluid shapes, instinct with fire,
Lead, as they dance, this gay creation,
Whose mild gradation
Of melting tints illudes the visual ray:
Dense earth in springing herbage lives,
Thence life and nurture gives
To sentient forms, that sink again to clay.

Ye maids and youths on fruitful plains,
Where LACSHMĪ revels and BHAVĀNĪ reigns,
Oh, haste! oh, bring your flow'ry treasures,
To rapid measures
Tripping at eve these hallow'd banks along:
The pow'r, in yon dim shrines ador'd,
To primal waves restor'd,
With many a smiling race shall bless your song.

Gītagōvinda:[1]
or,
The Songs of Jayadēva

'The firmament is obscured by clouds; the woodlands are black with *Tamāla*-trees; that youth, who roves in the forest, will be fearful in the gloom of night: go, my daughter; bring the wanderer home to my rustick mansion'. Such was the command of NANDA, the fortunate herdsman; and hence arose the love of RĀDHĀ and MĀDHAVA, who sported on the bank of *Yamunà*, or hastened eagerly to the secret bower.

If thy soul be delighted with the remembrance of HERI, or sensible to the raptures of love, listen to the voice of JAYADĒVA, whose notes are both sweet and brilliant. O THOU, who reclinest on the bosom of CAMALĀ; whose ears flame with gems, and whose locks are embellished with sylvan flowers; thou, from whom the day star derived his effulgence, who slewest the venom-breathing CĀLIYA, who beamedst, like a sun, on the tribe of YADU, that flourished like a lotos; thou, who sittest on the plumage of GARURA, who, by subduing demons, gavest exquisite joy to the assembly of immortals; thou, for whom the daughter of JANACA was decked in gay apparel, by whom DŪSHANA was overthrown; thou, whose eye sparkles like the water-lily, who calledst three worlds into existence; thou, by whom the rocks of *Mandar* were easily supported, who sippest nectar from the radiant lips of PEDMĀ, as the fluttering *Chacōra* drinks the moonbeams; *be victorious*, O HERI, *lord of conquest*.

RĀDHĀ sought him long in vain, and her thoughts were confounded by the fever of desire: she roved in the vernal morning among the twining *Vāsantis* covered with soft blossoms, when a damsel thus addressed her with youthful hilarity: 'The gale, that has wantoned round the beautiful clove-plants, breathes now from the hills of *Maylaya*; the circling arbours resound with the notes of the *Cōcil* and the murmurs of honey-making swarms. Now the hearts of damsels, whose lovers travel at a distance, are pierced with anguish; while the blossoms of *Bacul* are conspicuous among the flowrets covered with bees. The *Tamāla*, with leaves dark and odorous, claims a tribute from the musk, which it vanquishes; and the clustering flowers of the *Palāsa* resemble the nails of CĀMA, with which he rends the hearts of the young. The full-blown *Cēsara* gleams like the sceptre of the world's monarch, Love; and the pointed thyrse of the *Cētaca* resembles the darts, by which lovers are wounded. See the bunches of *Pātali*-flowers filled with bees, like the quiver of SMARA full of shafts; while the tender blossom of the *Caruna* smiles to see the whole world laying shame aside. The far-scented *Mādhavı* beautifies the trees, round which it twines; and the fresh *Mallicà* seduces with rich perfume even the hearts of hermits; while the *Amra*-tree with blooming tresses is embraced by the gay creeper *Atimucta*, and the blue streams of *Yamunà* wind round the groves of *Vrindāvan. In this charming season, which gives pain to separated lovers, young* HERI *sports and dances with a company of damsels.* A breeze, like the breath of love, from the fragrant flowers of the *Cētaca*, kindles every heart, whilst it perfumes the woods with the dust, which it shakes from the *Mallicà* with half-opened buds; and the *Cōcila* bursts into song, when he sees the blossoms glistening on the lovely *Rasāla*'.

The jealous RĀDHĀ gave no answer; and, soon after, her officious friend, perceiving the foe of MURA in the forest eager for the rapturous embraces of the herdsmen's daughters, with whom he was dancing, thus again addressed his forgotten mistress: 'With a garland of wild flowers descending even to the yellow mantle, that girds his azure limbs, distinguished by smiling cheeks and by ear-ring, that sparkle, as he plays, HERI *exults in the assemblage of amorous damsels.* One of them presses him with her swelling breast, while she warbles with exquisite melody. Another, affected by a glance from his eye, stands meditating on the lotos of his face. A third, on pretence of whispering a secret in her ear, approaches his temples, and kisses them with ardour. One seizes his mantle and draws him towards her, pointing to the bower on the banks of *Yamunà*, where elegant *Vanjulas* interweave their branches. He applauds another, who dances in the sportive circle, whilst her bracelets ring, as she

---

[1] [Jones's translation of Jayadeva's lovely, twelfth-century lyric drama *Gīta Govinda*, which celebrates the love of Krishna and Rādhā, moved him into major Sanskrit literature. Begun as an exercise to improve his reading skill in Sanskrit in 1789, it was polished and published in *Asiatick Researches* 3 (1792). The theme, which he interpreted as the attraction of the soul first by earthly and ultimately by heavenly love, intrigued him very much. He was unable to find Jayadeva's original musical scores interpolated into the drama, willingly sacrificing the lovely poetry in order to give Europe a literal and precise meaning through a prose translation.]

beats time with her palms. Now he caresses one, and kisses another, smiling on a third with complacency; and now he chases her, whose beauty has most allured him. Thus the wanton HERI frolicks, in the season of sweets, among the maids of *Vraja*, who rush to his embraces, as if he were Pleasure itself assuming a human form; and one of them, under a pretext of hymning his divine perfections, whispers in his ear: "Thy lips, my beloved, are nectar"'.

RĀDHĀ remains in the forest; but resenting the promiscuous passion of HERI, and his neglect of her beauty, which he once thought superiour, she retires to a bower of twining plants, the summit of which resounds with the humming of swarms engaged in their sweet labours; and there, falling languid on the ground, she thus addresses her female companion. '*Though he take recreation in my absence, and smile on all around him, yet my soul remembers him*, whose beguiling reed modulates a tune sweetened by the nectar of his quivering lip, while his ear sparkles with gems, and his eye darts amorous glances; Him, whose locks are decked with the plumes of peacocks resplendent with many-coloured moons,[2] and whose mantle gleams like a dark blue cloud illumined with rain-bows; Him, whose graceful smile gives new lustre to his lips, brilliant and soft as a dewy leaf, sweet and ruddy as the blossom of *Bandhujīva*, while they tremble with eagerness to kiss the daughters of the herdsmen; Him, who disperses the gloom with beams from the jewels, which decorate his bosom, his wrists, and his ankles, on whose forehead shines a circlet of sandalwood, which makes even the moon contemptible, when it sails through irradiated clouds; Him, whose ear-rings, are formed of entire gems in the shape of the fish *Macar* on the banners of Love; even the yellow-robed God, whose attendants are the chiefs of deities, of holy men, and of demons; Him, who reclines under a gay *Cadamba*-tree; who formerly delighted me, while he gracefully waved in the dance, and all his soul sparkled in his eye. My weak mind thus enumerates his qualities; and, though offended, strives to banish offence. What else can it do? It cannot part with its affection for CRISHNA, whose love is excited by other damsels, and who sports in the absence of RĀDHĀ. *Bring, O friend*, that vanquisher of the demon CĒSI, *to sport with* me, who am repairing to a secret

bower, who look timidly on all sides, who meditate with amorous fancy on his divine transfiguration. Bring him, whose discourse was once composed of the gentlest words, to converse with me, who am bashful on his first approach, and express my thoughts with a smile sweet as honey. Bring him, who formerly slept on my bosom, to recline with me on a green bed of leaves just gathered, while his lip sheds dew, and my arms enfold him. Bring him, who has attained the perfection of skill in love's art, whose hand used to press these firm and delicate spheres, to play with me, whose voice rivals that of the *Cōcil*, and whose tresses are bound with waving blossoms. Bring him, who formerly drew me by the locks to his embrace, to repose with me, whose feet tinkle, as they move, with rings of gold and of gems, whose loosened zone sounds, as it falls; and whose limbs are slender and flexible as the creeping plant. That God, whose cheeks are beautified by the nectar of his smiles, whose pipe drops in his ecstasy, I saw in the grove encircled by the damsels of *Vraja*, who gazed on him askance from the corners of their eyes: I saw him in the grove with happier damsels, yet the sight of him delighted me. Soft is the gale, which breathes over yon clear pool, and expands the clustering blossoms of the voluble *Asōca*; soft, yet grievous to me in the absence of the foe of MADHU. Delightful are the flowers of *Amra*-trees on the mountain-top, while the murmuring bees pursue their voluptuous toil; delightful, yet afflicting to me, O friend, in the absence of the youthful CĒSAVA'.

Meantime, the destroyer of CANSA, having brought to his remembrance the amiable RĀDHĀ, forsook the beautiful damsels of *Vraja*: he sought her in all parts of the forest; his old wound from love's arrow bled again; he repented of his levity, and, seated in a bower near the bank of *Yamunà*, the blue daughter of the sun, thus poured forth his lamentation.

'She is departed—she saw me, no doubt, surrounded by the wanton shepherdesses; yet, conscious of my fault, I durst not intercept her flight. *Wo is me! she feels a sense of injured honour, and is departed in wrath*. How will she conduct herself? How will she express her pain in so long a separation? What is wealth to me? What are numerous attendants? What are the pleasures of the world? What joy can I receive from a heavenly abode? I seem to behold her face with eye-brows contracting themselves through her just resentment: it resembles a fresh lotos, over which two black bees are fluttering: I seem, so present is she to my imagination, even now to caress her with eagerness. Why then do I seek her in this forest? Why do I lament without cause? O slender damsel, anger, I know, has torn thy soft bosom; but whither thou art retired, I know not.

---

[2] [Tennyson acknowledged borrowing this metaphor and other materials for his 'Thou Camest to Thy Bower, My Love', in *Poems by Two Brothers* (London, 1827). See Pachori, 'Tennyson's Early Poems and Their Hindu Imagery', *Literature East & West* 9 (1978-79): 132-34.]

How can I invite thee to return? Thou art seen by me, indeed, in a vision; thou seemest to move before me. Ah! why dost thou not rush, as before, to my embrace? Do but forgive me: never again will I commit a similar offence. Grant me but a sight of thee, O lovely RĀDHICĀ, for my passion torments me. I am not the terrible MAHĒSA: a garland of water-lilies · with subtil threads decks my shoulders; not serpents with twisted folds: the blue petals of the lotos glitter on my neck; not the azure gleam of poison: powdered sandal-wood is sprinkled on my limbs; not pale ashes: O God of Love, mistake me not for MAHĀDĒVA. Wound me not again; approach me not in anger; I love already but too passionately; yet I have lost my beloved. Hold not in thy hand that shaft barbed with an *Amra*-flower! Brace not thy bow, thou conqueror of the world! Is it valour to slay one who faints? My heart is already pierced by arrows from RĀDHĀ's eyes, black and keen as those of an antelope; yet mine eyes are not gratified with her presence. Her eyes are full of shafts; her eye-brows are bows; and the tips of her ears are silken strings: thus armed by ANANGA, the God of Desire, she marches, herself a goddess, to ensure his triumph over the vanquished universe. I meditate on her delightful embrace, on the ravishing glances darted from her eye, on the fragrant lotos of her mouth, on her nectar-dropping speech; on her lips ruddy as the berries of the *Bimba*; yet even my fixed meditation on such an assemblage of charms encreases, instead of alleviating, the misery of separation'.

The damsel, commissioned by RĀDHĀ, found the disconsolate God under an arbour of spreading *Vāniras* by the side of *Yamunà*; where, presenting herself gracefully before him, she thus described the affliction of his beloved:

'She despises essence of sandal-wood, and even by moon-light sits brooding over her gloomy sorrow; she declares the gale of *Malaya* to be venom, and the sandal-trees, through which it has breathed, to have been the haunt of serpents. *Thus*, O MĀDHAVA, *is she afflicted in thy absence with the pain, which love's dart has occasioned: her soul is fixed on thee*. Fresh arrows of desire are continually assailing her, and she forms a net of lotos-leaves as armour for her heart, which thou alone shouldst fortify. She makes her own bed of the arrows darted by the flowery-shafted God; but, when she hoped for thy embrace, she had formed for thee a couch of soft blossoms. Her face is like a water-lily, veiled in the dew of tears, and her eyes appear like moons eclipsed, which let fall their gathered nectar through pain caused by the tooth of the furious dragon. She draws thy image with

musk in the character of the Deity with five shafts, having subdued the *Macar*, or horned shark, and holding an arrow tipped with an *Amra*-flower; thus she draws thy picture, and worships it. At the close of every sentence, 'O MĀDHAVA', she exclaims, 'at thy feet am I fallen, and in thy absence even the moon, though it be a vase full of nectar, inflames my limbs'. Then, by the power of imagination, she figures thee standing before her; thee, who art not easily attained: she sighs, she smiles, she mourns, she weeps, she moves from side to side, she laments and rejoices by turns. Her abode is a forest; the circle of her female companions is a net; her sighs are flames of fire kindled in a thicket; herself (alas! through thy absence) is become a timid roe; and Love is the tiger, who springs on her like YAMA, the Genius of Death. So emaciated is her beautiful body, that even the light garland, which waves over her bosom, she thinks a load. *Such, O bright-haired God, is* RĀDHĀ *when thou art absent*. If powder of sandal-wood finely levigated be moistened and applied to her breasts, she starts, and mistakes it for poison. Her sighs form a breeze long extended, and burn her like the flame, which reduced CANDARPA to ashes. She throws around her eyes, like blue water-lilies with broken stalks, dropping lucid streams. Even her bed of tender leaves appear in her sight like a kindled fire. The palm of her hand supports her aching temple, motionless as the crescent rising at eve. 'HERI, HERI', thus in silence she meditates on thy name, as if her wish were gratified, and she were dying through thy absence. She rends her locks; she pants; she laments inarticulately; she trembles; she pines; she muses; she moves from place to place; she closes her eyes; she falls; she rises again; she faints: in such a fever of love, she may live, O celestial physician, if thou administer the remedy; but, shouldst Thou be unkind, her malady will be desperate. Thus, O divine healer, by the nectar of thy love must RĀDHĀ be restored to health; and, if thou refuse it, thy heart must be harder than the thunder-stone. Long has her soul pined, and long has she been heated with sandal-wood, moon-light, and water-lilies, with which others are cooled; yet she patiently and in secret meditates on Thee, who alone canst relieve her. Shouldst thou be inconstant, how can she, wasted as she is to a shadow, support life a single moment? How can she, who lately could not endure thy absence even an instant, forbear sighing now, when she looks with half-closed eyes on the *Rasāla* with bloomy branches, which remind her of the vernal season, when she first beheld thee with rapture?

'Here have I chosen my abode: go quickly to RĀDHĀ; soothe her with my message, and conduct her hither'. So spoke the foe of MADHU to the anxious damsel, who hastened back, and thus addressed her companion: 'Whilst a sweet breeze from the hills of *Malaya* comes wafting on his plumes the young God of Desire; while many a flower points his extended petals to pierce the bosom of separated lovers, the *Deity crowned with sylvan blossoms, laments, O friend, in thy absence.* Even the dewy rays of the moon burn him; and, as the shaft of love is descending, he mourns inarticulately with increasing distraction. When the bees murmur softly, he covers his ears; misery sits fixed in his heart, and every returning night adds anguish to anguish. He quits his radiant palace for the wild forest, where he sinks on a bed of cold clay, and frequently mutters thy name. In yon bower, to which the pilgrims of love are used to repair, he meditates on thy form, repeating in silence some enchanting word, which once dropped from thy lips, and thirsting for the nectar which they alone can supply. Delay not, O loveliest of women; follow the lord of thy heart: behold, he seeks the appointed shade, bright with the ornaments of love, and confident of the promised bliss. *Having bound his locks with forest-flowers, he hastens to yon arbour, where a soft gale breathes over the banks of Yamunà:* there, again pronouncing thy name, he modulates his divine reed. Oh! with what rapture doth he gaze on the golden dust, which the breeze shakes from expanded blossoms; the breeze, which has kissed thy cheek! With a mind, languid as a dropping wing, feeble as a trembling leaf, he doubtfully expects thy approach, and timidly looks on the path which thou must tread. Leave behind thee, O friend, the ring which tinkles on thy delicate ankle, when thou sportest in the dance; hastily cast over thee thy azure mantle, and run to the gloomy bower. The reward of thy speed, O thou who sparklest like lightning, will be to shine on the blue bosom of MURĀRI, which resembles a vernal cloud, decked with a string of pearls like a flock of white water-birds fluttering in the air. Disappoint not, O thou lotos-eyed, the vanquisher of MADHU; accomplish his desire; but go quickly: it is night; and the night also will quickly depart. Again and again he sighs; he looks around; he re-enters the arbour; he can scarce articulate thy sweet name; he again smooths his flowery couch; he looks wild; he becomes frantick: thy beloved will perish through desire. The bright-beamed God sinks in the west, and thy pain of separation may also be removed: the blackness of the night is increased, and the passionate imagination of GŌVINDA has acquired additional gloom. My address to thee has equalled in length and in sweetness the song of the Cocila: delay will make thee miserable, O my beauti-

ful friend. Seize the moment of delight in the place of assignation with the son of DĒVACĪ, who descended from heaven to remove the burdens of the universe;[3] he is a blue gem on the forehead of the three worlds, and longs to sip honey, like the bee, from the fragrant lotos of thy cheek'.

But the solicitous maid, perceiving that RĀDHĀ was unable, through debility, to move from her arbour of flowery creepers, returned to GŌVINDA, who was himself disordered with love, and thus described her situation.

'*She mourns, O sovereign of the world, in her verdant bower;* she looks eagerly on all sides in hope of thy approach; then, gaining strength from the delightful idea of the proposed meeting, she advances a few steps, and falls languid on the ground. When she rises, she weaves bracelets of fresh leaves; she dresses herself like her beloved, and, looking at herself in sport, exclaims, 'Behold the vanquisher of MADHU!' Then she repeats again and again the name of HERI, and, catching at a dark blue cloud, strives to embrace it, saying: 'It is my beloved who approaches'. Thus, while thou art dilatory, she lies expecting thee; she mourns; she weeps; she puts on her gayest ornaments to receive her lord; she compresses her deep sighs within her bosom; and then, meditating on thee, O cruel, she is drowned in a sea of rapturous imaginations. If a leaf but quiver, she supposes thee arrived; she spreads her couch; she forms in her mind a hundred modes of delight: yet, if thou go not to her bower, she must die this night through excessive anguish'.

By this time the moon spread a net of beams over the groves of *Vrindāvan*, and looked like a drop of liquid sandal on the face of the sky, which smiled like a beautiful damsel; while its orb with many spots betrayed, as it were, a consciousness of guilt, in having often attended amorous maids to the loss of their family honour. The moon, with a black fawn couched on its disc, advanced in its nightly course; but MĀDHAVA had not advanced to the bower of RĀDHĀ, who thus bewailed his delay with notes of varied lamentation.

'The appointed moment is come; but HERI, alas! comes not to the grove. Must the season of my unblemished youth pass thus idly away? *Oh! what refuge can I seek, deluded as I am by the guile of my female adviser?* The God with five arrows has wounded my heart; and I am deserted by Him, for whose sake I have sought at night the darkest recess of the forest. Since my best beloved

---

[3] [Devakī, King Vasudeva's wife, gave birth to Krishna in the prison of King Kansa of Mathura, who was her cruel brother. Krishna eventually fought and killed him.]

friends have deceived me, it is my wish to die: since my senses are disordered, and my bosom is on fire, why stay I longer in this world? The coolness of this vernal night gives me pain, instead of refreshment: some happier damsel enjoys my beloved; whilst I, alas! am looking at the gems in my bracelets, which are blackened by the flames of my passion. My neck, more delicate than the tenderest blossom, is hurt by the garland, that encircles it: flowers, are, indeed, the arrows of Love, and he plays with them cruelly. I make this wood my dwelling: I regard not the roughness of the *Vētas*-trees; but the destroyer of MADHU holds me not in his remembrance! Why comes he not to the bower of bloomy *Vanjulas*, assigned for our meeting? Some ardent rival, no doubt, keeps him locked in her embrace: or have his companions detained him with mirthful recreations? Else why roams he not through the cool shades? Perhaps, the heart-sick lover is unable through weakness to advance even a step!'—So saying, she raised her eyes; and, seeing her damsel return silent and mournful, unaccompanied by MĀDHAVA, she was alarmed even to phrensy; and, as if she actually beheld him in the arms of a rival, she thus described the vision which overpowered her intellect.

'Yes; in habiliments becoming the war of love, and with tresses waving like flowery banners, *a damsel, more alluring than* RĀDHĀ, *enjoys the conqueror of* MADHU. Her form is transfigured by the touch of her divine lover; her garland quivers over her swelling bosom; her face like the moon is graced with clouds of dark hair, and trembles, while she quaffs the nectareous dew of his lip; her bright ear-rings dance over her cheeks, which they irradiate; and the small bells on her girdle tinkle as she moves. Bashful at first, she smiles at length on her embracer, and expresses her joy with inarticulate murmurs; while she floats on the waves of desire, and closes her eyes dazzled with the blaze of approaching CĀMA: and now this heroine in love's warfare falls exhausted and vanquished by the resistless MURĀRI, but alas! in my bosom prevails the flame of jealousy, and yon moon, which dispels the sorrow of others, increases mine. See again, where the *foe of* MURA, *sports in yon grove on the bank of the Yamunà!* See, how he kisses the lip of my rival, and imprints on her forehead an ornament of pure musk, black as the young antelope on the lunar orb! Now, like the husband of RETI, he fixes white blossoms on her dark locks, where they gleam like flashes of lightning among the curled clouds. On her breasts, like two firmaments, he places a string of gems like a radiant constellation: he binds on her arms, graceful as the stalks of the water-lily, and adorned with hands glowing like the petals of its

flower, a bracelet of sapphires, which resemble a cluster of bees. Ah! see, how he ties round her waist a rich girdle illumined with golden bells, which seem to laugh, as they tinkle, at the inferior brightness of the leafy garlands, which lovers hang on their bowers to propitiate the God of Desire. He places her soft foot, as he reclines by her side, on his ardent bosom, and stains it with the ruddy hue of *Yāvaca*. Say, my friend, why pass I my nights in this tangled forest without joy, and without hope, while the faithless brother of HALADHERA clasps my rival in his arms? Yet why, my companion, shouldst thou mourn, though my perfidious youth has disappointed me? What offence is it of thine, if he sport with a crowd of damsels happier than I? Mark, how my soul, attracted by his irresistible charms, bursts from its mortal frame, and rushes to mix with its beloved. *She whom the God enjoys, crowned with sylvan flowers,* sits carelessly on a bed of leaves with Him, whose wanton eyes resemble blue water-lilies agitated by the breeze. She feels no flame from the gales of *Malaya* with Him, whose words are sweeter than the water of life. She derides the shafts of soul-born CĀMA, with Him, whose lips are like a red lotos in full bloom. She is cooled by the moon's dewy beams, while she reclines with Him, whose hands and feet glow like vernal flowers. No female companion deludes her, while she sports with Him, whose vesture blazes like tried gold. She faints not through excess of passion, while she caresses that youth, who surpasses in beauty the inhabitants of all worlds. O gale, scented with sandal, who breathest love from the regions of the south, be propitious but for a moment: when thou hast brought my beloved before my eyes, thou mayest freely waft away my soul. Love, with eyes like blue water-lilies, again assails me and triumphs; and, while the perfidy of my beloved rends my heart, my female friend is my foe, the cool breeze scorches me like a flame, and the nectar-dropping moon is my poison. Bring disease and death, O gale of *Malaya!* Seize my spirit, O God with five arrows! I ask not mercy from thee: no more will I dwell in the cottage of my father. Receive me in thy azure waves, O sister of YAMA, that the ardour of my heart may be allayed!'

Pierced by the arrows of love, she passed the night in the agonies of despair, and at early dawn thus rebuked her lover, whom she saw lying prostrate before her and imploring her forgiveness.

'Alas! *alas! Go,* MĀDHAVA, *depart,* O CĒSAVA; *speak not the language of guile; follow her,* O lotos-eyed God, *follow her, who dispels thy care.* Look at his eye half-opened, red with continued waking through the pleasurable night, yet smiling still with affection for my rival!

Thy teeth, O cerulean youth, are azure as thy complexion from the kisses, which thou hast imprinted on the beautiful eyes of thy darling graced with dark blue powder; and thy limbs marked with punctures in love's warfare, exhibit a letter of conquest written on polished sapphires with liquid gold. That broad bosom, stained by the bright lotos of her foot, displays a vesture of ruddy leaves over the tree of thy heart, which trembles within it. The pressure of her lip on thine wounds me to the soul. Ah! how canst thou assert, that we are one, since our sensations differ thus widely? Thy soul, O dark-limbed god, shows its blackness externally. How couldst thou deceive a girl who relied on thee; a girl who burned in the fever of love? Thou rovest in woods, and females are thy prey: what wonder? Even thy childish heart was malignant; and thou gavest death to the nurse, who would have given thee milk.[4] Since thy tenderness for me, of which these forests used to talk, has now vanished, and since thy breast, reddened by the feet of my rival, glows as if thy ardent passion for her were bursting from it, the sight of thee, O deceiver, makes me (ah! must I say it?) blush at my own affection'.

Having thus inveighed against her beloved, she sat overwhelmed in grief, and silently meditated on his charms; when her damsel softly addressed her.

'He is gone: the light air has wafted him away. What pleasure now, my beloved, remains in thy mansion? *Continue not, resentful woman, thy indignation against the beautiful* MĀDHAVA. Why shouldst thou render vain those round smooth vases, ample and ripe as the sweet fruit of yon *Tāla*-tree? How often and how recently have I said: 'forsake not the blooming HERI?' Why sittest thou so mournful? Why weepest thou with distraction, when the damsels are laughing around thee? Thou hast formed a couch of soft lotos-leaves: let thy darling charm thy sight, while he reposes on it. Afflict not thy soul with extreme anguish; but attend to my words, which conceal no guile. Suffer CĒSAVA to approach: let him speak with exquisite sweetness, and dissipate all thy sorrows. If thou art harsh to him, who is amiable; if thou art proudly silent, when he deprecates thy wrath with lowly prostrations; if thou showest aversion to him, who loves thee passionately; if, when he bends before thee, thy face be turned contemptuously away; by the same rule of contrariety, the dust of sandal-wood, which thou hast sprinkled, may become poison; the moon, with cool beams, a scorching

sun; the fresh dew, a consuming flame; and the sports of love be changed into agony'.

MĀDHAVA was not absent long: he returned to his beloved; whose cheeks were heated by the sultry gale of her sighs. Her anger was diminished, not wholly abated; but she secretly rejoiced at his return, while the shades of night also were approaching, she looked abashed at her damsel, while He, with faultering accents, implored her forgiveness.

'Speak but one mild word, and the rays of thy sparkling teeth will dispel the gloom of my fears. My trembling lips, like thirsty *Chacōras*, long to drink the moonbeams of thy cheek. *O my darling, who art naturally so tender-hearted, abandon thy causeless indignation. At this moment the flame of desire consumes my heart: Oh! grant me a draught of honey from the lotos of thy mouth.* Or, if thou beest inexorable, grant me death from the arrows of thy keen eyes; make thy arms my chains; and punish me according to thy pleasure. Thou art my life; thou art my ornament; thou art a pearl in the ocean of my mortal birth: oh! be favourable now, and my heart shall eternally be grateful. Thine eyes, which nature formed like blue water-lilies, are become, through thy resentment, like petals of the crimson lotos: oh! tinge with their effulgence these my dark limbs, that they may glow like the shafts of Love tipped with flowers. Place on my head that foot like a fresh leaf, and shade me from the sun of my passion, whose beams I am unable to bear. Spread a string of gems on those two soft globes; let the golden bells of thy zone tinkle, and proclaim the mild edict of love. Say, O damsel with delicate speech, shall I dye red with the juice of *alactaca* those beautiful feet, which will make the full-blown land-lotos blush with shame? Abandon thy doubts of my heart, now indeed fluttering through fear of thy displeasure, but hereafter to be fixed wholly on thee; a heart, which has no room in it for another: none else can enter it, but Love, the bodiless God. Let him wing his arrows; let him wound me mortally; decline not, O cruel, the pleasure of seeing me expire. Thy face is bright as the moon, though its beams drop the venom of maddening desire: let thy nectareous lip be the charmer, who alone has power to lull the serpent, or supply an antidote for his poison. Thy silence afflicts me: oh! speak with the voice of musick, and let thy sweet accents allay my ardour. Abandon thy wrath, but abandon not a lover, who surpasses in beauty the sons of men, and who kneels before thee, O thou most beautiful among women. Thy lips are a *Bandhujiva*-flower; the lustre the *Madhuca* beams on thy cheek; thine eye outshines the blue lotos; thy nose is a bud of the *Tila*; the *Cunda*-blossom yields

---

[4] [A reference to Putanā, a female demon sent by Kaṇsa to kill the baby Krishna by nursing him from her poisoned breasts, but Krishna killed her.]

to thy teeth: thus the flowery-shafted God borrows from thee the points of his darts, and subdues the universe. Surely, thou descendest from heaven, O slender damsel, attended by a company of youthful goddesses; and all their beauties are collected in thee'.

He spake; and, seeing her appeased by his homage, flew to his bower, clad in a gay mantle. The night now veiled all visible objects; and the damsel thus exhorted RĀDHĀ, while she decked her with beaming ornaments.

'Follow, *gentle* RĀDHICĀ, *follow the foe of* MADHU: his discourse was elegantly composed of sweet phrases; he prostrated himself at thy feet; and he now hastens to his delightful couch by yon grove of branching *Vanjulas.* Bind round thy ankle rings beaming with gems; and advance with mincing steps, like the pearl-fed *Marāla.* Drink with ravished ears the soft accents of HERI; and feast on love, while the warbling *Cōcilas* obey the mild ordinance of the flower-darting God. Abandon delay: see, the whole assembly of slender plants, pointing to the bower with fingers of young leaves agitated by the gale, make signals for thy departure. Ask those two round hillocks, which receive pure dew-drops from the garland playing on thy neck, and the buds on whose top start aloft with the thought of thy darling; ask, and they will tell, that thy soul is intent on the warfare of love; advance, fervid warrior, advance with alacrity, while the sound of thy tinkling waist-bells shall represent martial musick. Lead with thee some favoured maid; grasp her hand with thine, whose fingers are long and smooth as love's arrows: march; and, with the noise of thy bracelets, proclaim thy approach to the youth, who will own himself thy slave: 'She will come; she will exult on beholding me; she will pour accents of delight; she will enfold me with eager arms; she will melt with affection': Such are his thoughts at this moment; and, thus thinking, he looks through the long avenue; he trembles; he rejoices; he burns; he moves from place to place; he faints, when he sees thee not coming, and falls in his gloomy bower. The night now dresses in habiliments fit for secrecy, the many damsels, who hasten to their places of assignation: she sets off with blackness their beautiful eyes; fixes dark *Tamāla*-leaves behind their ears; decks their locks with the deep azure of water-lilies, and sprinkles musk on their panting bosoms. The nocturnal sky, black as the touchstone, tries now the gold of their affection, and is marked with rich lines from the flashes of their beauty, in which they surpass the brightest *Cashmirians*'.

RĀDHĀ, thus incited, tripped through the forest; but shame overpowered her, when, by the light of innumerable gems, on the arms, the feet, and the neck of her beloved, she saw him at the door of his flowery mansion: then her damsel again addressed her with ardent exultation.

'Enter, sweet RĀDHĀ the bower of HERI: seek delight, O thou, whose bosom laughs with the foretaste of happiness. Enter, sweet RĀDHĀ, the bower graced with a bed of *Asōca*-leaves: seek delight, O thou, whose garland leaps with joy on thy breast. Enter, sweet RĀDHĀ, the bower illumined with gay blossoms; seek delight, O thou, whose limbs far excel them in softness. Enter, O RĀDHĀ, the bower made cool and fragrant by gales from the woods of *Malaya*: seek delight, O thou, whose amorous lays are softer than breezes. Enter, O RĀDHĀ, the bower spread with leaves of twining creepers: seek delight, O thou, whose arms have been long inflexible. Enter, O RĀDHĀ, the bower, which resounds with the murmur of honey-making bees: seek delight, O thou, whose embrace yields more exquisite sweetness. Enter, O RĀDHĀ, the bower attuned by the melodious band of Cocilas: seek delight, O thou, whose lips, which outshine the grains of the pomegranate, are embellished, when thou speakest, by the brightness of thy teeth. Long has he borne thee in his mind; and now, in an agony of desire, he pants to taste nectar from thy lip. Deign to restore thy slave, who will bend before the lotos of thy foot, and press it to his irradiated bosom; a slave, who acknowledges himself bought by thee for a single glance from thy eye, and a toss of thy disdainful eye-brow'.

She ended; and RĀDHĀ with timid joy, darting her eyes on GŌVINDA, while she musically sounded the rings of her ankles and the bells of her zone, entered the mystick bower for her only beloved. *There she beheld her* MĀDHAVA, *who delighted in her alone; who so long had sighed for her embrace; and whose countenance then gleamed with excessive rapture*: his heart was agitated by her sight, as the waves of the deep are affected by the lunar orb. His azure breast glittered with pearls of unblemished lustre, like the full bed of the cerulean *Yamunà,* interspersed with curls of white foam. From his graceful waist, flowed a pale yellow robe, which resembled the golden dust of the water-lily, scattered over its blue petals. His passion was inflamed by the glances of her eyes, which played like a pair of water-birds with azure plumage, that sport near a full-blown lotos on a pool in the season of dew. Bright ear-rings, like two suns, displayed in full expansion the flowers of his cheeks and lips, which glistened with the liquid radiance of smiles. His locks, interwoven with blossoms, were like a cloud variegated with moon-beams; and on his forehead shone a circle of odorous oil, extracted from the sandal of

Malaya, like the moon just appearing on the dusky horizon; while his whole body seemed in a flame from the blaze of unnumbered gems. Tears of transport gushed in a stream from the full eyes of RĀDHĀ, and their watery glances beamed on her best beloved. Even shame, which before had taken its abode in their dark pupils, was itself ashamed and departed, when the fawn-eyed RĀDHĀ gazed on the brightened face of CRISHNA, while she passed by the soft edge of his couch, and the bevy of his attendant nymphs, pretending to strike the gnats from their cheeks in order to conceal their smiles, warily retired from his bower.

GŌVINDA, seeing his beloved cheerful and serene, her lips sparkling with smiles, and her eye speaking desire, thus eagerly addressed her; while she carelessly reclined on the leafy bed strewn with soft blossoms.

'Set the lotos of thy foot on this azure bosom; and let this couch be victorious over all, who rebel against love. *Give short rapture, sweet* RĀDHĀ, *to* NĀRĀYAN, *thy adorer*. I do thee homage; I press with my blooming palms thy feet, weary with so long a walk. O that I were the golden ring, that plays round thy ankle! Speak but one gentle word; bid nectar drop from the bright moon of thy mouth. Since the pain of absence is removed, let me thus remove the thin vest that enviously hides thy charms. Blest should I be, if those raised globes were fixed on my bosom, and the ardour of my passion allayed. O! suffer me to quaff the liquid bliss of those lips; restore with their water of life thy slave, who has long been lifeless, whom the fire of separation has consumed. Long have these ears been afflicted, in thy absence, by the notes of the *Cōcila*: relieve them with the sound of thy tinkling waistbells, which yield musick, almost equal to the melody of thy voice. Why are those eyes half closed? Are they ahamed of seeing a youth, to whom thy careless resentment gave anguish? Oh! let affliction cease: and let ecstasy drown the remembrance of sorrow'.

In the morning she rose disarrayed, and her eyes betrayed a night without slumber; when the yellow-robed God, who gazed on her with transport, thus meditated on her charms in his heavenly mind: 'Though her locks be diffused at random, though the lustre of her lips be faded, though her garland and zone be fallen from their enchanting stations, and though she hide their places with her hands, looking toward me with bashful silence, yet even thus disarranged, she fills me with extatic delight'. But RĀDHĀ, preparing to array herself, before the company of nymphs could see her confusion, spake thus with exultation to her obsequious lover.

'Place, O son of YADU, with fingers cooler than sandal-wood, place a circlet of musk on this breast, which resembles a vase of consecrated water, crowned with fresh leaves, and fixed near a vernal bower, to propitiate the God of Love. Place, my darling, the glossy powder, which would make the blackest bee envious, on this eye, whose glances are keener than arrows darted by the husband of RETI.[5] Fix, O accomplished youth, the two gems, which form part of love's chain, in these ears, whence the antelopes of thine eyes may run downwards and sport at pleasure. Place now a fresh circle of musk, black as the lunar spots, on the moon of my forehead; and mix gay flowers on my tresses with a peacock's feathers, in graceful order, that they may wave like the banners of CĀMA. Now replace, O tender hearted, the loose ornaments of my vesture; and refix the golden bells of my girdle on their destined station, which resembles those hills, where the God with five shafts, who destroyed SAMBAR, keeps his elephant ready for battle'.

While she spake, the heart of YADAVA triumphed; and, obeying her sportful behests, he placed musky spots on her bosom and forehead, dyed her temples with radiant hues, embellished her eyes with additional blackness, decked her braided hair and her neck with fresh garlands, and tied on her wrists the loosened bracelets, on her ankles the beamy rings, and round her waist the zone of bells, that sounded with ravishing melody.

Whatever is delightful in the modes of musick, whatever is divine in meditations on VISHNU, whatever is exquisite in the sweet art of love, whatever is greaceful in the fine strains of poetry, all that let the happy and wise learn from the songs of JAYADĒVA, whose soul is united with the foot of NĀRĀYAN. May that HERI be your support, who expanded himself into an infinity of bright forms, when, eager to gaze with myriads of eyes on the daughter of the ocean, he displayed his great character of the all-pervading deity, by the multiplied reflections of his divine person in the numberless gems on the many heads of the king of serpents, whom he chose for his couch; that HERI, who removing the lucid veil from the bosom of PEDMĀ, and fixing his eyes on the delicious buds, that grew on it, diverted her attention by declaring that, when she had chosen him as her bridegroom near the sea of milk, the disappointed husband of PERVATI drank in despair the venom, which dyed his neck azure![6]

[5] [Kama, the love god, whom Jones celebrated in 'A Hymn to Camdeo'.]

[6] [A reference to the ocean-churning episode in the *Bhāgavat*, when Lakshmi (Padmā) was presented to Vishnu as a bride. In despair, according to Vishnu's joking account here, Siva drank poison.]

SACONTALĀ;
OR,
THE FATAL RING:
AN
INDIAN DRAMA
BY CĀLIDĀS¹

Translated From The Original Sanscrit
And Pracrit

¹ [Jones first came to know about Kālidāsa's *Sakuntalā* in the late summer of 1787. While in Europe he had heard about Indian Natakas known as Brahminical histories mixed with fables, but his inquiries in Calcutta among the Brahmins convinced him that these were popular works, not histories, and most probably discourses on dancing, music, and poetry. Then a Brāhmin, Pandit Rādhākānt, informed him that the Nātakas were like the English plays staged in Calcutta during the rainy season. When Jones asked for the best specimen of such a play, he was given the *Sakuntalā* in the padded Bengali recension, which was the only one then available in authentic manuscript.

[In a year's time he read the play with the help of Pandit Rāmlochan, a Vaidya of 65 or 66 and a Sanskrit teacher at the University of Nadia, who further assisted him with the interlinear Latin translation of the *Sakuntalā*. Jones rendered it literally into an English version and had Joseph Cooper publish it in Calcutta in the summer of 1789. Thus Jones introduced Kālidāsa, whom he called the Indian Shakespeare, and the best of Sanskrit drama to a startled Europe, dramatically stimulating Western interest in the Golden Age of the Gupta dynasty in Indian history. Kālidāsa was one of the 'Nava-ratna' (Nine Jewels) that adorned the court of Emperor Vikramāditya (Chandra Gupta II). By this translation, Jones ushered in an Age of Indian Renaissance and added Indian drama to world literature.

[A year later, there was a London edition of the *Sakuntalā* by Edwards in 1790, and then another in 1792. Several translations in other European languages were quickly produced: George Forster's famous German translation from Jones in 1791; Nikolai M. Karamsin's Russian, from Forster, in 1792; Hans West's Danish, from Jones, in 1793; A Bruguière's French, from Jones, in 1803; and Doria's Italian, from Bruguière, in 1815.

[Jones's *Sakuntalā* had raised few doubts in Europe about the authenticity of Kālidāsa's poetic-dramatic genius, as most reviewers—previously unaware of the treasures of Sanskrit literature—were enthralled by the playwright's delicate poetry and imagination. Yet they were living in a century that had produced Thomas Chatterton and James Macpherson, whose 'medieval' works had been proven to be forgeries. In his letter to Sir Joseph Banks (18 October 1791), Jones asserted: 'I can assure you that the translation is as literal as possible; but I am not sure, that my own errors or inattention may not have occasioned mistakes' (*Letters*) 2: 894).

[Henry F. Cary later wrote in 'Sir William Jones', *Lives of English Poets* (London, 1846)): 'The doubts suggested by the critics in England, concerning the authenticity of this work, he considered as scarcely deserving of a serious reply' (p. 377).]

Preface

In one of the letters which bear the title of EDIFYING, though most of them swarm with ridiculous errours, and all must be consulted with extreme diffidence, I met, some years ago, with the following passage: 'In the north of India there are many books, called Nātac, which, as the Brahmens assert, contain a large portion of ancient history without any mixture of fable'; and having an eager desire to know the real state of this empire before the conquest of it by the Savages of the North, I was very solicitous, on my arrival in Bengal, to procure access to those books, either by the help of translations, if they had been translated, or by learning the language in which they were originally composed, and which I had yet a stronger inducement to learn from its connection with the administration of justice to the Hindûs; but when I was able to converse with the Brāhmens, they assured me that the Natacs were not histories, and abounded with fables; that they were extremely popular works, and consisted of conversations in prose and verse, held before ancient Rājās in their publick assemblies, on an infinite variety of subjects, and in various dialects of India: this definition gave me no very distinct idea; but I concluded that they were dialogues on moral or literary topicks; whilst other Europeans, whom I consulted, had understood from the natives that they were discourses on dancing, musick, or poetry. At length a very sensible Brāhmen, named Rādhācānt, who had long been attentive to English manners, removed all my doubts, and gave me no less delight than surprise, by telling me that our nation had compositions of the same sort, which were publickly represented at Calcutta in the cold season, and bore the name, as he had been informed, of plays. Resolving at my leisure to read the best of them, I asked which of their Nātacs was most universally esteemed; and he answered without hesitation, Sacontalā, supporting his opinion, as usual among the Pandits, by a couplet to this effect: 'The ring of Sacontalā, in which the fourth act, and four stanzas of that act, are eminently brilliant, displays all the rich exuberance of Cālidāsa's genius'. I soon procured a correct copy of it; and, assisted by my teacher Rāmalōchan, began with translating it verbally into Latin, which bears so great a resemblance to Sanscrit, that it is more convenient than any modern language for a scrupulous interlineary ver-

sion: I then turned it word for word into English, and afterwards, without adding or suppressing any material sentence, disengaged it from the stiffness of a foreign idiom, and prepared the faithful translation of the Indian drama, which I now present to the publick as a most pleasing and authentick picture of old Hindû manners, and one of the greatest curiosities that the literature of Asia has yet brought to light.

Dramatick poetry must have been immemorially ancient in the Indian empire: the invention of it is commonly ascribed to Bheret, a sage believed to have been inspired, who invented also a system of musick which bears his name; but this opinion of its origin is rendered very doubtful by the universal belief, that the first Sanscrit verse ever heard by mortals was pronounced in a burst of resentment by the great Vālmic, who flourished in the silver age of the world, and was author of an Epick Poem on the war of his contemporary, Rāma, king of Ayōdhyà; so that no drama in verse could have been represented before his time; and the Indians have a wild story, that the first regular play, on the same subject with the Rāmāyan, was composed by Hanumat or Pāvan, who commanded an army of Satyrs or Mountaineers in Rāma's expedition against Lancā: they add, that he engraved it on a smooth rock, which, being dissatisfied with his composition, he hurled into the sea; and that, many years after, a learned prince ordered expert divers to take impressions of the poem on wax, by which means the drama was in great measure restored; and my Pandit assures me that he is in possession of it. By whomsoever or in whatever age this species of entertainment was invented, it is very certain, that it was carried to great perfection in its kind, when Vicramāditya, who reigned in the first century before Christ, gave encouragement to poets, philologers, and mathematicians, at a time when the Britons were as unlettered and unpolished as the army of Hanumat: nine men of genius, commonly called the nine gems, attended his court, and were splendidly supported by his bounty; and Cālidās is unanimously allowed to have been the brightest of them.—A modern epigram was lately repeated to me, which does so much honour to the author of Sacontalā, that I cannot forbear exhibiting a literal version of it: 'Poetry was the sportful daughter of Vālmic, and, having been educated by Vyāsa, she chose Cālidās for her bridegroom after the manner of Viderbha: she was the mother of Amara, Sundar, Sanc'ha, Dhanic; but now, old and decrepit, her beauty faded, and her unadorned feet slipping as she walks, in whose cottage does she disdain to take shelter?'

All the other works of our illustrious poet, the Shakespeare of India, that have yet come to my knowledge, are a second play, in five acts, entitled Urvasī; an heroic poem, or rather a series of poems in one book, on the Children of the Sun; another, with perfect unity of action, on the Birth of Cumāra, god of war; two or three love tales in verse; and an excellent little work on Sanscrit Metre, precisely in the manner of Terentianus; but he is believed by some to have revised the works of Vālmic and Vyāsa, and to have corrected the perfect editions of them which are now current: this at least is admitted by all, that he stands next in reputation to those venerable bards; and we must regret, that he has left only two dramatick poems, especially as the stories in his Raghu-vansa would have supplied him with a number of excellent subjects.—Some of his contemporaries, and other Hindû poets even to our own times, have composed so many tragedies, comedies, farces, and musical pieces, that the Indian theatre would fill as many volumes as that of any nation in ancient or modern Europe: all the Pandits assert that their plays are innumerable; and, on my first inquiries concerning them, I had notice of more than thirty, which they consider as the flower of their Nātacs, among which the Malignant Child, the Rape of Ushā, the Taming of Durvāsas, the Seizure of the Lock, Mālati and Mādhava, with five or six dramas on the adventures of their incarnate gods, are the most admired after those of Cālidās. They are all in verse, where the dialogue is elevated; and in prose, where it is familiar; the men of rank and learning are represented speaking pure Sanscrit, and the women Pracrit, which is little more than the language of the Brāhmens melted down by a delicate articulation to the softness of Italian; while the low persons of the drama speak the vulgar dialects of the several provinces which they are supposed to inhabit.

The play of Sacontalā must have been very popular when it was first represented; for the Indian empire was then in full vigour, and the national vanity must have been highly flattered by the magnificent introduction of those kings and heroes in whom the Hindûs gloried; the scenery must have been splendid and beautiful; and there is good reason to believe, that the court at Avanti[2] was equal in brilliancy during the reign of Vicramāditya, to that of any monarch in any age or country.—Dushmanta, the hero of the piece, appears in the chronological tables of the Brāhmens among the Children of the Moon, and in the twenty-first generation after the flood; so that, if we can at all rely on the chronology of the Hindûs, he was nearly

contemporary with Obed, or Jeffe; and Puru, his most celebrated ancestor, was the fifth in descent from Budha, or Mercury, who married, they say, a daughter of the pious king, whom Vishnu preserved in an ark from the universal deluge: his eldest son Bheret was the illustrious progenitor of Curu, from whom Pāndu was lineally descended, and in whose family the Indian Apollo became incarnate; whence the poem, next in fame to the Rāmāyan, is called Mahābhārat.

As to the machinery of the drama, it is taken from the system of mythology, which prevails to this day, and which it would require a large volume to explain; but we cannot help remarking, that the deities introduced in the Fatal Ring are clearly allegorical personages. Marīchi, the first production of Brahmā, or the Creative Power, signifies light, that subtil fluid which was created before its reservoir, the sun, as water was created before the sea; Casyapa, the offspring of Marīchi, seems to be a personification of infinite space, comprehending innumerable worlds; and his children by Aditi, or his active power (unless Aditi mean the primeval day, and Diti, his other wife, the night), are Indra, or the visible firmament, and the twelve Adityas, or suns, presiding over as many months.

On the characters and conduct of the play I shall offer no criticism; because I am convinced that the tastes of men differ as much as their sentiments and passions, and that, in feeling the beauties of art, as in smelling flowers, tasting fruits, viewing prospects, and hearing melody, every individual must be guided by his own sensations and the incommunicable associations of his own ideas. This only I may add, that if Sacontalā should ever be acted in India, where alone it could be acted with perfect knowledge of Indian dresses, manners, and scenery, the piece might easily be reduced to five acts of a moderate length, by throwing the third act into the second, and the sixth into the fifth; for it must be confessed that the whole of Dushmanta's conversation with his buffoon, and great part of his courtship in the hermitage, might be omitted without any injury to the drama.[3]

It is my anxious wish that others may take the pains to learn Sanscrit, and may be persuaded to translate the works of Cālidās: I shall hardly again employ my leisure in a task so foreign to my professional (which are, in truth, my favourite) studies; and have no intention of translating any other book from any language, except the Law Tract of Menu, and the new Digest of Indian and Arabian laws;

---

[2] [The capital city of King Vikramāditya, near Ujjain in central India.]

[3] [Thus Jones intuitively realized that his Sanskrit manuscript was padded.]

but, to show, that the Brāhmens, at least, do not think polite literature incompatible with jurisprudence, I cannot avoid mentioning, that the venerable compiler of the Hindû Digest, who is now in his eighty-sixth year, has the whole play of Sacontala by heart; as he proved when I last conversed with him, to my entire conviction. Lest, however, I should hereafter seem to have changed a resolution which I mean to keep inviolate, I think it proper to say, that I have already translated four of five other books, and among them the *Hitōpadēsa*, which I undertook, merely as an exercise in learning Sanscrit, three years before I knew that Mr. Wilkins, without whose aid I should never have learnt it, had any thought of giving the same work to the publick.

Persons of the Drama

Dushmanta, Emperor of India
Sacontalā, the Heroine of the Piece
Anusūyā, }
Priyamvadā, } Damsels attendant on her
Mādhavya, the Emperor's Buffoon
Gautamī, an old female Hermit
Sārngarava, }
Sāradwata, } two Brāhmens
Canna, Foster-father of Sacontalā
Cumbhīlaca, a Fisherman
Misracēsī, a Nymph
Mātali, Charioteer of Indra
A little Boy
Casyapa, }
Aditi, } Deities, Parents of Indra

Officers of State and Police, Brāhmens, Damsels, Hermits, Pupils, Chamberlains Warders of the Palace, Messengers, and Attendants

The Prologue

A Brāhmen *pronounces the benediction.*

Water was the first work of the Creator; and Fire receives the oblations ordained by law; the Sacrifice is performed with solemnity; the Two Lights of heaven distinguish time; the subtil Ether, which is the vehicle of sound, pervades the universe; the Earth is the natural parent of all increase; and by Air all things breathing are animated; may ĪSA, the God of Nature, apparent in these eight forms, bless and sustain you!

The Manager *enters.*

Man. What occasion is there for a long speech?—[*Looking towards the dressing room.*]—When your decorations, Madam, are completed, be pleased to come forward.

An Actress *enters.*

Actr. I attend, Sir.—What are your commands?
Man. This, Madam, is the numerous and polite assembly of the famed Hero, our king Vicramāditya, the patron of every delightful art; and before this audience we must do justice to a new production of Cālidās, a dramatick piece, entitled Sacontalā, or, The Fatal Ring: it is requested, therefore, that all will be attentive.
Actr. Who, Sir, could be inattentive to an entertainment so well intended?
Man. [*Smiling*]. I will speak, Madam, without reserve.— As far as an enlightened audience receive pleasure from our theatrical talents, and express it, so far, and no farther, I set a value on them; but my own mind is diffident of its powers, how strongly soever exerted.
Actr. You judge rightly in measuring your own merit by the degree of pleasure which this assembly may receive; but its value, I trust, will presently appear.—Have you any farther commands?
Man. What better can you do, since you are now on the stage, than exhilarate the souls, and gratify the sense, of our auditory with a song?
Actr. Shall I sing the description of a season? and which of the seasons do you chuse to hear described?
Man. No finer season could be selected than the summer, which is actually begun, and abounds with delights. How sweet is the close of a summer day, which invites our youth to bathe in pure streams, and induces gentle slumber under the shades refreshed by sylvan breezes, which have passed over the blooming Pātalis and stolen their fragrance!
Actr. [*Singing.*] 'Mark how the soft blossoms of the Nāgacēsar are lightly kissed by the bees! Mark how the damsels delicately place behind their ears the flowers of Sirīsha!'[4]
Man. A charming strain! the whole company sparkles, as it were, with admiration; and the musical mode to which the words are adapted, has filled their souls with rapture.

---

[4] [Jones's interest in botany and zoology was further increased by Kālidāsa's abundant references to Indian flowers, flora, and fauna in the play. He provided Linnaean identifications such as *Mimosa odoratissima* for the 'Sirisha,' *Mangisera* for the 'Amra,' *Nyctanthes Zambak* for the 'Mallica,' and *Banisteria* for the 'Mādhavi creeper' (see *Letters* 2: 894-95).]

By what other performance can we ensure a continuance of their favour?

Actr. Oh! by none better than by the Fatal Ring, which you have just announced.

Man. How could I forget it! In that moment I was lulled to distraction by the melody of thy voice, which allured my heart, as the king Dushmanta is now allured by the swift antelope. [*They both go out*].

Sacontalā;
or,
The Fatal Ring

Act I[5]

Scene—A Forest.

Dushmanta, *in a car, pursuing an antelope, with a bow and quiver, attended by his* Charioteer.

Char. [*Looking at the antelope, and then at the king.*] WHEN I cast my eye on that black antelope, and on thee, O king, with thy braced bow, I see before me, as it were, the God Mahēsa chasing a hart, with his bow, named pināca, braced in his left hand.

Dushm. The fleet animal has given us a long chase. Oh! there he runs, with his neck bent gracefully, looking back, from time to time, at the car which follows him. Now, through fear of a descending shaft, he contracts his forehand, and extends his flexible haunches; and now, through fatigue, he pauses to nibble the grass in his path with his mouth half opened. See how he springs and bounds with long steps, lightly skimming the ground, and rising high in the air! And now so rapid is his flight, that he is scarce discernible!

Char. The ground was uneven, and the horses were checked in their course. He has taken advantage of our delay. It is level now, and we may easily overtake him.

Dushm. Loosen the reins.

Char. As the king commands.—[*He drives the car first at full speed, and then gently.*]—He could not escape. The horses were not even touched by the clouds of dust which they raised; they tossed their manes, erected their ears, and rather glided than galloped over the smooth plain.

Dushm. They soon outran the swift antelope.—Objects which, from their distance, appeared minute, presently became larger: what was really divided, seemed united,

[5] [See Jones's earlier summary of the story probably from the *Mahābhārata*, in his letter to the second Earl Spencer (4 September 1787): 'The dramatick piece, which is neither Tragedy nor Comedy, but like many of Shakespeare's fairy-pieces, is called *Sacontalā*, and the story is this' (*Letters* 2: 766, 767-68, 792). The translated version is slightly different from the summary including the name of the character Dushyanta.]

as we passed; and what was in truth bent, seemed straight. So swift was the motion of the wheels, that nothing, for many moments, was either distant or near. [*He fixes an arrow in his bowstring.*]

[*Behind the scenes*]. He must not be slain. This antelope, O king, has an asylum in our forest: he must not be slain.

Char. [*Listening and looking.*] Just as the animal presents a fair mark for your arrow, two hermits are advancing to interrupt your aim.

Dushm. Then stop the car.

Char. The king is obeyed. [*He draws in the reins.*]

       Enter a Hermit *and his* Pupil.

Herm [*Raising his hands.*] Slay not, O mighty sovereign, slay not a poor fawn, who has found a place of refuge. No, surely, no; he must not be hurt. An arrow in the delicate body of a deer would be like fire in a bale of cotton. Compared with thy keen shafts, how weak must be the tender hide of a young antelope! Replace quickly, oh! replace the arrow which thou hast aimed. The weapons of you kings and warriors are destined for the relief of the oppressed, not for the destruction of the guiltless.

Dushm. [*Saluting them.*] It is replaced. [*He places the arrow in his quiver.*]

Herm. [*With joy.*] Worthy is that act of thee, most illustrious of monarchs; worthy, indeed, of a prince descended from Puru. Mayst thou have a son adorned with virtues, a sovereign of the world!

Pup. [*Elevating both his hands.*] Oh! by all means, may thy son be adorned with every virtue, a sovereign of the world!

Dushm. [*Bowing to them.*] My head bears with reverence the order of a Brāhmen.

Herm. Great king, we came hither to collect wood for a solemn sacrifice; and this forest, on the banks of the Malinì, affords an asylum to the wild animals protected by Sacontalā, whom our holy preceptor Canna has received as a sacred deposit. If you have no other avocation, enter yon grove, and let the rights of hospitality be duly performed. Having seen with your own eyes the virtuous behaviour of those whose only wealth is their piety, but whose worldly cares are now at an end, you will then exclaim, 'How many good-subjects are defended by this arm, which the bowstring has made callous!'

Dushm. Is the master of your family at home?

Herm. Our preceptor is gone to Sōmatìrtha, in hopes of deprecating some calamity, with which destiny threatens the irreproachable Sacontalā; and he has charged her, in his absence, to receive all guests with due honour.

Dushm. Holy man, I will attend her; and she, having

observed my devotion, will report it favourably to the venerable sage.

Both. Be it so; and we depart on our own business. [*The* Hermit *and his* Pupil *go out.*]

Dushm. Drive on the car. By visiting the abode of holiness, we shall purify our souls.

Char. As the king (may his life be long!) commands. [*He drives on.*]

Dushm. [*Looking on all sides.*] That we are near the dwelling-place of pious hermits, would clearly have appeared, even if it had not been told.

Char. By what marks?

Dushm. Do you not observe them? See under yon trees the hallowed grains which have been scattered on the ground, while the tender female parrots were feeding their unfledged young in their pendent nests. Mark in other places the shining pieces of polished stone which have bruised the oily fruit of the sacred Ingudì. Look at the young fawns, which, having acquired confidence in man, and accustomed themselves to the sound of his voice, frisk at pleasure, without varying their course. Even the surface of the river is reddened with lines of consecrated bark, which float down its stream. Look again; the roots of yon trees are bathed in the waters of holy pools, which quiver as the breeze plays upon them; and the glowing lustre of yon fresh leaves is obscured, for a time, by smoke that rises from oblations of clarified butter. See too, where the young roes graze, without apprehension from our approach, on the lawn before yonder garden, where the tops of the sacrificial grass, cut for some religious rite, are sprinkled around.

Char. I now observe all those marks of some holy habitation.

Dushm. [*Turning aside.*] This awful sanctuary, my friend, must not be violated. Here, therefore, stop the car; that I may descend.

Char. I hold in the reins. The king may descend at his pleasure.

Dushm. [*Having descended, and looking at his own dress.*] Groves devoted to religion must be entered in humbler habiliments. Take these regal ornaments;—[*the* Charioteer *receives them*]—and, whilst I am observing those who inhabit this retreat, let the horses be watered and dressed.

Char. Be it as you direct! [*He goes out.*]

Dushm. [*Walking round and looking.*] Now then I enter the sanctuary.—[*He enters the grove.*]—Oh! this place must be holy, my right arm throbs.[6]— [*Pausing and*

---

6 [The playwright suggests an old Indian superstition

considering.]—What new acquisition does this omen promise in a sequestered grove? But the gates of predestined events are in all places open.

[*Behind the scenes.*] Come hither, my beloved companions; Oh! come hither.

Dushm. [*Listening.*] Hah! I hear female voices to the right of yon arbour. I am resolved to know who are conversing.— [*He walks round and looks.*]—There are some damsels, I see, belonging to the hermit's family who carry water-pots of different sizes proportioned to their strength, and are going to water the delicate plants. Oh! how charmingly they look! If the beauty of maids who dwell in woodland retreats cannot easily be found in the recesses of a palace, the garden flowers must make room for the blossoms of the forest, which excel them in colour and fragrance. [*He stands gazing at them.*]

    Enter Sacontalā, Anusūyā, *and* Priyamvadā.

Anu. O my Sacontalā, it is in thy society that the trees of our father Canna seem to me delightful: it well becomes thee, who art soft as the fresh-blown Mallicà, to fill with water the canals which have been dug round these tender shrubs.

Sac. It is not only in obedience to our father that I thus employ myself, though that were a sufficient motive, but I really feel the affection of a sister for these young plants. [*Watering them.*]

Pri. My beloved friend, the shrubs which you have watered flower in the summer, which is now begun: let us give water to those which have passed their flowering time; for our virtue will be the greater when it is wholly disinterested.

Sac. Excellent advice! [*Watering other plants.*]

Dushm. [*Aside in transport.*] How! is that Canna's daughter, Sacontalā?—[*With surprise*]—The venerable sage must have an unfeeling heart, since he has allotted a mean employment to so lovely a girl, and has dressed her in a course mantle of woven bark. He, who could wish that so beautiful a creature, who at first sight ravishes my soul, should endure the hardships of his austere devotion, would attempt, I suppose, to cleave the hard wood Samì with a leaf of the blue lotos. Let me retire behind this tree,

that I may gaze on her charms without diminishing her confidence. [*He retires.*]

Sac. My friend Priyamvadā has tied this mantle of bark so closely over my bosom that it gives me pain: Anusūyā, I request you to untie it. [Anusūyā, *unties the mantle.*]

Pri [*Laughing.*] Well, my sweet friend, enjoy, while you may, that youthful prime which gives your bosom so beautiful a swell.

Dushm. [*Aside.*] Admirably spoken, Priyamvada! No; her charms cannot be hidden, even though a robe of intertwisted fibres be thrown over her shoulders, and conceal a part of her bosom, like a veil of yellow leaves enfolding a radiant flower. The water lily, though dark moss may settle on its head, is nevertheless beautiful; and the moon with dewy beams is rendered yet brighter by its black spots. The bark itself acquires elegance from the features of a girl with antelope's eyes, and rather augments than diminishes my ardour. Many are the rough stalks which support the water lily; but many and exquisite are the blossoms which hang on them.

Sac. [*Looking before her*]. Yon Amra tree, my friends, points with the finger of its leaves, which the gale gently agitates, and seem inclined to whisper some secret. I will go near it. [*They all approach the tree.*]

Pri. O my Sacontalā, let us remain some time in this shade.

Sac. Why here particularly?

Pri. Because the Amra tree seems wedded to you, who are graceful as the blooming creeper which twines round it.

Sac. Properly are you named Priyamvadā, or speaking kindly.

Dushm. [*Aside.*] She speaks truly. Yes; her lip glows like the tender leaflet; her arms resemble two flexible stalks; and youthful beauty shines, like a blossom, in all her lineaments.

Anu. See, my Sacontalā, how yon fresh Mallicà, which you have surnamed Vanàdòsinì, or Delight of the Grove, has chosen the sweet Amra for her bridegroom.

Sac. [*Approaching, and looking at it with pleasure.*] How charming is the season, when the nuptials even of plants are thus publickly celebrated! [*She stands admiring it.*]

Pri. [*Smiling.*] Do you know, my Anusūyā, why Sacontalā gazes on the plants with such rapture?

Anu. No, indeed: I was trying to guess. Pray, tell me.

Pri. 'As the Grove's Delight is united to a suitable tree, thus I too hope for a bridegroom to my mind'.—That is her private thought at this moment.

Sac. Such are the flights of your own imagination. [*Inverting the water-pot.*]

---

prevalent even among the royalty in those days that the throbbing of any right organ—arm, eye, or leg—is a propitious omen for men, and the left for women, and a bad omen if vice versa. King Dushyanta senses a welcome predestination of events which might grace him in the pious hermitage. Conversely, Śakuntalā gets scared in Dushyanta's court when her right eye twitches (Act V), and she then is rejected by him.]

Anu. Here is a plant, Sacontalā, which you have forgotten, though it has grown up, like yourself, under the fostering care of our father Canna.

Sac. Then I shall forget myself.—O wonderful!—[*approaching the plant.*]—O Priyamvadā! [*looking at it with joy*] I have delightful tidings for you.

Pri. What tidings, my beloved, for me?

Sac. This Mādhavi-creeper, though it be not the usual time for flowering, is covered with gay blossoms from its root to its top.

Both. [*Approaching it hastily.*] Is it really so, sweet friend?

Sac. Is it so? Look yourselves.

Pri. [*With eagerness.*] From this omen, Sacontalā, I announce you an excellent husband, who will very soon take you by the hand. [*Both girls look at* Sacontalā.]

Sac. [*Displeased.*] A strange fancy of yours!

Pri. Indeed, my beloved, I speak not jestingly. I heard something from our father Canna. Your nurture of these plants has prospered; and thence it is, that I foretel your approaching nuptials.

Anu. It is thence, my Priyamvadā, that she has watered them with so much alacrity.

Sac. The Mādhavi plant is my sister; can I do otherwise than cherish her? [*Pouring water on it.*]

Dushm. [*Aside.*] I fear she is of the same religious order with her foster-father. Or has a mistaken apprehension risen in my mind? My warm heart is so attached to her, that she cannot but be a fit match for a man of the military class. The doubts which awhile perplex the good, are soon removed by the prevalence of their strong inclinations. I am enamoured of her, and she cannot, therefore, be the daughter of a Brahmen, whom I could not marry.

Sac. [*Moving her head.*] Alas! a bee has left the blossom of this Mallicá, and is fluttering round my face.[7] [*She expresses uneasiness.*]

Dushm. [*Aside, with affection.*] How often have I seen our court damsels affectedly turn their heads aside from some roving insect, merely to display their graces! but this rural charmer knits her brows, and gracefully moves her eyes through fear only, without art or affectation. Oh! happy bee, who touchest the corner of that eye beautifully trembling; who, approaching the tip of that ear, murmurest as softly as if thou wert whispering a secret of love; and who sippest nectar, while she waves her graceful hand, from that lip, which contains all the treasures of

delight! Whilst I am solicitous to know in what family she was born, thou art enjoying bliss, which to me would be supreme felicity.

Sac. Disengage me, I entreat, from this importunate insect, which quite baffles my efforts.

Pri. What power have we to deliver you? The king Dushmanta is the sole defender of our consecrated groves.

Dushm. [*Aside.*] This is a good occasion for me to discover myself—[*advancing a little.*]—I must not, I will not, fear. Yet—[*checking himself and retiring*]—my royal character will thus abruptly be known to them. No; I will appear as a simple stranger, and claim the duties of hospitality.

Sac. This impudent bee will not rest. I will remove to another place. [*Stepping aside and looking round.*]—Away! away! He follows me wherever I go. Deliver me, oh! deliver me from this distress.

Dushm. [*Advancing hastily.*] Ah! While the race of Puru govern the world, and restrain even the most profligate, by good laws well administered, has any man the audacity to molest the lovely daughters of pious hermits? [*They look at him with emotion.*]

Anu. Sir, no man is here audacious; but this damsel, our beloved friend, was teased by a fluttering bee. [*Both girls look at* Sacontalā.]

Dushm. [*Approaching her.*] Damsel, may thy devotion prosper! [Sacontalā *looks on the ground, bashful and silent.*]

Anu. Our guest must be received with due honours.

Pri. Stranger, you are welcome. Go, my Sacontalā; bring from the cottage a basket of fruit and flowers. This river will, in the mean time, supply water for his feet. [*Looking at the water-pots.*]

Dushm. Holy maid, the gentleness of thy speech does me sufficient honour.

Anu. Sit down awhile on this bank of earth, spread with the leaves of Septaperna: the shade is refreshing, and our lord must want repose after his journey.

Dushm. You too must all be fatigued by your hospitable attentions: rest yourselves, therefore, with me.

Pri. [*Aside to* Sacontalā.] Come, let us all be seated: our guest is contented with our reception of him. [*They all seat themselves.*]

Sac. [*Aside.*] At the sight of this youth I feel an emotion scarce consistent with a grove devoted to piety.

Dushm. [*Gazing at them alternately.*] How well your friendship agrees, holy damsels, with the charming equality of your ages and of your beauties!

Pri. [*Aside to Anusūyā.*] Who can this be, my Anusūyā? The union of delicacy with robustness in his form, and of

---

[7] [The bee's fluttering round Śakuntalā's soft lovely face, mistaking it for a flower, is an old analogy in Sanskrit literature when a virgin's face is compared with a juicy flower.]

sweetness with dignity in his discourse, indicate a character fit for ample dominion.

Anu. [*Aside to* Priyamvadā.] I too have been admiring him. I must ask him a few questions.—[*Aloud.*] Your sweet speech, Sir, gives me confidence. What imperial family is embellished by our noble guest? What is his native country? Surely it must be afflicted by his absence from it. What, I pray, could induce you to humiliate that exalted form of yours by visiting a forest peopled only by simple anchorites?

Sac. [*Aside.*] Perplex not thyself, oh my heart! let the faithful Anusūyā direct with her counsel the thoughts which rise in thee.

Dushm. [*Aside.*] How shall I reveal, or how shall I disguise myself?—[*Musing.*]—Be it so.—[*Aloud to* Anusūyā.] Excellent lady, I am a student of the Vēda, dwelling in the city of our king, descended from Puru; and, being occupied in the discharge of religious and moral duties, am come hither to behold the sanctuary of virtue.

Anu. Holy men, employed like you, are our lords and masters.

[Sacontalā *looks modest, yet with affection; while her companions gaze alternately at her and at the king* .]

Anu. [*Aside to* Sacontalā.] Oh! if our venerable father were present—

Sac. What if he were?

Anu. He would entertain our guest with a variety of refreshments.

Sac. [*Pretending displeasure.*] Go too; you had some other idea in your head; I will not listen to you. [*She sits apart.*]

Dushm [*Aside to* Anusūyā *and* Priyamvadā.] In my turn, holy damsels, allow me to ask one question concerning your lovely friend.

Both The request, Sir, does us honour.

Dushm. The sage Canna, I know, is ever intent upon the great Being; and must have declined all earthly connections. How then can this damsel be, as it is said, his daughter?

Anu. Let our lord hear. There is, in the family of Cusa, a pious prince of extensive power, eminent in devotion and in arms.

Dushm. You speak, no doubt, of Causica, the sage and monarch.

Anu. Know, Sir, that he is in truth her father; while Canna bears that reverend name, because he brought her up, since she was left an infant.

Dushm. Left? the word excites my curiosity; and raises in me a desire of knowing her whole story.

Anu. You shall hear it, Sir, in few words.—When that sage king had begun to gather the fruits of his austere devotion, the gods of Swerga became apprehensive of his increasing power, and sent the nymph Mēnacà to frustrate, by her allurements, the full effect of his piety.

Dushm. Is a mortal's piety so tremendous to the inferior deities? What was the event?

Anu. In the bloom of the vernal season, Causica, beholding the beauty of the celestial nymph, and wafted by the gale of desire— [*She stops and looks modest.*]

Dushm. I now see the whole. Sacontalā then is the daughter of a king, by a nymph of the lower heaven.

Anu. Even so.

Dushm. [*Aside.*] The desire of my heart is gratified.—[*Aloud.*] How, indeed, could her transcendent beauty be the portion of mortal birth? Yon light, that sparkles with tremulous beams, proceeds not from a terrestrial cavern. [Sacontalā *sits modestly, with her eyes on the ground.*]

Dushm. [*Again aside.*] Happy man that I am! Now has my fancy an ample range. Yet, having heard the pleasantry of her companions on the subject of her nuptials, I am divided with anxious doubt, whether she be not wholly destined for a religious life.

Pri. [*Smiling, and looking first at* Sacontalā, *then at the king.*] Our lord seems desirous of asking other questions. [Sacontalā *rebukes* Priyamvadā *with her hand.*]

Dushm. You know my very heart. I am, indeed, eager to learn the whole of this charmer's life; and must put one question more.

Pri. Why should you muse on it so long?—[*Aside.*] One would think this religious man was forbidden by his vows to court a pretty woman.

Dushm. This I ask. Is the strict rule of a hermit so far to be observed by Canna, that he cannot dispose of his daughter in marriage, but must check the natural impulse of juvenile love? Can she (oh preposterous fate!) be destined to reside for life among her favourite antelopes, the black lustre of whose eyes is far surpassed by hers?

Pri. Hitherto, Sir, our friend has lived happy in this consecrated forest, the abode of her spiritual father; but it is now his intention to unite her with a bridegroom equal to herself.

Dushm. [*Aside, with ecstasy.*] Exult, oh my heart, exult. All doubt is removed; and what before thou wouldst have dreaded as a flame, may now be approached as a gem inestimable.

Sac. [*Seeming angry.*] Anusūyā, I will stay here no longer.

Anu. Why so, I pray?

Sac. I will go to the holy matron Gautamī, and let her

know how impertinently our Priyamvadā has been prattling. [*She rises.*]

Anu. It will not be decent, my love, for an inhabitant of this hallowed wood to retire before a guest has received complete honour. [Sacontalā, *giving no answer, offers to go.*]

Dushm. [*Aside.*] Is she then departing?—[*He rises, as if going to stop her, but checks himself.*]—The actions of a passionate lover are as precipitate as his mind is agitated. Thus I, whose passion impelled me to follow the hermit's daughter, am restrained by a sense of duty.

Pri. [*Going up to* Sacontalā.] My angry friend, you must not retire.

Sac. [*Stepping back and frowning.*] What should detain me?

Pri. You owe me the labour, according to our agreement, of watering two more shrubs. Pay me first, to acquit your conscience, and then depart, if you please. [*Holding her.*]

Dushm. The damsel is fatigued, I imagine, by pouring so much water on the cherished plants. Her arms, graced with palms like fresh blossoms, hang carelessly down; her bosom heaves with strong breathing; and now her dishevelled locks, from which the string has dropped, are held by one of her lovely hands. Suffer me, therefore, thus to discharge the debt.— [*Giving his ring to* Priyamvadā. *Both damsels, reading the name* Dushmanta, *inscribed on the ring, look with surprise at each other.*]— It is a toy unworthy of your fixed attention; but I value it as a gift from the king.

Pri. Then you ought not, Sir, to part with it. Her debt is from this moment discharged on your word only. [*She returns the ring.*]

Anu. You are now released, Sacontalā, by this benevolent lord—or favoured, perhaps, by a monarch himself. To what place will you now retire?

Sac. [*Aside.*] Must I not wonder at all this if I preserve my senses?

Pri. Are not you going, Sacontalā?

Sac. Am I your subject? I shall go when it pleases me.

Dushm. [*Aside, looking at* Sacontalā.] Either she is affected towards me, as I am towards her, or I am distracted with joy. She mingles not her discourse with mine; yet, when I speak, she listens attentively. She commands not her actions in my presence; and her eyes are engaged on me alone.

[*Behind the scenes.*] Oh pious hermits, preserve the animals of this hallowed forest! The king Dushmanta is hunting in it. The dust raised by the hoofs of his horses, which pound the pebbles ruddy as early dawn, falls like a swarm of blighting insects on the consecrated boughs

which sustain your mantles of woven bark, moist with the water of the stream in which you have bathed.

Dushm. [*Aside.*] Alas! my officers, who are searching for me, have indiscreetly disturbed this holy retreat.

[*Again behind the scenes.*] Beware, ye hermits, of yon elephant, who comes overturning all that oppose him; now he fixes his trunk with violence on a lofty branch that obstructs his way; and now he is entangled in the twining stalks of the Vratati. How are our sacred rites interrupted! How are the protected herds dispersed! The wild elephant, alarmed at the new appearance of a car, lays our forest waste.

Dushm. [*Aside.*] How unwillingly am I offending the devout foresters! Yes; I must go to them instantly.

Pri. Noble stranger, we are confounded with dread of the enraged elephant. With your permission, therefore, we retire to the hermit's cottage.

Anu. O Sacontalā, the venerable matron will be much distressed on your account. Come quickly, that we may be all safe together.

Sac. [*Walking slowly.*] I am stopped, alas! by a sudden pain in my side.[8]

Dushm. Be not alarmed, amiable damsels. It shall be my care that no disturbance happen in your groves.

Pri. Excellent stranger, we were wholly unacquainted with your station; and you will forgive us, we hope, for the offence of intermitting awhile the honours due to you: but we humbly request that you will give us once more the pleasure of seeing you, though you have not now been received with perfect hospitality.

Dushm. You depreciate your own merits. The sight of you, sweet damsels, has sufficiently honoured me.

Sac. My foot, O Anusūyā, is hurt by this pointed blade of Cusa grass; and now my loose vest of bark is caught by a branch of the Curuvaca. Help me to disentangle myself, and support me.— [*She goes out, looking from time to time at* Dushmanta, *and supported by the damsels.*]

Dushm. [*Sighing.*] They are all departed; and I too, alas! must depart. For how short a moment have I been blessed with a sight of the incomparable Sacontalā! I will send my attendants to the city, and take my station at no great distance from this forest. I cannot, in truth, divert my

---

8 [Kālidāsa demonstrates here his astute knowledge of female physiology, pointing out a possibility that Śankutalā, before the kings' arrival in the hermitage, had had her menstruation; and the pain felt in her lower abdomen, on the right side, is an indicator of the time of her ovulation. According to current gynecological research, mittelschmerz is the best indicator of ovulation. Hence Śakuntalā's pregnancy takes place in no time.]

mind from the sweet occupation of gazing on her. How, indeed, should I otherwise occupy it? My body moves onward; but my restless heart runs back to her; like a light flag borne on a staff against the wind, and fluttering in an opposite direction. [*He goes out.*]

Act II

Scene—*A Plain, with royal pavilions on the skirt of the forest.* Mādhavya. [*Sighing and lamenting.*] Strange recreation this?—Ah me! I am wearied to death —My royal friend has an unaccountable taste.—What can I think of a king so passionately fond of chasing unprofitable quadrupeds?—'Here runs an antelope! there goes a boar!'—Such is our only conversation.—Even at noon, in excessive heat, when not a tree in the forest has a shadow under it, we must be skipping and prancing about, like the beasts whom we follow.—Are we thirsty? We have nothing to drink but the waters of mountain torrents, which taste of burned stones and mawkish leaves—Are we hungry? We must greedily devour lean venison, and that commonly roasted to a stick.—Have I a moment's repose at night?—My slumber is disturbed by the din of horses and elephants, or by the sons of slave-girls hollooing out, 'More venison, more venison!' —Then comes a cry that pierces my ear, 'Away to the forest, away!'—Nor are these my only grievances: fresh pain is now added to the smart of my first wounds; for, while we were separated from our king, who was chasing a foolish deer, he entered, I find, yon lonely place, and there, to my infinite grief, saw a certain girl, called Sacontalā, the daughter of a hermit: from that moment not a word of returning to the city!—These distressing thoughts have kept my eyes open the whole night.—Alas! when shall we return?—I cannot set eyes on my beloved friend Dushmanta since he set his heart on taking another wife.—[*Stepping aside and looking*]—Oh! there he is.— How changed!—He carries a bow, indeed, but wears for his diadem a garland of wood-flowers.—He is advancing: I must begin my operations.—[*He stands leaning on a staff.*]—Let me thus take a moment's rest.—[*Aloud.*] Dushmanta *enters, as described.*

Dushm. [*Aside, sighing*] My darling is not so easily attainable; yet my heart assumes confidence from the manner in which she seemed affected: surely, though our love has not hitherto prospered, yet the inclinations of us both are fixed on our union.—[*Smiling.*]—Thus do lovers agreeably beguile themselves, when all the powers of their souls are intent on the objects of their desire!—But

am I beguiled? No; when she cast her eyes even on her companions, they sparkled with tenderness; when she moved her graceful arms, they dropped, as if languid with love; when her friend remonstrated against her departure, she spoke angrily.—All this was, no doubt, on my account—Oh! how quick-sighted is love in discerning his own advantages!

Mādh. [*Bending downward, as before.*] Great prince! my hands are unable to move; and it is with my lips only that I can mutter a blessing on you. May the king be victorious!

Dushm. [*Looking at him smiling.*] Ah! what has crippled thee, friend Mādhavya?

Mādh. You strike my eye with your own hand, and then ask what makes it weep.

Dushm. Speak intelligibly. I know not what you mean.

Mādh. Look at yon Vētas tree bent double in the river. Is it crooked, I pray, by its own act, or by the force of the stream?

Dushm. It is bent, I suppose, by the current.

Mādh. So am I by your Majesty.

Dushm. How so, Mādhavya?

Mādh. Does it become you, I pray, to leave the great affairs of your empire, and so charming a mansion as your palace, for the sake of living here like a forester? Can you hold a council in a wood? I, who am a reverend Brahmen, have no longer the use of my hands and feet: they are put out of joint by my running all day long after dogs and wild beasts. Favour me, I entreat, with your permission to repose but a single day.

Dushm. [*Aside.*] Such are this poor fellow's complaints; whilst I, when I think of Canna's daughter, have as little relish for hunting as he. How can I brace this bow, and fix a shaft in the string, to shoot at those beautiful deer who dwell in the same groves with my beloved, and whose eyes derive lustre from hers?

Mādh. [*Looking stedfastly at the king.*] What scheme is your royal mind contriving? I have been crying, I find, in a wilderness.

Dushm. I think of nothing but the gratification of my old friend's wishes.

Mādh. [*Joyfully.*] Then may the king live long! [*Rising, but counterfeiting feebleness.*]

Dushm. Stay; and listen to me attentively.

Mādh. Let the king command.

Dushm. When you have taken repose, I shall want your assistance in another business, that will give you no fatigue.

Mādh. Oh! what can that be, unless it be eating rice-pudding?

Dushm You shall know in due time.

Mādh. I shall be delighted to hear it.

Dushm. Hola! who is there?

        The Chamberlain *enters.*

Cham. Let my sovereign command me.

Dushm. Raivataca, bid the General attend.

Cham. I obey.—[*He goes out, and returns with the* General.]—Come quickly, Sir, the king stands expecting you.

Gen. [*Aside looking at* Dushmanta.] How comes it that hunting, which moralists reckon a vice,[9] should be a virtue in the eyes of a king? Thence it is, no doubt, that our emperor, occupied in perpetual toil, and inured to constant heat, is become so lean, that the sunbeams hardly affect him; while he is so tall, that he looks to us little men, like an elephant grazing on a mountain: he seems all soul.— [*Aloud, approaching the king.*]—May our monarch ever be victorious!—This forest, O king, is infested by beasts of prey: we see the traces of their huge feet in every path.—What orders is it your pleasure to give?

Dushm. Bhadrasēna, this moralizing Mādhavya has put a stop to our recreation by forbidding the pleasures of the chase.

Gen. [*Aside to* Mādhavya.] Be firm to your word, my friend; whilst I sound the king's real inclinations.— [*Aloud.*] Oh! Sir, the fool talks idly. Consider the delights of hunting. The body, it is true, becomes emaciated, but it is light and fit for exercise. Mark how the wild beasts of various kinds are variously affected by fear and by rage! What pleasure equals that of a proud archer, when his arrow hits the mark as it flies?—Can hunting be justly called a vice? No recreation, surely, can be compared with it.

Mādh. [*Angrily.*] Away, thou false flatterer! The king, indeed, follows his natural bent, and is excusable; but thou, son of a slave girl, hast no excuse.—Away to the wood!—How I wish thou hadst been seized by a tiger or an old bear, who was prowling for a skakàl, like thyself! *Dushm.* We are now, Bhadrasēna, encamped near a sacred hermitage; and I cannot at present applaud your panegyrick on hunting. This day, therefore, let the wild buffalos roll undisturbed in the shallow water, or toss up the sand with their horns; let the herd of antelopes,

assembled under the thick shade, ruminate without fear; let the large boars root up the herbage on the brink of yon pool; and let this my bow take repose with a slackened string.

Gen. As our lord commands.

Dushm. Recall the archers who have advanced before me, and forbid the officers to go very far from this hallowed grove. Let them beware of irritating the pious: holy men are eminent for patient virtues, yet conceal within their bosoms a scorching flame; as carbuncles are naturally cool to the touch; but, if the rays of the sun have been imbibed by them, they burn the hand.

Mādh. Away now, and triumph on the delights of hunting.

Gen. The king's orders are obeyed. [*He goes out.*]

Dushm. [*To his attendants.*] Put off your hunting apparel; and thou, Raivataca, continue in waiting at a little distance.

Cham. I shall obey. [*Goes out.*]

Mādh. So! you have cleared the stage: not even a fly is left on it. Sit down, I pray, on this pavement of smooth pebbles, and the shade of this tree shall be your canopy: I will sit by you; for I am impatient to know what will give me no fatigue.

Dushm. Go first, and seat thyself.

Mādh. Come, my royal friend. [*They both sit under a tree.*]

Dushm. Friend Mādhavya, your eyes have not been gratified with an object which best deserves to be seen.

Mādh. Yes, truly; for a king is before them.

Dushm. All men are apt, indeed, to think favourably of themselves; but I meant Sacontalā, the brightest ornament of these woods.

Mādh. [*Aside.*] I must not foment this passion.—[*Aloud.*] What can you gain by seeing her? She is a Brahmen's daughter, and consequently no match for you!

Dushm. What! Do people gaze at the new moon, with uplifted heads and fixed eyes, from a hope of possessing it? But you must know, that the heart of Dushmanta is not fixed on an object which he must for ever despair of attaining.

Mādh. Tell me how.

Dushm. She is the daughter of a pious prince and warriour, by a celestial nymph; and, her mother having left her on earth, she has been fostered by Canna, even as a fresh blossom of Malati, which drops on its pendent stalk, is raised and expanded by the sun's light.

Mādh. [*Laughing.*] Your desire to possess this rustick girl, when you have women bright as gems in your palace already, is like the fancy of a man, who has lost his relish for dates, and longs for the sour tamarind.

---

[9] [A probable allusion to Buddhism and Jainism, existing two centuries earlier than Kālidāsa, when hunting was considered an immoral act full of vice (which gave birth to the doctrine of nonviolence, of 'Ahimsa', as preached by Mahatma Gandhi).]

Dushm. Did you know her, you would not talk so wildly.

Mādh. Oh! certainly, whatever a king admires must be superlatively charming.

Dushm. [*Smiling.*] What need is there of long description? When I meditate on the power of Brahmā, and on her lineaments, the creation of so transcendent a jewel outshines, in my apprehension, all his other works: she was formed and moulded in the eternal mind, which had raised with its utmost exertion, the ideas of perfect shapes, and thence made an assemblage of all abstract beauties.

Mādh. She must render, then, all other handsome women contemptible.

Dushm. In my mind she really does. I know not yet what blessed inhabitant of this world will be the possessor of that faultless beauty, which now resembles a blossom whose fragrance has not been diffused; a fresh leaf, which no hand has torn from its stalk; a pure diamond, which no polisher has handled; new honey, whose sweetness is yet untasted; or rather the celestial fruit of collected virtues, to the perfection of which nothing can be added.

Mādh. Make haste, then, or the fruit of all virtues will drop into the hand of some devout rustick, whose hair shines with oil of Ingudì.

Dushm. She is not her own mistress; and her foster-father is at a distance.

Mādh. How is she disposed towards you?

Dushm. My friend, the damsels in a hermit's family are naturally reserved: yet she did look at me, wishing to be unperceived; then she smiled, and started a new subject of conversation. Love is by nature averse to a sudden communication, and hitherto neither fully displays, nor wholly conceals, himself in her demeanour towards me.

Mādh. [*Laughing.*] Has she thus taken possession of your heart on so transient a view?

Dushm. When she walked about with her female friends, I saw her yet more distinctly, and my passion was greatly augmented. She said sweetly, but untruly, 'My foot is hurt by the points of the Cusa grass:' then she stopped; but soon, advancing a few paces, turned back her face, pretending a wish to disentangle her vest of woven bark from the branches in which it had not really been caught.

Mādh. You began with chasing an antelope, and have now started new game: thence it is, I presume, that you are grown so fond of a consecrated forest.

Dushm. Now the business for you, which I mentioned, is this: you, who are a Brāhmen, must find some expedient for my second entrance into that asylum of virtue.

Mādh. And the advice which I give is this: remember that you are a king.

Dushm What then?

Mādh. 'Hola! bid the hermits bring my sixth part of their grain'. Say this, and enter the grove without scruple.

Dushm. No, Mādhavya: they pay a different tribute, who, having abandoned all the gems and gold of this world, possess riches far superior. The wealth of princes, collected from the four orders of their subjects, is perishable; but pious men give us a sixth part of the fruits of their piety; fruits which will never perish.

[*Behind the scenes.*] Happy men that we are! we have now attained the object of our desire.

Dushm. Hah! I hear the voices of some religious anchorites.

The Chamberlain *enters.*

Cham. May the king be victorious!—Two young men, sons of a hermit, are waiting at my station, and soliciting an audience.

Dushm. Introduce them without delay.

Cham. As the king commands.—[*He goes out, and reenters with two* Brāhmens.]—Come on; come this way.

First Brāhm. [*Looking at the king.*] Oh! what confidence is inspired by his brilliant appearance!—Or proceeds it rather from his disposition to virtue and holiness.— Whence comes it, that my fear vanishes?—He now has taken his abode in a wood which supplies us with every enjoyment; and with all his exertions for our safety, his devotion increases from day to day.—The praise of a monarch who has conquered his passions ascends even to heaven: inspired bards are continually singing, 'Behold a virtuous prince!' but with us the royal name stands first: 'Behold, among kings, a sage!'

Second Brāhm. Is this, my friend, the truly virtuous Dushmanta?

First Brāhm. Even he.

Second Brāhm. It is not then wonderful, that he alone, whose arm is lofty and strong as the main bar of his city gate, possesses the whole earth, which forms a dark boundary to the ocean; or that the gods of Swerga, who fiercely contend in battle with evil powers, proclaim victory gained by his braced bow, not by the thunderbolt of INDRA.

Both. [*Approaching him.*] O king, be victorious!

Dushm. [*Rising.*] I humbly salute you both.

Both. Blessings on thee!

Dushm. [*Respectfully.*] May I know the cause of this visit?

First Brāhm. Our sovereign is hailed by the pious inhabitants of these woods; and they implore—

Dushm. What is their command?

First Brāhm. In the absence of our spiritual guide, Canna, some evil demons are disturbing our holy retreat. Deign, therefore, accompanied by thy charioteer, to be master of our asylum, if it be only for a few short days.

Dushm [*Eagerly.*] I am highly favoured by your invitation.

Mādh [*Aside.*] Excellent promoters of your design! They draw you by the neck, but not against your will.

Dushm. Raivataca, bid my charioteer bring my car, with my bow and quiver.

Cham. I obey. [*He goes out.*]

First Brāhm. Such condescension well becomes thee, who art an universal guardian.

Second Brāhm. Thus do the descendants of Puru perform their engagement to deliver their subjects from fear of danger.

Dushm. Go first, holy men: I will follow instantly.

Both. Be ever victorious! [*They go out.*]

Dushm. Shall you not be delighted, friend Mādhavya, to see my Sacontalā?

Mādh. At first I should have had no objection; but I have a considerable one since the story of the demons.

Dushm. Oh! fear nothing: you will be near me.

Mādh. And you, I hope, will have leisure to protect me from them.

The Chamberlain *re-enters.*

Cham. May our lord be victorious!—The imperial car is ready; and all are expecting your triumphant approach. Carabba too, a messenger from the queen-mother, is just arrived from the city.

Dushm. Is he really come from the venerable queen?

Cham. There can be no doubt of it.

Dushm. Let him appear before me.

[*The* Chamberlain *goes out, and returns with the* Messenger.]

Cham. There stands the king—O Carabba, approach him with reverence.

Mess. [*Prostrating himself.*] May the king be ever victorious!—The royal mother sends this message—

Dushm. Declare her command.

Mess. Four days hence the usual fast for the advancement of her son will be kept with solemnity; and the presence of the king (may his life be prolonged!) will then be required.

Dushm. On one hand is a commission from holy Brāhmens; on the other, a command from my revered parent: both duties are sacred, and neither must be neglected.

Mādh [*Laughing.*] Stay suspended between them both, like king Trisancu between heaven and earth; when the pious men said, 'Rise!' and the gods of Swerga said,

'Fall!'

Dushm. In truth I am greatly perplexed. My mind is principally distracted by the distance of the two places where the two duties are to be performed; as the stream of a river is divided by rocks in the middle of its bed.— [*Musing.*]—Friend Mādhavya, my mother brought you up as her own son, to be my playfellow, and to divert me in my childhood. You may very properly act my part in the queen's devotions. Return then to the city, and give an account of my distress through the commission of these reverend foresters.

Mādh. That I will;—but you could not really suppose that I was afraid of demons!

Dushm. How come you, who are an egregious Brāhmen, to be so bold on a sudden?

Mādh. Oh! I am now a young king.

Dushm. Yes, certainly; and I will dispatch my whole train to attend your highness, whilst I put an end to the disturbance in this hermitage.

Mādh. [*Strutting.*] See, I am a prince regnant.

Dushm. [*Aside.*] This buffoon of a Brāhmen has a slippery genius. He will perhaps disclose my present pursuit to the women in the palace. I must try to deceive him.—[*Taking* Mādhavya *by the hand.*]—I shall enter the forest, be assured, only through respect for its pious inhabitants; not from any inclination for the daughter of a hermit. How far am I raised above a girl educated among antelopes; a girl, whose heart must ever be a stranger to love!—The tale was invented for my diversion.

Mādh. Yes, to be sure; only for your diversion.

Dushm. Then farewel, my friend; execute my commission faithfully, whilst I proceed—to defend the anchorites. [*All go out.*]

Act III

Scene—*The* Hermitage *in a Grove.*
The Hermit's Pupil *bearing consecrated grass.*
Pupil [*Meditating with wonder.*]

How great is the power of Dushmanta!—The monarch and his charioteer had no sooner entered the grove than we continued our holy rites without interruption.—What words can describe him?—By his barely aiming a shaft, by the mere sound of his bow-string, by the simple murmur of his vibrating bow, he disperses at once our calamities.—Now then I deliver to the priests this bundle of fresh Cusa grass to be scattered round the place of sacrifice.—[*Looking behind the scenes.*]—Ah! Priyamvadā, for whom are you carrying that ointment of Usīra root, and those leaves of water lilies?—[*Listening atten-tively.*]—What say you?—That Sacontalā is extremely disordered by the sun's heat, and that you have procured for her a cooling medicine! Let her, my Priyamvadā, be diligently attended; for she is the darling of our venerable father Canna.—I will administer, by the hand of Gautamī, some healing water consecrated in the ceremony called Vaitāna. [*He goes out*]

Dushmanta *enters, expressing the distraction of a lover.*

Dushm. I well know the power of her devotion: that she will suffer none to dispose of her but Canna, I too well know. Yet my heart can no more return to its former placid state, than water can reascend the steep, down which it has fallen.—O God of Love, how can thy darts be so keen, since they are pointed with flowers?—Yes, I discover the reason of their keenness. They are tipped with the flames which the wrath of Hara[10] kindled, and which bláze at this moment, like the Bārava fire under the waves: how else couldst thou, who wast consumed even to ashes, be still the inflamer of our souls? By thee and by the moon, though each of you seems worthy of con-fidence, we lovers are cruelly deceived. They who love as I do, ascribe flowery shafts to thee, and cool beams to the moon, with equal impropriety; for the moon sheds fire on them with her dewy rays, and thou pointest with sharp diamonds those arrows which seem to be barbed with blossoms. Yet this god, who bears a fish on his banners, and who wounds me to the soul, will give me real delight, if he destroy me with the aid of my beloved, whose eyes are large and beautiful as those of a roe.—O powerful divinity, even when I thus adore thy attributes, hast thou no compassion? Thy fire, O Love, is fanned into a blaze by a hundred of my vain thoughts.—Does it become thee to draw thy bow even to thy ear, that the shaft, aimed at my bosom, may inflict a deeper wound? Where now can I recreate my afflicted soul by the permission of those pious men whose uneasiness I have removed by dismiss-ing my train?—[*Sighing*]—I can have no relief but from a sight of my beloved.—[*Looking up.*]—This intensely hot noon must, no doubt, be passed by Sacontalā with her damsels on the banks of this river over-shadowed with Tamālas. —It must be so:—I will advance thither.—[*Walking round and looking.*]—My sweet friend has, I guess, been lately walking under that row of young trees; for I see the stalks of some flowers, which probably she gathered, still unshrivelled; and some fresh leaves, newly

10 [Another name of Siva, who extinguished the physical identity of Kāma, the god of love.]

plucked, still dropping milk.—[*Feeling a breeze.*]—Ah! this bank has a delightful air!—Here may the gale em-brace me, wafting odours from the water lilies, and cool my breast, inflamed by the bodiless god, with the liquid particles which it catches from the waves of the Mālinì.—[*Looking down.*]—Happy lover! Sacontalā must be somewhere in this grove of flowering creepers; for I discern on the yellow sand at the door of yon arbour some recent footsteps, raised a little before, and depressed behind by the weight of her elegant limbs.—I shall have a better view from behind this thick foliage.—[*He con-ceals himself, looking vigilantly.*]—Now are my eyes fully gratified.—The darling of my heart, with her two faithful attendants, reposes on a smooth rock strown with fresh flowers.—These branches will hide me, whilst I hear their charming conversation. [*He stands concealed, and gazes.*]

Sacontalā *and her two Damsels discovered.*

Both. [*Fanning her.*] Say, beloved Sacontalā, does the breeze, raised by our fans of broad lotos leaves, refresh you?

Sac. [*Mournfully.*] Why, alas, do my dear friends take this trouble? [*Both look sorrowfully at each other.*]

Dushm. [*Aside.*] Ah! she seems much indisposed. What can have been the fatal cause of so violent a fever?—Is it what my heart suggests? Or— [*Musing*—I am perplexed with doubts.—The medicine extracted from the balmy Usīra has been applied, I see, to her bosom: her only bracelet is made of thin filaments from the stalks of a water lily, and even that is loosely bound on her arm. Yet, even thus disordered, she is exquisitely beautiful.—Such are the hearts of the young! Love and the sun equally inflame us; but the scorching heat of summer leads not equally to happiness with the ardour of youthful desires.

Pri [*Aside to* Anusūyā.] Did you not observe how the heart of Sacontalā was affected by the first sight of our pious monarch? My suspicion is, that her malady has no other cause.

Anu. [*Aside to* Priyamvadā.] The same suspicion had risen in my mind. I will ask her at once.—[*Aloud*]—My sweet Sacontalā, let me put one question to you. What has really occasioned your indisposition?

Dushm. [*Aside.*] She must now declare it. Ah! though her bracelets of lotos are bright as moon beams, yet they are marked, I see, with black spots from internal ardour.

Sac. [*Half raising herself.*] Oh! say what you suspect to have occasioned it.

Anu. Sacontalā, we must necessarily be ignorant of what is passing in your breast; but I suspect your case to be that which we have often heard related in tales of love. Tell us openly what causes your illness. A physician, without knowing the cause of a disorder, cannot even begin to apply a remedy.

Dushm. [*Aside.*] I flatter myself with the same suspicion.

Sac. [*Aside.*] My pain is intolerable; yet I cannot hastily disclose the occasion of it.

Pri. My sweet friend, Anusūyā, speaks rationally. Consider the violence of your indisposition. Every day you will be more and more emaciated, though your exquisite beauty has not yet forsaken you.

Dushm. [*Aside.*] Most true. Her forehead is parched; her neck droops; her waist is more slender than before; her shoulders languidly fall; her complection is wan; she resembles a Mādhavī creeper, whose leaves are dried by a sultry gale: yet, even thus transformed, she is lovely, and charms my soul.

Sac. [*Sighing.*] What more can I say? Ah! why should I be the occasion of your sorrow?

Pri. For that very reason, my beloved, we are solicitous to know your secret; since, when each of us has a share of your uneasiness, you will bear more easily your own portion of it.

Dushm. [*Aside.*] Thus urged by two friends, who share her pains as well as her pleasures, she cannot fail to disclose the hidden cause of her malady; whilst I, on whom she looked at our first interview with marked affection, am filled with anxious desire to hear her answer.

Sac. From the very instant when the accomplished prince, who has just given repose to our hallowed forest, met my eye—[*She breaks off, and looks modest.*]

Both. Speak on, beloved Sacontalā.

Sac. From that instant my affection was unalterably fixed on him—and thence I am reduced to my present languor.

Anu. Fortunately your affection is placed on a man worthy of yourself.

Pri. Oh! could a fine river have deserted the sea and flowed into a lake?

Dushm. [*Joyfully.*] That which I was eager to know, her own lips have told. Love was the cause of my distemper, and love has healed it; as a summer's day, grown black with clouds, relieves all animals from the heat which itself had caused.

Sac. If it be no disagreeable task, contrive, I entreat you, some means by which I may find favour in the king's eyes.

Dushm. [*Aside.*] That request banishes all my cares, and gives me rapture even in my present uneasy situation.

Pri. [*Aside to* Anusūyā.] A remedy for her, my friend, will scarce be attainable. Exert all the powers of your mind;

for her illness admits of no delay.

Anu. [*Aside to* Priyamvadā.] By what expedient can her cure be both accelerated and kept secret?

Pri. [*As before.*] Oh! to keep it secret will be easy; but to attain it soon, almost insuperably difficult.

Anu. [*As before.*] How so?

Pri. The young king seemed, I admit, by his tender glances, to be enamoured of her at first sight; and he has been observed, within these few days, to be pale and thin, as if his passion had kept him long awake.

Dushm. [*Aside.*] So it has.—This golden bracelet, sullied by the flame which preys on me, and which no dew mitigates, but the tears gushing nightly from these eyes, has fallen again and again on my wrist, and has been replaced on my emaciated arm.

Pri. [*Aloud.*] I have a thought, Anusūyā.—Let us write a love letter, which I will conceal in a flower, and, under the pretext of making a respectful offering, deliver it myself into the king's hand.

Anu. An excellent contrivance! It pleases me highly;—but what says our beloved Sacontalā?

Sac. I must consider, my friend, the possible consequences of such a step.

Pri Think also of a verse or two, which may suit your passion, and be consistent with the character of a lovely girl born in an exalted family.

Sac. I will think of them in due time; but my heart flutters with the apprehension of being rejected.

Dushm. [*Aside.*] Here stands the man supremely blessed in thy presence, from whom, O timid girl, thou art apprehensive of a refusal! Here stands the man, from whom, O beautiful maid, thou fearest rejection, though he loves thee distractedly. He who shall possess thee will seek no brighter gem; and thou art the gem which I am eager to possess.

Anu. You depreciate, Sacontalā, your own incomparable merits. What man in his senses would intercept with an umbrella the moonlight of autumn, which alone can allay the fever caused by the heat of the noon?

Sac. [*Smiling.*] I am engaged in thought. [*She meditates.*]

Dushm. Thus then I fix my eyes on the lovely poetess, without closing them a moment, while she measures the feet of her verse: her forehead is gracefully moved in cadence, and her whole aspect indicates pure affection.

Sac. I have thought of a couplet; but we have no writing implements.

Pri. Let us hear the words; and then I will mark them with my nail on this lotos leaf, soft and green as the breast of a young paroquet: it may easily be cut into the form of a letter.—Repeat the verses.

Sac. 'Thy heart, indeed, I know not: but mine, oh! cruel, love warms by day and by night; and all my faculties are centred on thee'.

Dushm. [*Hastily advancing, and pronouncing a verse in the same measure.*]

'Thee, O slender maid, love only warms; but me he burns; as the day-star only stifles the fragrance of the night-flower, but quenches the very orb of the moon'.

Anu. [*Looking at him joyfully.*] Welcome, great king: the fruit of my friend's imagination has ripened without delay. [Sacontalā *expresses an inclination to rise.*]

Dushm. Give yourself no pain. Those delicate limbs, which repose on a couch of flowers, those arms, whose bracelets of lotos are disarranged by a slight pressure, and that sweet frame, which the hot noon seems to have disordered, must not be fatigued by ceremony.

Sac. [*Aside.*] O my heart, canst thou not rest at length after all thy sufferings?

Anu. Let our sovereign take for his seat a part of the rock on which she reposes. [Sacontalā *makes a little room.*]

Dushm. [*Seating himself.*] Priyamvadā, is not the fever of your charming friend in some degree abated?

Pri. [*Smiling.*] She has just taken a salutary medicine, and will soon be restored to health. But, O mighty prince, as I am favoured by you and by her, my friendship for Sacontalā prompts me to converse with you for a few moments.

Dushm. Excellent damsel, speak openly; and suppress nothing.

Pri. Our lord shall hear.

Dushm. I am attentive.

Pri. By dispelling the alarms of our pious hermits, you have discharged the duty of a great monarch.

Dushm. Oh! talk a little on other subjects.

Pri. Then I must inform you that our beloved companion is enamoured of you, and has been reduced to her present languor by the resistless divinity, love. You only can preserve her inestimable life.

Dushm. Sweet Priyamvadā, our passion is reciprocal; but it is I who am honoured.

Sac. [*Smiling, with a mixture of affection and resentment.*] Why should you detain the virtuous monarch, who must be afflicted by so long an absence from the secret apartments of his palace?

Dushm. This heart of mine, oh thou who art of all things the dearest to it, will have no object but thee, whose eyes enchant me with their black splendour, if thou wilt but speak in a milder strain. I, who was nearly slain by love's arrow, am destroyed by thy speech.

Anu. [*Laughing.*] Princes are said to have many favourite consorts. You must assure us, therefore, that our beloved friend shall not be exposed to affliction through your conduct.

Dushm. What need is there of many words? Let there be ever so many women in my palace, I will have only two objects of perfect regard; the seagirt earth, which I govern, and your sweet friend, whom I love.

Both. Our anxiety is dissipated. [Sacontalā *strives in vain to conceal her joy.*]

Pri. [*Aside to* Anusūyā.] See how our friend recovers her spirits by little and little, as the peahen, oppressed by the summer heat, is refreshed by a soft gale and a gentle shower.

Sac. [*To the damsels.*] Forgive, I pray, my offence in having used unmeaning words: they were uttered only for your amusement in return for your tender care of me.

Pri. They were the occasion, indeed, of our serious advice. But it is the king who must forgive: who else is offended?

Sac. The great monarch will, I trust, excuse what has been said either before him or in his absence.—[*Aside to the damsels.*] Intercede with him, I entreat you.

Dushm. [*Smiling.*] I would cheerfully forgive any offence, lovely Sacontalā, if you, who have dominion over my heart, would allow me full room to sit by you, and recover from my fatigue, on this flowery couch pressed by your delicate limbs.

Pri. Allow him room; it will appease him, and make him happy.

Sac. [*Pretending anger, aside to* Priyamvadā.] Be quiet, thou mischief-making girl! Dost thou sport with me in my present weak state?

Anu. [*Looking behind the scenes.*] Oh! my Priyamvadā, there is our favourite young antelope running wildly and turning his eyes on all sides: he is, no doubt, seeking his mother, who has rambled in the wide forest. I must go and assist his search.

Pri. He is very nimble; and you alone will never be able to confine him in one place. I must accompany you. [*Both going out.*]

Sac. Alas! I cannot consent to your going far: I shall be left alone.

Both. [*Smiling.*] Alone! with the sovereign of the world by your side! [*They go out.*]

Sac. How could my companions both leave me?

Dushm. Sweet maid, give yourself no concern. Am not I, who humbly solicit your favour, present in the room of them? — [*Aside*] — I must declare my passion.— [*Aloud.*]—Why should not I, like them, wave this fan of lotos leaves, to raise cool breezes and dissipate your uneasiness? Why should not I, like them, lay softly in my lap those feet, red as water lilies, and press them, O my charmer, to relieve your pain?

Sac. I should offend against myself, by receiving homage from a person entitled to my respect. [*She rises, and walks slowly through weakness.*]

Dushm. The noon, my love, is not yet passed; and your sweet limbs are weak. Having left that couch where fresh flowers covered your bosom, you can ill sustain this intense heat with so languid a frame. [*He gently draws her back.*]

Sac. Leave me, oh leave me. I am not, indeed, my own mistress, or— the two damsels were only appointed to attend me. What can I do at present?

Dushm. [*Aside.*] Fear of displeasing her makes me bashful.

Sac. [*Overhearing him.*] The king cannot give offence. It is my unhappy fate only that I accuse.

Dushm. Why should you accuse so favourable a destiny?

Sac. How rather can I help blaming it, since it has permitted my heart to be affected by amiable qualities, without having left me at my own disposal?

Dushm. [*Aside.*] One would imagine that the charming sex, instead of being, like us, tormented with love, kept love himself within their hearts, to torment him with delay. [*Sacontalā going out.*]

Dushm. [*Aside.*] How! must I then fail of attaining felicity? [*Following her and catching the skirt of her mantle.*]

Sac. [*Turning back.*] Son of Puru, preserve thy reason; oh! preserve it.—The hermits are busy on all sides of the grove.

Dushm. My charmer, your fear of them is vain. Canna himself, who is deeply versed in the science of law, will be no obstacle to our union. Many daughters of the holiest men have been married by the ceremony called Gandharva,[11] as it is practised by Indra's band, and even their fathers have approved them.—[*Looking round.*]—What say you? are you still inflexible? Alas! I must then depart. [*Going from her a few paces, then looking back.*]

Sac. [*Moving also a few steps, and then turning back her face.*] Though I have refused compliance, and have only allowed you to converse with me for a moment, yet, O son of Puru—let not Sacontalā be wholly forgotten.

Dushm. Enchanting girl, should you be removed to the ends of the world, you will be fixed in this heart, as the shade of a lofty tree remains with it even when the day is departed.

[11] [A mythical marital rite observed only among gods, nymphs, and kings, where no priestly supervision or parental approval was needed.]

Sac. [*Going out, aside.*] Since I have heard his protestations, my feet move, indeed, but without advancing. I will conceal myself behind those flowering Curuvacas, and thence I shall see the result of his passion. [*She hides herself behind the shrubs.*]

Dushm. [*Aside.*] Can you leave me, beloved Sacontalā; me who am all affection? Could you not have tarried a single moment? Soft is your beautiful frame, and indicates a benevolent soul; yet your heart is obdurate: as the tender Sirīsha hangs on a hard stalk.

Sac. [*Aside.*] I really have now lost the power of departing.

Dushm. [*Aside.*] What can I do in this retreat since my darling has left it?—[*Musing and looking round.*]—Ah! my departure is happily delayed.—Here lies her bracelet of flowers, exquisitely perfumed by the root of Usīra which had been spread on her bosom: it has fallen from her delicate wrist, and is become a new chain for my heart. [*Taking up the bracelet with reverence.*]

Sac. [*Aside, looking at her hand.*] Ah me! such was my languor, that the filaments of lotos stalks which bound my arm dropped on the ground unperceived by me.

Dushm. [*Aside, placing it in his bosom.*] Oh! how delightful to the touch!—From this ornament of your lovely arm, O my darling, though it be inanimate and senseless, your unhappy lover has regained confidence—a bliss which you refused to confer.

Sac. [*Aside.*] I can stay here no longer. By this pretext I may return. [*Going slowly towards him.*]

Dushm. [*With rapture.*] Ah! the empress of my soul again blesses these eyes. After all my misery I was destined to be favoured by indulgent heaven.—The bird Chatac, whose throat was parched with thirst, supplicated for a drop of water, and suddenly a cool stream poured into his bill from the bounty of a fresh cloud.

Sac. Mighty king, when I had gone half way to the cottage, I perceived that my bracelet of thin stalks had fallen from my wrist; and I return because my heart is almost convinced that you must have seen and taken it. Restore it, I humbly entreat, lest you expose both yourself and me to the censure of the hermits.

Dushm. Yes, on one condition I will return it.

Sac. On what condition? Speak—

Dushm. That I may replace it on the wrist to which it belongs.

Sac. [*Aside.*] I have no alternative. [*Approaching him.*]

Dushm. But in order to replace it, we must both be seated on that smooth rock. [*Both sit down.*]

Dushm. [*Taking her hand.*] O exquisite softness! This hand has regained its native strength and beauty, like a young shoot of Cāmalatà: or it resembles rather the god of love himself, when, having been consumed by the fire of Hara's wrath, he was restored to life by a shower of nectar sprinkled by the immortals.

Sac. [*Pressing his hand.*] Let the son of my lord make haste to tie on the bracelet.

Dushm. [*Aside, with rapture.*] Now I am truly blessed —That phrase, the son of my lord, is applied only to a husband.—[*Aloud.*]—My charmer, the clasp of this bracelet is not easily loosened: it must be made to fit you better.

Sac. [*Smiling.*] As you please.

Dushm. [*Quitting her hand.*] Look, my darling: this is the new moon which left the firmament in honour of superior beauty, and, having descended on your enchanting wrist, has joined both its horns round it in the shape of a bracelet.

Sac. I really see nothing like a moon: the breeze, I suppose, has shaken some dust from the lotos flower behind my ears, and that has obscured my sight.

Dushm. [*Smiling.*] If you permit me, I will blow the fragrant dust from your eye.

Sac. It would be a kindness; but I cannot trust you.

Dushm. Oh! fear not, fear not. A new servant never transgresses the command of his mistress.

Sac. But a servant over-assiduous deserves no confidence.

Dushm. [*Aside.*] I will not let slip this charming occasion.—[*Attempting to raise her head—Sacontalā faintly repels him, but sits still.*]—O damsel with an antelope's eyes, be not apprehensive of my indiscretion.—[*Sacontalā looks up for a moment, and then bashfully drops her head.*—Dushmanta, *aside, gently raising her head.*]— That lip, the softness of which is imagined, not proved, seems to pronounce, with a delightful tremour, its permission for me to allay my thirst.

Sac. The son of my lord seems inclined to break his promise.

Dushm. Beloved, I was deceived by the proximity of the lotos to that eye which equals it in brightness. [*He blows gently on her eye.*]

Sac. Well: now I see a prince who keeps his word as it becomes his imperial character. Yet I am really ashamed that no desert of mine entitles me to the kind service of my lord's son.

Dushm. What reward can I desire, except that which I consider as the greatest, the fragrance of your delicious lip?

Sac. Will that content you?

Dushm. The bee is contented with the mere odour of the water lily.

Sac. If he were not, he would get no remedy.

Dushm. Yes, this and this— [*Kissing her eagerly.*]

Behind the scenes. Hark! the Chacravāca is calling her mate on the bank of the Mālinì: the night is beginning to spread her shades.

Sac. [*Listening alarmed.*] O son of my lord, the matron Gautamī approaches to enquire my health. Hide yourself, I entreat, behind yon trees.

Dushm. I yield to necessity. [*He retires.*]

Gautamī *enters with a vase in her hand.*

Gaut. [*Looking anxiously at* Sacontalā.] My child, here is holy water for thee.—What ! hast thou no companion here but the invisible gods; thou who art so much indisposed?

Sac. Both Priyamvadā and Anusūyā are just gone down to the river.

Gaut. [*Sprinkling her.*] Is thy fever, my child, a little abated? [*Feeling her hand.*]

Sac. Venerable matron, there is a change for the better.

Gaut. Then thou art in no danger. Mayst thou live many years! The day is departing: let us both go to the cottage.

Sac. [*Aside, rising slowly.*] O my heart, no sooner hadst thou begun to taste happiness, than the occasion slipped away!—[*She advances a few steps, and returns to the arbour.*]—O bower of twining plants, by whom my sorrows have been dispelled, on thee I call; ardently hoping to be once more happy under thy shade. [*She goes out with* Gautamī.]

Dushm. [*Returning to the bower,* and sighing.] How, alas, have my desires been obstructed!—Could I do less than kiss the lips of my charmer, though her modest cheeks were half averted; lips, whose sweetness had enchanted me, even when they pronounced a denial?—Whither now can I go?—I will remain a while in this arbour of creepers, which my darling's presence has illuminated.—[*Looking round.*]—Yes; this is her seat on the rock, spread with blossoms, which have been pressed by her delicate limbs. —Here lies her exquisite love letter on the leaf of a water lily; here lay her bracelet of tender filaments which had fallen from her sweet wrist.—Though the bower of twining Vetasas be now desolate, since my charmer has left it, while my eyes are fixed on all these delightful memorials of her, I am unable to depart.—[*Musing*]—Ah! how imperfectly has this affair been conducted by a lover, like me, who, with his darling by his side, has let the occasion slip.—Should Sacontalā visit once more this calm retreat, the opportunity shall not pass again unimproved: the pleasures of youth are by nature transitory.—Thus my foolish heart forms resolutions, while it is distracted by the sudden interruption of its happiness.

Why did it ever allow me to quit without effect the presence of my beloved?

Behind the scenes. O king, while we are beginning our evening sacrifice, the figures of blood-thirsty demons, embrowned by clouds collected at the departure of day, glide over the sacred hearth, and spread consternation around.

Dushm. Fear not, holy men.—Your king will protect you. [*He goes out.*]

Act IV

Scene—A lawn *before the Cottage.*
The two damsels are discovered gathering flowers.
Anusūyā.

O my Priyamvadā, though our sweet friend has been happily married, according to the rites of Gandharvas, to a bridegroom equal in rank and accomplishments, yet my affectionate heart is not wholly free from care; and one doubt gives me particular uneasiness.

Pri. What doubt, my Anusūyā?

Anu. This morning the pious prince was dismissed with gratitude by our hermits, who had then completed their mystick rites: he is now gone to his capital, Hastināpura, where, surrounded by a hundred women in the recesses of his palace, it may be doubted whether he will remember his charming bride.

Pri. In that respect you may be quite easy. Men, so well informed and well educated as he, can never be utterly destitute of honour.—We have another thing to consider. When our father Canna shall return from his pilgrimage, and shall hear what has passed, I cannot tell how he may receive the intelligence.

Anu. If you ask my opinion, he will, I think, approve of the marriage.

Pri. Why do you think so?

Anu. Because he could desire nothing better, than that a husband so accomplished and so exalted should take Sacontalā by the hand. It was, you know, the declared object of his heart, that she might be suitably married; and, since heaven has done for him what he most wished to do, how can he possibly be dissatisfied?

Pri. You reason well; but—[*Looking at her basket.*]—My friend, we have plucked a sufficient store of flowers to scatter over the place of sacrifice.[12]

12 [Sprinkling of flowers over the place of sacrifice—in this case the spot where the Gandharva marriage of Śakuntalā and Dushyanta occurred—was an ancient Hindu custom. It was especially popular during the time of Kālidāsa.]

Anu. Let us gather more to decorate the temples of the goddesses who have procured for Sacontalā so much good fortune. [*They both gather more flowers.*]

Behind the scenes. It is I.—Hola!

Anu [*Listening.*] I hear the voice, as it seems, of a guest arrived in the hermitage.

Pri. Let us hasten thither. Sacontalā is now reposing; but though we may, when she wakes, enjoy her presence, yet her mind will all day be absent with her departed lord.

Anu. Be it so; but we have occasion, you know, for all these flowers. [*They advance.*]

Again behind the scenes. How! Dost thou show no attention to a guest? Then hear my imprecations—'He on whom thou art meditating, on whom alone thy heart is now fixed, while thou neglectest a pure gem of devotion who demands hospitality, shall forget thee, when thou seest him next, as a man restored to sobriety forgets the words which he uttered in a state of intoxication'. [*Both damsels look at each other with affliction.*]

Pri. Wo is me! Dreadful calamity! Our beloved friend has, through mere absence of mind, provoked by her neglect, some holy man who expected reverence.

Anu. [*Looking.*] It must be so; for the cholerick Durvāsas is going hastily back.

Pri. Who else has power to consume, like raging fire, whatever offends him? Go, my Anusūyā; fall at his feet, and persuade him, if possible, to return: in the mean time I will prepare water and refreshments for him.

Anu. I go with eagerness. [*She goes out.*]

Pri [*Advancing hastily, her foot slips.*] Ah! through my eager haste I have let the basket fall; and my religious duties must not be postponed. [*She gathers fresh flowers.*]

Anusūyā *re-enters.*

Anu. His wrath, my beloved, passes all bounds.—Who living could now appease him by the humblest prostrations or entreaties? yet at last he a little relented.

Pri. That little is a great deal for him.—But inform me how you soothed him in any degree.

Anu. When he positively refused to come back, I threw myself at his feet, and thus addressed him: 'Holy sage, forgive, I entreat, the offence of an amiable girl, who has the highest veneration for you, but was ignorant, through distraction of mind, how exalted a personage was calling to her'.

Pri. What then? What said he?

Anu. He answered thus: 'My word must not be recalled ; but the spell which it has raised shall be wholly removed when her lord shall see his ring'. Saying this, he disappeared.

Pri. We may now have confidence; for before the monarch departed, he fixed with his own hand on the finger of Sacontalā the ring, on which we saw the name of Dushmanta engraved, and which we will instantly recognize. On him therefore alone will depend the remedy for our misfortune.

Anu. Come, let us now proceed to the shrines of the goddesses, and implore their succour. [*Both advance.*]

Pri. [*Looking.*] See, my Anusūyā, where our beloved friend sits, motionless as a picture, supporting her languid head with her left hand. With a mind so intent on one subject, she can pay no attention to herself, much less to a stranger.

Anu. Let the horrid imprecation, Priyamvadā, remain a secret between us two: we must spare the feelings of our beloved, who is naturally susceptible of quick emotions.

Pri. Who would pour boiling water on the blossom of a tender Mallicà? [*Both go out.*]

A Pupil of Canna *enters.*

Pup. I am ordered by the venerable Canna, who is returned from the place of his pilgrimage, to observe the time of the night, and am, therefore, come forth to see how much remains of it.—[*Walking round, and observing the heavens.*]—On one side, the moon, who kindles the flowers of the Oshadhì, has nearly sunk in his western bed; and, on the other, the sun, seated behind his charioteer Arun, is beginning his course: the lustre of them both is conspicuous, when they rise and when they set; and by their example should men be equally firm in prosperous and in adverse fortune.—The moon has now disappeared, and the night flower pleases no more: it leaves only a remembrance of its odour, and languishes like a tender bride whose pain is intolerable in the absence of her beloved.—The ruddy morn impurples the dew drops on the branches of yonder Vadarī; the peacock, shaking off sleep, hastens from the cottages of hermits interwoven with holy grass; and yonder antelope, springing hastily from the place of sacrifice, which is marked with his hoofs, raises himself on high, and stretches his graceful limbs.—How is the moon fallen from the sky with diminished beams! the moon who had set his foot on the head of Sumēru, king of mountains, and had climbed, scattering the rear of darkness, even to the central palace of Vishnu!—Thus do the great men of this world ascend with extreme labour to the summit of ambition, but easily and quickly descend from it.

Anusūyā enters meditating.

Anu. [*Aside.*] Such has been the affection of Sacontalā, though she was bred in austere devotion, averse from sensual enjoyments!—How unkind was the king to leave her!

Pup. [*Aside.* The proper time is come for performing the hōma: I must apprise our preceptor of it. [*He goes out.*]

Anu. The shades of night are dispersed; and I am hardly awake; but were I ever so perfectly in my senses, what could I now do? My hands move not readily to the usual occupations of the morning.—Let the blame be cast on love, on love only, by whom our friend has been reduced to her present condition, through a monarch who has broken his word.—Or does the imprecation of Durvāsas already prevail?—How else could a virtuous king, who made so solemn an engagement, have suffered so long a time to elapse without sending even a message?—Shall we convey the fatal ring to him?—Or what expedient can be suggested for the relief of this incomparable girl, who mourns without ceasing?—Yet what fault has she committed?—With all my zeal for her happiness, I cannot summon courage enough to inform our father Canna that she is pregnant.—What then, oh! what step can I take to relieve her anxiety?

Priyamvadā enters.

Pri. Come, Anusūyā, come quickly. They are making suitable preparations for conducting Sacontalā to her husband's palace.

Anu. [*With surprise.*] What say you, my friend?

Pri. Hear me. I went just now to Sacontalā, meaning only to ask if she had slept well—

Anu. What then? oh! what then?

Pri. She was sitting with her head bent on her knee, when our father Canna, entering her apartment, embraced and congratulated her.—'My sweet child', said he, 'there has been a happy omen: the young Brāhmen who officiated in our morning sacrifice, though his sight was impeded by clouds of smoke, dropped the clarified butter into the very centre of the adorable flame.—Now, since the pious act of my pupil has prospered, my foster child must not be suffered any longer to languish in sorrow; and this day I am determined to send thee from the cottage of the old hermit who bred thee up, to the palace of the monarch who has taken thee by the hand'.

Anu. My friend, who told Canna what passed in his absence?

Pri. When he entered the place where the holy fire was blazing, he heard a voice from heaven pronouncing divine measures.—

Anu. [*Amazed*]. Ah! you astonish me.

Pri. Hear the celestial verse:—'Know that thy adopted daughter, O pious Brāhmen, has received from Dushmanta a ray of glory destined to rule the world; as the wood Samì becomes pregnant with mysterious fire'.

Anu. [*Embracing* Priyamvadā.] I am delighted, my beloved; I am transported with joy. But—since they mean to deprive us of our friend so soon as to-day, I feel that my delight is at least equalled by my sorrow.

Pri. Oh! we must submit patiently to the anguish of parting. Our beloved friend will now be happy; and that should console us.

Anu. Let us now make haste to dress her in bridal array. I have already, for that purpose, filled the shell of a cocoa nut, which you see fixed on an Amra tree, with the fragrant dust of Nāgacēsaras: take it down, and keep it in a fresh lotos leaf, whilst I collect some Gōrāchana from the forehead of a sacred cow, some earth from consecrated ground, and some fresh Cusa grass, of which I will make a paste to ensure good fortune.

Pri. By all means. [*She takes down the perfume.—Anusūyā goes out.*]

Behind the scenes. O Gautamī, bid the two Misras, Sārngarava and Sāradwata, make ready to accompany my child Sacontalā.

Pri. [*Listening.*] Lose no time, Anusūyā, lose no time. Our father Canna is giving orders for the intended journey to Hastināpura.

Anusūyā *re-enters with the ingredients of her charm.*

Anu. I am here: let us go, my Priyamvadā. [*They both advance.*]

Pri. [*Looking.*] There stands our Sacontalā, after her bath at sunrise, while many holy women, who are congratulating her, carry baskets of hallowed grain.—Let us hasten to greet her.

Enter Sacontalā, Gautamī, *and female Hermits.*

Sac. I prostrate myself before the goddess.

Gaut. My child, thou canst not pronounce too often the word goddess: thus wilt thou procure great felicity for thy lord.

Herm. Mayst thou, O royal bride, be delivered of a hero! [*The* Hermits *go out.*]

Both damsels. [*Approaching* Sacontalā.] Beloved friend, was your bath pleasant?

Sac. O! my friends, you are welcome: let us sit a while together. [*They seat themselves.*]

Anu. Now you must be patient, whilst I bind on a charm to secure your happiness.

Sac. That is kind.—Much has been decided this day: and the pleasure of being thus attended by my sweet friends will not soon return. [*Wiping off her tears.*]

Pri. Beloved, it is unbecoming to weep at a time when you are going to be so happy.—[*Both damsels burst into tears as they dress her.*]—Your elegant person deserves richer apparel: it is now decorated with such rude flowers as we could procure in this forest.

Canna's Pupil *enters with rich clothes.*

Pup. Here is a complete dress. Let the queen wear it auspiciously; and may her life be long! [*The women look with astonishment.*]

Gaut. My son, Hārīta, whence came this apparel?

Pup. From the devotion of our father Canna.

Gaut. What dost thou mean?

Pup. Be attentive. The venerable sage gave this order: 'Bring fresh flowers for Sacontalā from the most beautiful trees'; and suddenly the wood-nymphs appeared, raising their hands, which rivalled new leaves in beauty and softness. Some of them wove a lower mantle bright as the moon, the presage of her felicity; another pressed the juice of Lācshà to stain her feet exquisitely red; the rest were busied in forming the gayest ornaments; and they eagerly showered their gifts on us.

Pri. [*Looking at* Sacontalā.] Thus it is, that even the bee, whose nest is within the hollow trunk, does homage to the honey of the lotos flower.

Gaut. The nymphs must have been commissioned by the goddess of the king's fortune, to predict the accession of brighter ornaments in his palace. [*Sacontalā looks modest.*]

Pup. I must hasten to Canna, who is gone to bathe in the Mālinì, and let him know the signal kindness of the woodnymphs. [*He goes out.*]

Anu. My sweet friend, I little expected so splendid a dress:—how shall I adjust it properly?—[*Considering.*] —Oh! my skill in painting will supply me with some hints; and I will dispose the drapery according to art.

Sac. I well know your affection for him.

Canna *enters meditating.*

Can. [*Aside.*] This day must Sacontalā depart: that is resolved; yet my soul is smitten with anguish.—My speech is interrupted by a torrent of tears, which my reason suppresses and turns inward: my very sight is dimmed.—Strange that the affliction of a forester,[13] retired from the haunts of men, should be so excessive! —Oh, with what pangs must they who are fathers of families, be afflicted on the departure of a daughter! [*He walks round musing.*]

Pri. Now, my Sacontalā, you are becomingly decorated: put on this lower vest, the gift of sylvan goddesses. [*Sacontalā rises and puts on the mantle*]

Gaut. My child, thy spiritual father, whose eyes overflow

13 [Kālidāsa imparts a moral lesson to the audience by suggesting a dichotomy between a worldly father and an ascetic father, as the sage Kanvà is in this situation. He ought not to be expected to feel the anguish at Śakuntalā's departure.]

with tears of joy, stands desiring to embrace thee. Hasten, therefore, to do him reverence. [Sacontala *modestly bows to him.*]

Can. Mayst thou be cherished by thy husband, as Sarmishthà was cherished by Yayàti! Mayst thou bring forth a sovereign of the world, as she brought forth Puru!

Gaut. This, my child, is not a mere benediction; it is a boon actually conferred.

Can. My best beloved, come and walk with me round the sacrificial fire.—[*They all advance.*]—May these fires preserve thee! Fires which spring to their appointed stations on the holy hearth, and consume the consecrated wood, while the fresh blades of mysterious Cusa lie scattered around them!—Sacramental fires, which destroy sin with the rising fumes of clarified butter!—[Sacontala *walks with solemnity round the hearth.*]—Now set out, my darling, on thy auspicious journey.—[*Looking round.*]—Where are the attendants, the two Misras?

Enter Sàrngarava *and* Sàradwata.

Both. Holy sage, we are here.

Can. My son, Sàrngarava, show thy sister her way.

Sàrn. Come, damsel.— [*They all advance.*]

Can. Hear, all ye trees of this hallowed forest; ye trees, in which the sylvan goddesses have their abode; hear, and proclaim, that Sacontala is going to the palace of her wedded lord; she who drank not, though thirsty, before you were watered; she who cropped not, through affection for you, one of your fresh leaves, though she would have been pleased with such an ornament for her locks; she whose chief delight was in the season when your branches are spangled with flowers!

CHORUS *of invisible* Woodnymphs.

May her way be attended with prosperity! May propitious breezes sprinkle, for her delight, the odoriferous dust of rich blossoms! May pools of clear water, green with the leaves of the lotos, refresh her as she walks! and may shady branches be her defence from the scorching sunbeams! [*All listen with admiration.*]

Sàrn. Was that the voice of the Còcila wishing a happy journey to Sacontala?—Or did the nymphs, who are allied to the pious inhabitants of these woods, repeat the warbling of the musical bird, and make its greeting their own?

Gaut. Daughter, the sylvan goddesses, who love their kindred hermits, have wished you prosperity, and are entitled to humble thanks.

[Sacontala *walks round, bowing to the nymphs.*]

Sac. [*Aside to* Priyamvadà.] Delighted as I am, O Priyamvadà, with the thought of seeing again the son of my lord, yet, on leaving this grove, my early asylum, I am scarce able to walk.

Pri. You lament not alone.—Mark the affliction of the forest itself when the time of your departure approaches! —The female antelope browses no more on the collected Cusa grass; and the peahen ceases to dance on the lawn: the very plants of the grove, whose pale leaves fall on the ground, lose their strength and their beauty.

Sac. Venerable father, suffer me to address this Màdhavì creeper, whose red blossoms inflame the grove.

Can. My child, I know thy affection for it.

Sac. [*Embracing the plant.*] O most radiant of twining plants, receive my embraces, and return them with thy flexible arms: from this day, though removed to a fatal distance, I shall for ever be thine.—O beloved father, consider this creeper as myself.

Can. My darling, thy amiable qualities have gained thee a husband equal to thyself: such an event has been long, for thy sake, the chief object of my heart; and now, since my solicitude for thy marriage is at an end, I will marry thy favourite plant to the bridegroom Amra, who sheds fragrance near her.—Proceed, my child, on thy journey.

Sac. [*Approaching the two damsels.*] Sweet friends, let this Màdhavì creeper be a precious deposit in your hands.

Anu. and Pri. Alas! in whose care shall we be left? [*They both weep.*]

Can. Tears are vain, Anusùyà: our Sacontala ought rather to be supported by your firmness, than weakened by your weeping. [*All advance.*]

Sec. Father! when yon female antelope, who now moves slowly from the weight of the young ones with which she is pregnant, shall be delivered of them, send me, I beg, a kind message with tidings of her safety.—Do not forget.

Can. My beloved, I will not forget it.

Sac. [*Advancing, then stopping.*] Ah! what is it that clings to the skirts of my robe, and detains me? [*She turns round, and looks.*]

Can. It is thy adopted child, the little fawn, whose mouth, when the sharp points of Cusa grass had wounded it, has been so often smeared by thy hand with the healing oil of Ingudì; who has been so often fed by thee with a handful of Syàmakà grains, and now will not leave the footsteps of his protectress.

Sac. Why dost thou weep, tender fawn, for me, who must leave our common dwelling-place?—As thou wast reared by me when thou hadst lost thy mother, who died soon after thy birth, so will my foster-father attend thee, when we are separated, with anxious care.—Return, poor thing, return—we must part. [*She bursts into tears.*]

Can. Thy tears, my child, ill suit the occasion: we shall

all meet again: be firm: see the direct road before thee, and follow it.—When the big tear lurks beneath thy beautiful eyelashes, let thy resolution check its first efforts to disengage itself.—In thy passage over this earth, where the paths are now high, now low, and the true path seldom distinguished, the traces of thy feet must needs be unequal; but virtue will press thee right onward.

Sārn. It is a sacred rule, holy sage, that a benevolent man should accompany a traveller till he meet with abundance of water; and that rule you have carefully observed: we are now near the brink of a large pool. Give us, therefore, your commands, and return.

Can. Let us rest a while under the shade of this Vata tree—[*They all go to the shade.*]—What message can I send with propriety to the noble Dushmanta? [*He meditates.*]

Anu. [*Aside to* Sacontalā.] My beloved friend, every heart in our asylum is fixed on you alone, and all are afflicted by your departure.—Look; the bird Chacravāca, called by his mate, who is almost hidden by water lilies, gives her no answer; but having dropped from his bill the fibres of lotos stalks which he had plucked, gazes on you with inexpressible tenderness.

Can. My son Sārngarava, remember, when thou shalt present Sacontalā to the king, to address him thus, in my name: 'Considering us hermits as virtuous, indeed, but rich only in devotion, and considering also thy own exalted birth, retain thy love for this girl, which arose in thy bosom without any interference of her kindred; and look on her among thy wives with the same kindness which they experience: more than that cannot be demanded; since particular affection must depend on the will of heaven'.

Sārn. Your message, venerable man, is deeply rooted in my remembrance.

Can. [*Looking tenderly at* Sacontalā.] Now, my darling, thou too must be gently admonished.—We, who are humble foresters, are yet acquainted with the world which we have forsaken.

Sārn. Nothing can be unknown to the wise.

Can. Hear, my daughter.—When thou art settled in the mansion of thy husband, show due reverence to him, and to those whom he reveres: though he have other wives, be rather an affectionate handmaid to them than a rival.—Should he displease thee, let not thy resentment lead thee to disobedience.—In thy conduct to thy domesticks be rigidly just and impartial; and seek not eagerly thy own gratifications.—By such behaviour young women become respectable; but perverse wives are the bane of a family.—What thinks Gautamī of this lesson?

Gaut. It is incomparable:—my child, be sure to remember it.

Can. Come, my beloved girl, give a parting embrace to me and to thy tender companions.

Sac. Must Anusūyā and Priyamvadā return to the hermitage?

Can. They too, my child, must be suitably married; and it would not be proper for them yet to visit the city; but Gautamī will accompany thee.

Sac. [*Embracing him.*] Removed from the bosom of my father, like a young sandal tree, rent from the hills of Malaya, how shall I exist in a strange soil?

Can. Be not so anxious. When thou shalt be mistress of a family, and consort of a king, thou mayst, indeed, be occasionally perplexed by the intricate affairs which arise from exuberance of wealth, but wilt then think lightly of this transient affliction, especially when thou shalt have a son (and a son thou wilt have) bright as the rising day-star.—Know also with certainty, that the body must necessarily, at the appointed moment, be separated from the soul: who, then, can be immoderately afflicted, when the weaker bounds of extrinsick relations are loosened, or even broken.

Sac. [*Falling at his feet.*] My father, I thus humbly declare my veneration for you.

Can. Excellent girl, may my effort for thy happiness prove successful.

Sac. [*Approaching her two companions.*] Come, then, my beloved friends, embrace me together. [*They embrace her.*]

Anu. My friend, if the virtuous monarch should not at once recollect you, only show him the ring on which his own name is engraved.

Sac. [*Starting.*] My heart flutters at the bare apprehension which you have raised.

Pri. Fear not, sweet Sacontalā: love always raises ideas of misery, which are seldom or never realised.

Sārn. Holy sage, the sun has risen to a considerable height: let the queen hasten her departure.

Sac. [*Again embracing* Canna.] When, my father, oh! when again shall I behold this asylum of virtue?

Can. Daughter, when thou shalt long have been wedded, like this fruitful earth, to the pious monarch, and shalt have borne him a son, whose car shall be matchless in battle, thy lord shall transfer to him the burden of empire, and thou, with thy Dushmanta, shalt again seek tranquillity, before thy final departure, in this loved and consecrated grove.

Gaut. My child, the proper time for our journey passes away rapidly: suffer thy father to return.—Go, venerable man, go back to thy mansion, from which she is doomed to be so long absent.

Can. Sweet child, this delay interrupts my religious duties.

Sac. You, my father, will perform them long without sorrow; but I, alas! am destined to bear affliction.

Can. O! my daughter, compel me not to neglect my daily devotions.—[*Sighing.*]—No, my sorrow will not be diminished.—Can it cease, my beloved, when the plants which rise luxuriantly from the hallowed grains which thy hand has strown before my cottage, are continually in my sight?—Go, may thy journey prosper.

[Sacontalā *goes out with* Gautamī *and the two Misras.*]

Both damsels. [*Looking after* Sacontalā *with anguish.*] Alas! alas! our beloved is hidden by the thick trees.

Can. My children, since your friend is at length departed, check your immoderate grief, and follow me. [*They all turn back.*]

Both. Holy father, the grove will be a perfect vacuity without Sacontalā.

Can. Your affection will certainly give it that appearance.—[*He walks round, meditating.*]—Ah me!—Yes; at last my weak mind has attained its due firmness after the departure of my Sacontalā.—In truth a daughter must sooner or later be the property of another; and, having now sent her to her lord, I find my soul clear and undisturbed, like that of a man who has restored to its owner an inestimable deposit which he long had kept with solicitude. [*They go out.*]

Act V

Scene—*The* Palace.
An old Chamberlain, *sighing.*
Old Chamberlain. [Pārvatāvana.]

Alas! what a decrepit old age have I attained!—This wand, which I first held for the discharge of my customary duties in the secret apartments of my prince, is now my support, whilst I walk feebly through the multitude of years which I have passed.—I must now mention to the king, as he goes through the palace, an event which concerns himself: it must not be delayed.—[*Advancing slowly.*]—What is it?—Oh! I recollect: the devout pupils of Canna desire an audience.—How strange a thing is human life!—The intellects of an old man seem at one time luminous, and then on a sudden are involved in darkness, like the flame of a lamp at the point of extinction.—[*He walks round and looks.*]—There is Dushmanta: he has been attending to his people, as to his own family; and now with a tranquil heart seeks a solitary chamber; as an elephant the chief of his herd, having grazed the whole morning, and being heated by the meridian sun, repairs to a cool station during the oppressive heats.—Since the king is just risen from his tribunal, and must be fatigued, I am almost afraid to inform him at present that Canna's pupils are arrived: yet how should they who support nations enjoy rest?—The sun yokes his bright steeds for the labour of many hours; the gale breathes by night and by day; the prince of serpents continually sustains the weight of this earth; and equally incessant is the toil of that man, whose revenue arises from a sixth part of his people's income. [*He walks about.*]

Enter Dushmanta, Mādhavya, *and Attendants.*

Dushm. [*Looking oppressed with business.*] Every petitioner having attained justice, is departed happy; but kings who perform their duties conscientiously are afflicted without end.—The anxiety of acquiring dominion gives extreme pain; and when it is firmly established, the cares of supporting the nation incessantly harass the sovereign; as a large umbrella, of which a man carries the staff in his own hand, fatigues while it shades him.

Behind the scenes. May the king be victorious!

Two Bards *repeat stanzas.*

First Bard. Thou seekest not thy own pleasure: no, it is for the people that thou art harassed from day to day. Such, when thou wast created, was the disposition implanted in thy soul! Thus a branchy tree bears on his head the scorching sunbeams, while his broad shade allays the fever of those who seek shelter under him.

Second Bard. When thou wieldest the rod of justice, thou bringest to order all those who have deviated from the path of virtue: thou biddest contention cease: thou wast formed for the preservation of thy people: thy kindred possess, indeed, considerable wealth; but so boundless is thy affection, that all thy subjects are considered by thee as thy kinsmen.

Dushm. [*Listening.*] That sweet poetry refreshes me after the toil of giving judgements and publick orders.

Mādh. Yes; as a tired bull is refreshed when the people say, 'There goes the lord of cattle'.

Dushm. [*Smiling.*] Oh! art thou here, my friend: let us take our seats together. [*The king and* Mādhavya *sit down.— Musick behind the scenes.*]

Mādh. Listen, my royal friend. I hear a well-tuned Vīnà sounding, as if it were in concert with the lutes of the gods, from yonder apartment.—The queen Hansamatì is preparing, I imagine, to greet you with a new song.

Dushm. Be silent, that I may listen.

Cham. [*Aside.*] The king's mind seems intent on some other business. I must wait his leisure. [*Retiring on one side.*]

SONG. [*Behind the scenes.*]

'Sweet bee, who, desirous of extracting fresh honey, wast wont to kiss the soft border of the new-blown Amra flower, how canst thou now be satisfied with the water lily, and forget the first object of thy love?'

Dushm. The ditty breathes a tender passion.

Mādh. Does the king know its meaning? It is too deep for me.

Dushm. [*Smiling.*] I was once in love with Hansamatì, and am now reproved for continuing so long absent from her.—Friend Mādhavya, inform the queen in my name that I feel the reproof.

Mādh. As the king commands; but—[*Rising slowly.*]— My friend, you are going to seize a sharp lance with another man's hand. I cannot relish your commission to an enraged woman.—A hermit cannot be happy till he has taken leave of all passions whatever.

Dushm. Go, my kind friend: the urbanity of thy discourse will appease her.

Mādh. What an errand! [*He goes out.*]

Dushm. [*Aside.*] Ah! what makes me so melancholy on hearing a mere song on absence, when I am not in fact separated from any real object of my affection?—Perhaps the sadness of men, otherwise happy, on seeing beautiful forms and listening to sweet melody, arises from some faint remembrance of past joys and the traces of connections in a former state of existence.[14] [*He sits pensive and sorrowful.*]

Old Cham. [*Advancing humbly.*] May our sovereign be victorious!—Two religious men, with some women, are come from their abode in a forest near the Snowy Mountains, and bring a message from Canna.—The king will command.

Dushm. [*Surprised.*] What! are pious hermits arrived in the company of women?

Cham. It is even so.

Dushm. Order the priest Sōmaratā, in my name, to shew them due reverence in the form appointed by the Vēda; and bid him attend me. I shall wait for my holy guests in a place fit for their reception.

---

[14] [A similar statement is made by Dante in the 'Inferno' on the remembrance of happy times: 'There is no greater sorrow than to remember the happy times when one is in misery'. See also Tennyson in 'Locksley Hall': 'That a sorrow's crown of sorrow is remembering happier things'.]

Cham. I obey. [*He goes out.*]

Dushm. Warder, point the way to the hearth of the consecrated fire.

Ward. This, O king, this is the way.—[*He walks before.*]—Here is the entrance of the hallowed enclosure; and there stands the venerable cow to be milked for the sacrifice, looking bright from the recent sprinkling of mystick water.—Let the king ascend.

[*Dushmanta is raised to the place of sacrifice on the shoulders of his Warders.*]

Dushm. What message can the pious Canna have sent me?—Has the devotion of his pupils been impeded by evil spirits—or by what other calamity?—Or has any harm, alas! befallen the poor herds who graze in the hallowed forest?—Or have the sins of the king tainted the flowers and fruits of the creepers planted by female hermits?—My mind is entangled in a labyrinth of confused apprehensions.

Ward. What our sovereign imagines, cannot possibly have happened; since the hermitage has been rendered secure from evil by the mere sound of his bowstring. The pious men, whom the king's benevolence has made happy, are come, I presume, to do him homage.

Enter Sārngarava, Sāradwata *and* Gautamī, *leading* Sacontalā *by the hand*; *and before them the old* Chamberlain *and the* Priest.

Cham. This way, respectable strangers; come this way.

Sārn. My friend Sāradwata, there sits the king of men, who has felicity at command, yet shows equal respect to all: here no subject, even of the lowest class, is received with contempt. Nevertheless, my soul having ever been free from attachment to worldly things, I consider this hearth, although a crowd now surround it, as the station merely of consecrated fire.

Sārad. I was not less confounded than yourself on entering the populous city; but now I look on it, as a man just bathed in pure water, on a man smeared with oil and dust, as the pure on the impure, as the waking on the sleeping, as the free man on the captive, as the independent on the slave.

Priest. Thence it is, that men, like you two, are so elevated above other mortals.

Sac. [*Perceiving a bad omen.*] Venerable mother, I feel my right eye throb! What means this involuntary motion?

Gaut. Heaven avert the omen, my sweet child! May every delight attend thee! [*They all advance.*]

Priest. [*Shewing the king to them.*] There, holy men, is the protector of the people; who has taken his seat, and expects you.

Sārn. This is what we wished; yet we have no private

interest in the business. It is ever thus: trees are bent by the abundance of their fruit; clouds are brought low, when they teem with salubrious rain; and the real benefactors of mankind are not elated by riches.

Ward. O king, the holy guests appear before you with placid looks, indicating their affection.

Dushm. [*Gazing at* Sacontalā.] Ah! what damsel is that, whose mantle conceals the far greater part of her beautiful form?—She looks, among the hermits, like a fresh green bud among faded and yellow leaves.

Ward. This at least, O king, is apparent; that she has a form which deserves to be seen more distinctly.

Dushm. Let her still be covered: she seems pregnant; and the wife of another must not be seen even by me.

Sac. [*Aside, with her hand to her bosom.*] O my heart, why dost thou palpitate?—Remember the beginning of thy lord's affection, and be tranquil.

Priest. May the king prosper! The respectable guests have been honoured as the law ordains; and they have now a message to deliver from their spiritual guide: let the king deign to hear it.

Dushm. [*With reverence.*] I am attentive.

Both Misras. [*Extending their hands.*] Victory attend thy banners!

Dushm. I respectfully greet you both.

Both. Blessings on our sovereign!

Dushm. Has your devotion been uninterrupted?

Sārn. How should our rites be disturbed, when thou art the preserver of all creatures? How, when the bright sun blazes, should darkness cover the world?

Dushm. [*Aside.*] The name of royalty produces, I suppose, all worldly advantages!—[*Aloud.*]—Does the holy Canna then prosper?

Sārn. O king, they who gather the fruits of devotion may command prosperity. He first inquires affectionately whether thy arms are successful, and then addresses thee in these words:—

Dushm. What are his orders?

Sārn. 'The contract of marriage, reciprocally made between thee and this girl, my daughter, I confirm with tender regard; since thou art celebrated as the most honourable of men, and my Sacontalā is Virtue herself in a human form, no blasphemous complaint will henceforth be made against Brahmā for suffering discordant matches: he has now united a bride and bridegroom with qualities equally transcendent.—Since, therefore, she is pregnant by thee, receive her in thy palace, that she may perform, in conjunction with thee, the duties prescribed by religion'.

Gaut. Great king, thou hast a mild aspect; and I wish to address thee in few words.

Dushm. [*Smiling.*] Speak, venerable matron.

Gaut. She waited not the return of her spiritual father; nor were thy kindred consulted by thee. You two only were present, when your nuptials were solemnized: now, therefore, converse freely together in the absence of all others.

Sac. [*Aside.*] What will my lord say?

Dushm. [*Aside, perplexed.*] How strange an adventure!

Sac. [*Aside.*] Ah me! how disdainfully he seems to receive the message!

Sārn. [*Aside.*] What means that phrase which I overheard, 'How strange an adventure?'—[*Aloud.*]—Monarch, thou knowest the hearts of men. Let a wife behave ever so discreetly, the world will think ill of her, if she live only with her paternal kinsmen; and a lawful wife now requests, as her kindred also humbly entreat, that whether she be loved or not, she may pass her days in the mansion of her husband.

Dushm. What sayest thou!—Am I the lady's husband?

Sac. [*Aside, with anguish.*] O my heart, thy fears have proved just.

Sārn. Does it become a magnificent prince to depart from the rules of religion and honour, merely because he repents of his engagements?

Dushm. With what hope of success could this groundless fable have been invented?

Sārn. [*Angrily.*] The minds of those whom power intoxicates are perpetually changing.

Dushm. I am reproved with too great severity.

Gaut. [*To* Sacontalā.] Be not ashamed, my sweet child: let me take off thy mantle, that the king may recollect thee. [*She unveils her.*]

Dushm. [*Aside, looking at* Sacontalā.] While I am doubtful whether this unblemished beauty which is displayed before me has not been possessed by another, I resemble a bee fluttering at the close of night over a blossom filled with dew; and in this state of mind, I neither can enjoy nor forsake her.

Ward. [*Aside to* Dushmanta.] The king best knows his rights and his duties: but who would hesitate when a woman, bright as a gem, brings lustre to the apartments of his palace?

Sārn. What, O king, does thy strange silence import?

Dushm. Holy man, I have been meditating again and again, but have no recollection of my marriage with this lady. How then can I lay aside all consideration of my military tribe, and admit into my palace a young woman who is pregnant by another husband?

Sac. [*Aside.*] Ah! wo is me.—Can there be a doubt even

of our nuptials?—The tree of my hope, which had risen so luxuriantly, is at once broken down.

Sārn. Beware, lest the godlike sage, who would have bestowed on thee, as a free gift, his inestimable treasure, which thou hadst taken, like a base robber, should now cease to think of thee, who art lawfully married to his daughter, and should confine all his thoughts to her whom thy perfidy disgraces.

Sārad. Rest a while, my Sārngarava; and thou, Sacontalā, take thy turn to speak; since thy lord has declared his forgetfulness.

Sac. [*Aside.*] If his affection has ceased, of what use will it be to recall his remembrance of me?—Yet, if my soul must endure torment, be it so: I will speak to him.— [*Aloud to* Dushmanta.]—O my husband!—[*Pausing.*]— Or (if the just application of that sacred word be still doubted by thee) O son of Puru, is it becoming, that, having been once enamoured of me in the consecrated forest, and having shown the excess of thy passion, thou shouldst this day deny me with bitter expressions?

Dushm. [*Covering his ears.*] Be the crime removed from my soul!—Thou hast been instructed for some base purpose to vilify me, and make me fall from the dignity which I have hitherto supported; as a river which has burst its banks and altered its placid current, overthrows the trees that had risen aloft on them.

Sac. If thou sayst this merely from want of recollection, I will restore thy memory by producing thy own ring, with thy name engraved on it!

Dushm. A capital invention!

Sac. [*Looking at her finger.*] Ah me! I have no ring. [*She fixes her eyes with anguish on* Gautamī.]

Gaut. The fatal ring must have dropped, my child, from thy hand, when thou tookest up water to pour on thy head in the pool of Sachītīrtha, near the station of Sacrāvatāra.

Dushm. [*Smiling.*] So skilful are women in finding ready excuses!

Sac. The power of Brahmā must prevail: I will yet mention one circumstance.

Dushm. I must submit to hear the tale.

Sac. One day, in a grove of Vētasas, thou tookest water in thy hand from its natural vase of lotos leaves—

Dushm. What followed?

Sac. At that instant a little fawn, which I had reared as my own child, approached thee; and thou saidst with benevolence: 'Drink thou first, gentle fawn'. He would not drink from the hand of a stranger, but received water eagerly from mine; when thou saidst, with increasing affection: 'Thus every creature loves its companions; you are both foresters alike, and both alike amiable'.

Dushm. By such interested and honied falsehoods are the souls of voluptuaries ensnared!

Gaut. Forbear, illustrious prince, to speak harshly. She was bred in a sacred grove where she learned no guile.

Dushm. Pious matron, the dexterity of females, even when they are untaught, appears in those of a species different from our own.—What would it be if they were duly instructed!—The female Cŏcilas, before they fly towards the firmament, leave their eggs to be hatched, and their young fed, by birds who have no relation to them.

Sac. [*With anger.*] Oh! void of honour, thou measurest all the world by thy own bad heart. What prince ever resembled, or ever will resemble, thee, who wearest the garb of religion and virtue, but in truth art a base deceiver; like a deep well whose mouth is covered with smiling plants!

Dushm. [*Aside.*] The rusticity of her education makes her speak thus angrily and inconsistently with female decorum.—She looks indignant; her eye glows; and her speech, formed of harsh terms, faulters as she utters them. Her lip, ruddy as the Bimba fruit, quivers as if it were nipped with frost; and her eyebrows, naturally smooth and equal, are at once irregularly contracted.—Thus having failed in circumventing me by the apparent lustre of simplicity, she has recourse to wrath, and snaps in two the bow of Cāma, which, if she had not belonged to another, might have wounded me.—[*Aloud.*]—The heart of Dushmanta, young woman, is known to all; and thine is betrayed by thy present demeanor.

Sac. [*Ironically* You kings are in all cases to be credited implicitly: you perfectly know the respect which is due to virtue and to mankind; while females, however modest, however virtuous, know nothing, and speak nothing truly.—In a happy hour I came hither to seek the object of my affection: in a happy moment I received the hand of a prince descended from Puru; a prince who had won my confidence by the honey of his words, whilst his heart concealed the weapon that was to pierce mine. [*She hides her face and weeps.*]

Sārn. This insufferable mutability of the king's temper kindles my wrath.—Henceforth let all be circumspect before they form secret connections: a friendship hastily contracted, when both hearts are not perfectly known, must ere long become enmity.

Dushm. Wouldst thou force me then to commit an enormous crime, relying solely on her smooth speeches?

Sārn. [*Scornfully.*] Thou has heard an answer.—The words of an incomparable girl, who never learned what iniquity was, are here to receive no credit; while they, whose learning consists in accusing others, and inquiring into crimes, are the only persons who speak truth!

Dushm. O man of unimpeached veracity, I certainly am what thou describest; but what would be gained by accusing thy female associate?

Sārn. Eternal misery.

Dushm. No; misery will never be the portion of Puru's descendants.

Sārn. What avails our altercation?—O king, we have obeyed the commands of our preceptor, and now return. Sacontalā is by law thy wife, whether thou desert or acknowledge her; and the dominion of a husband is absolute.—Go before us, Gautamī. [*The two Misras and* Gautamī *returning.*]

Sac. I have been deceived by this perfidious man; but will you, my friends, will you also forsake me? [*Following them.*]

Gaut. [*Looking back.*] My son, Sacontalā follows us with affectionate supplications. What can she do here with a faithless husband; she who is all tenderness?

Sārn. [*Angrily to* Sacontalā.] O wife, who seest the faults of thy lord, dost thou desire independence? [*Sacontalā stops, and trembles.*]

Sārad. Let the queen hear. If thou beest what the king proclaims thee, what right hast thou to complain? But if thou knowest the purity of thy own soul, it will become thee to wait as a handmaid in the mansion of thy lord. Stay, then, where thou art: we must return to Canna.

Dushm. Deceive her not, holy men, with vain expectations. The moon opens the night flower; and the sun makes the water lily blossom: each is confined to its own object: and thus a virtuous man abstains from any connection with the wife of another.

Sārn. Yet thou, O king, who fearest to offend religion and virtue, art not afraid to desert thy wedded wife; pretending that the variety of thy publick affairs has made thee forget thy private contract.

Dushm. [*To his Priest.*] I really have no remembrance of any such engagement; and I ask thee, my spiritual counsellor, whether of the two offences be the greater, to forsake my own wife, or to have an intercourse with the wife of another?

Priest. [*After some deliberation.*] We may adopt an expedient between both.

Dushm. Let my venerable guide command.

Priest. The young woman may dwell till her delivery in my house.

Dushm. For what purpose?

Priest. Wise astrologers have assured the king, that he will be the father of an illustrious prince, whose dominion will be bounded by the western and eastern seas: now, if the holy man's daughter shall bring forth a son whose hands and feet bear the marks of extensive sovereignty, I will do homage to her as my queen, and conduct her to the royal apartments; if not, she shall return in due time to her father.

Dushm. Be it as you judge proper.

Priest. [*To* Sacontalā.] This way, my daughter, follow me.

Sac. O earth! mild goddess, give me a place within thy bosom![15] [*She goes out weeping with the Priest; while the two Misras go out by a different way with* Gautamī— Dushmanta *stands meditating on the beauty of* Sacontalā; *but the imprecation still clouds his memory.*]

Behind the scenes. Oh! miraculous event!

Dushm. [*Listening.*] What can have happened!

The Priest *re-enters.*

Priest. Hear, O king, the stupendous event. When Canna's pupils had departed, Sacontalā, bewailing her adverse fortune, extended her arms and wept; when—

Dushm. What then?

Priest. A body of light, in a female shape, descended near Apsarastīrtha, where the nymphs of heaven are worshiped; and having caught her hastily in her bosom, disappeared. [*All express astonishment.*]

Dushm. I suspected from the beginning some work of sorcery.—The business is over; and it is needless to reason more on it.—Let thy mind, Sōmarāta, be at rest.

Priest. May the king be victorious. [*He goes out.*]

Dushm. Chamberlain, I have been greatly harassed; and thou, Warder, go before me to a place of repose.

Ward. This way; let the king come this way.

Dushm. [*Advancing, aside.*] I cannot with all my efforts recollect my nuptials with the daughter of the hermit; yet so agitated is my heart, that it almost induces me to believe her story. [*All go out.*]

---

[15] [Kālidāsa reminds the audience of a similar appeal made by Sita in the Rāmāyana at the time of her ordeal after Rāma's triumphal return from Lanka. Her request, since she was Lakshmi, was granted, but not that of Śakuntalā, who was only half-divine.]

Act VI

Scene—*A* Street.

*Enter a Superintendent of Police with two Officers,
leading a man with his hands bound.
First Officer, Striking the prisoner.*

Take that, Cumbhīlaca, if Cumhīlaca be thy name; and
tell us now where thou gottest this ring, bright with a large
gem, on which the king's name is engraved.

Cumbh. [*Trembling.*] Spare me, I entreat your honours to
spare me: I am not guilty of so great a crime as you
suspect.

First Off. O distinguished Brāhmen, didst thou then receive it from the king as a reward of some important
service?

Cumbh. Only hear me: I am a poor fisherman dwelling at
Sacrāvatāra—

Second Off. Did we ask, thou thief, about thy tribe or thy
dwelling place?

Sup. O Sūchaca, let the fellow tell his own story.—Now
conceal nothing, sirrah.

First Off. Dost thou hear? Do as our master commands.

Cumbh. I am a man who support my family by catching
fish in nets, or with hooks, and by various other con-
trivances.

Sup. [*Langhing.*] A virtuous way of gaining a livelihood!

Cumbh. Blame me not, master. The occupation of our
forefathers, how low soever, must not be forsaken; and a
man who kills animals for sale may have a tender heart
though his act be cruel.

Sup. Go on, go on.

Cumbh. One day having caught a large Rōhita fish, I cut
it open, and saw this bright ring in its stomach; but when
I offered to sell it, I was apprehended by your honours.
So far only am I guilty of taking the ring. Will you now
continue beating and bruising me to death?

Sup. [*Smelling the ring.*] It is certain, Jāluca, that this gem
has been in the body of a fish. The case requires consideration; and I will mention it to some of the king's
household.

Both Off. Come on, cutpurse. [*They advance.*]

Sup. Stand here, Sūchaca, at the great gate of the city, and
wait for me, while I speak to some of the officers in the
palace.

Both Off. Go, Rājayucta. May the king favour thee. [*The
Superintendent goes out.*]

Second Off. Our master will stay, I fear, a long while.

First Off. Yes; access to kings can only be had at their
leisure.

Second Off. The tips of my fingers itch, my friend Jāluca,

to kill this cutpurse.

Cumbh. You would put to death an innocent man.

First Off. [*Looking.*] Here comes our master.—The king
has decided quickly. Now, Cumbhīlaca, you will either
see your companions again, or be the food of shakàls and
vultures.

The Superintendent re-enters.

Sup. Let the fisherman immediately—

Cumbh. [*In an agony.*] Oh! I am a dead man.

Sup.—be discharged.—Hola! set him at liberty. The king
says he knows his innocence; and his story is true.

Second Off. As our master commands.—The fellow is
brought back from the mansion of Yama, to which he was
hastening. [*Unbinding the fisherman.*]

Cumbh. [*Bowing.*] My lord, I owe my life to your kindness.

Sup. Rise, friend; and hear with delight that the king gives
thee a sum of money equal to the full value of the ring: it
is a fortune to a man in thy station. [*Giving him the
money.*]

Cumbh. [*With rapture.*] I am transported with joy.

First Off. This vagabond seems to be taken down from
the stake, and set on the back of a state elephant.

Second Off. The king, I suppose, has a great affection for
his gem.

Sup. Not for its intrinsick value; but I guessed the cause
of his ecstasy when he saw it.

Both Off. What could occasion it?

Sup. I suspect that it called to his memory some person
who has a place in his heart; for though his mind be
naturally firm, yet, from the moment when he beheld the
ring, he was for some minutes excessively agitated.

Second Off. Our master has given the king extreme pleasure.

First Off. Yes; and by the means of this fish-catcher.
[*Looking fiercely at him.*]

Cumbh. Be not angry—Half the money shall be divided
between you to purchase wine.

First Off. Oh! now thou art our beloved friend.—Good
wine is the first object of our affection.—Let us go
together to the vintners. [*They all go out.*]

Scene—*The* Garden *of the* Palace,

The Nymph Misracēsī *appears in the air.*

Misr. My first task was duly performed when I went
to bathe in the Nymphs' pool; and I now must see with
my own eyes how the virtuous king is afflicted.—Sacontalā is dear to this heart, because she is the daughter of
my beloved Mēnacà, from whom I received both commissions.—[*She looks round.*]—Ah! on a day full of
delights the monarch's family seem oppressed with some

new sorrow.—By exerting my supernatural power I could know what has passed; but respect must be shown to the desire of Mēnacà. I will retire, therefore, among those plants, and observe what is done without being visible. [*She descends, and takes her station.*]

Enter two Damsels, attending on the God of Love.

First Dams. [*Looking at an Amra flower.*] The blossoms of yon Amra, waving on the green stalk, are fresh and light as the breath of this vernal month. I must present the goddess Retī with a basket of them.

Second Dams. Why, my Parabhriticà, dost thou mean to present it alone?

First Dams. O my friend Madhucaricà, when a female Cōcilà, which my name implies, sees a blooming Amra, she becomes entranced, and loses her recollection.

Second Dams. [*With transport.*] What! is the season of sweets actually returned?

First Dams. Yes; the season in which we must sing of nothing but wine and love.

Second Dams. Support me, then, while I climb up this tree, and strip it of its fragrant gems, which we will carry as an offering to Cāma.

First Dams. If I assist, I must have a moiety of the reward which the god will bestow.

Second Dams. To be sure, and wihout any previous bargain. We are only one soul, you know, though Brahmā has given it two bodies.—[*She climbs up, and gathers the flowers.*]—Ah! the buds are hardly opened.—Here is one a little expanded, which diffuses a charming odour—[*Taking a handful of buds.*]—This flower is sacred to the god who bears a fish on his banner.—O sweet blossom, which I now consecrate, thou well deservest to point the sixth arrow of Cāmadēva, who now takes his bow to pierce myriads of youthful hearts. [*She throws down a blossom.*]

The old Chamberlain *enters.*

Cham. [*Angrily.*] Desist from breaking off those half-opened buds: there will be no jubilee this year; our king has forbidden it.

Both Dams. Oh! pardon us. We really knew not the prohibition.

Cham. You knew it not!—Even the trees which the spring was decking, and the birds who perch on them, sympathize with our monarch. Thence it is, that yon buds, which have long appeared, shed not yet their prolifick dust; and the flower of the Curuvaca, though perfectly formed, remains veiled in a closed chalice; while the voice of the Cōcila, though the cold dews fall no more, is fixed within his throat; and even Smara, the god of desire, replaces the shaft half-drawn from his quiver.

Misr. [*Aside.*] The king, no doubt, is constant and tender-hearted.

First Dams.] A few days ago, Mitravasu, the governor of our province, dispatched us to kiss the feet of the king, and we come to decorate his groves and gardens with various emblems: thence it is, that we heard nothing of his interdict.

Cham. Beware then of reiterating your offence.

Second Dams. To obey our lord will certainly be our delight; but if we are permitted to hear the story, tell us, we pray, what has induced our sovereign to forbid the usual festivity.

Misr. [*Aside.*] Kings are generally fond of gay entertainments; and there must be some weighty reason for the prohibition.

Cham. [*Aside.*] The affair is publick: why should I not satisfy them?—[*Aloud.*]—Has not the calamitous desertion of Sacontalā reached your ears?

First Dams. We heard her tale from the governor, as far as the sight of the fatal ring.

Cham. Then I have little to add.—When the king's memory was restored, by the sight of his gem, he instantly exclaimed: 'Yes, the incomparable Sacontalā is my lawful wife; and when I rejected her, I had lost my reason'.— He showed strong marks of extreme affliction and penitence; and from that moment he has abhorred the pleasures of life. No longer does he exert his respectable talents from day to day for the good of his people: he prolongs his nights wihout closing his eyes, perpetually rolling on the edge of his couch; and when he rises, he pronounces not one sentence aptly; mistaking the names of the women in his apartments, and through distraction, calling each of them Sacontalā: then he sits abashed, with his head long bent on his knees.

Misr. [*Aside.*] This is pleasing to me, very pleasing.

Cham. By reason of the deep sorrow which now prevails in his heart, the vernal jubilee has been interdicted.

Both Dams. The prohibition is highly proper.

Behind the scenes. Make way! The king is passing.

Cham. [*Listening.*] Here comes the monarch: depart therefore, damsels, to your own province. [*The two Damsels go out.*]

Dushmanta *enters in penitential weeds, preceded by a Warder, and attended by* Mādhavya.

Cham. [*Looking at the king.*] Ah! how majestic are noble forms in every habiliment!—Our prince, even in the garb of afflication, is a venerable object.—Though he has abandoned pleasure, ornaments, and business; though he is become so thin, that his golden bracelet falls loosened even down to his wrist; though his lips are parched with the heat of his sighs, and his eyes are fixed open by long sorrow and want of sleep, yet am I dazzled by the blaze of virtue which beams in his countenance like a diamond exquisitely polished.

Misr. [*Aside, gazing on* Dushmanta.] With good reason is my beloved Sacontalā, though disgraced and rejected, heavily oppressed with grief through the absence of this youth.

Dushm. [*Advancing slowly in deep meditation.*] When my darling with an antelope's eyes would have reminded me of our love, I was assuredly slumbering; but excess of misery has awakened me.

Misr. [*Aside.*] The charming girl will at last be happy.

Mādh. [*Aside.*] This monarch of ours is caught again in the gale of affection; and I hardly know a remedy for his illness.

Cham. [*Approaching Dushmanta.*] May the king be victorious!—Let him survey yon fine woodland, these cool walks, and this blooming garden; where he may repose with pleasure on banks of delight.

Dushm. [*Not attending to him.*] Warder, inform the chief minister in my name, that having resolved on a long absence from the city, I do not mean to sit for some time in the tribunal; but let him write and dispatch to me all the cases that may arise among my subjects.

Ward. As the king commands. [*He goes out.*]

Dushm. [*To the Chamberlain.*] And thou, Pārvatāyana, neglect not thy stated business.

Cham. By no means. [*He goes out.*]

Mādh. You have not left a fly in the garden.—Amuse yourself now in this retreat, which seems pleased with the departure of the dewy season.

Dushm. O Mādhavya, when persons accused of great offences prove wholly innocent, see how their accusers are punished!—A phrensy obstructed my remembrance of any former love for the daughter of the sage; and now the heart-born god, who delights in giving pain, has fixed in his bow-string a new shaft pointed with the blossom of an Amra.—The fatal ring having restored my memory, see me deplore with tears of repentance the loss of my best beloved, whom I rejected without cause; see me overwhelmed with sorrow, even while the return of spring fills the hearts of all others with pleasure.

Mādh. Be still, my friend, whilst I break Love's arrows with my staff. [*He strikes off some flowers from an Amra tree.*]

Dushm. [*Meditating.*] Yes, I acknowledge the supreme power of Brahmā.—[*To* Mādhavya.] Where now, my friend, shall I sit and recreate my sight with the slender shrubs which bear a faint resemblance to the shape of Sacontalā?

Mādh. You will soon see the damsel skilled in painting, whom you informed that you would spend the forenoon in yon bower of Mādhavì creepers; and she will bring the queen's picture which you commanded her to draw.

Dushm. My soul will be delighted even by her picture.— Show the way to the bower.

Mādh. This way, my friend.—[*They both advance,* Misracēsì *following them.*]—The arbour of twining Mādhavìs, embellished with fragments of stone like bright gems, appears by its pleasantness, though without a voice, to bid thee welcome.—Let us enter it, and be seated.[*They both sit down in the bower.*]

Misr. [*Aside.*] From behind these branchy shrubs I shall behold the picture of my Sacontalā.—I will afterwards hasten to report the sincere affection of her husband. [*She conceals herself.*]

Dushm. [*Sighing.*] O my approved friend, the whole adventure of the hermitage is now fresh in my memory.— I informed you how deeply I was affected by the first sight of the damsel; but when she was rejected by me you were not present.—Her name was often repeated by me (how, indeed, should it not?) in our conversation.— What! hast thou forgotten, as I had, the whole story?

Misr. [*Aside.*] The sovereigns of the world must not, I find, be left an instant without the objects of their love.

Mādh. Oh, no: I have not forgotten it; but at the end of our discourse you assured me that your love tale was invented solely for your diversion; and this, in the simplicity of my heart, I believed.—Some great event seems in all this affair to be predestined in heaven.

Misr. [*Aside.*] Nothing is more true.

Dushm. [*Having meditated.*] O! my friend, suggest relief for my torment.

Mādh. What new pain torments you? Virtuous men should never be thus afflicted: the most violent wind shakes not mountains.

Dushm. When I reflect on the situation of your friend Sacontalā, who must now be greatly affected by my desertion of her, I am without comfort.—She made an attempt to follow the Brāhmens and the matron: Stay, said the sage's pupil, who was revered as the sage himself; Stay, said he, with a loud voice. Then once more she fixed on me, who had betrayed her, that celestial face, then bedewed with gushing tears; and the bare idea of her pain burns me like an envenomed javelin.

Misr. [*Aside.*] How he afflicts himself! I really sympathize with him.

Mādh. Surely some inhabitant of the heavens must have wafted her to his mansion.

Dushm. No; what male divinity would have taken the pains to carry off a wife so firmly attached to her lord? Mēnacà, the nymph of Swerga, gave her birth; and some of her attendant nymphs have, I imagine, concealed her at the desire of her mother.

Misr. [*Aside.*] To reject Sacontalā was, no doubt, the effect of a delirium, not the act of a waking man.

Mādh. If it be thus, you will soon meet her again.

Dushm. Alas! why do you think so?

Mādh. Because no father and mother can long endure to see their daughter deprived of her husband.

Dushm. Was it sleep that impaired my memory? Was it delusion? Was it an error of my judgement? Or was it the destined reward of my bad actions? Whatever it was, I am sensible that, until Sacontalā return to these arms, I shall be plunged in the abyss of affliction.

Mādh. Do not despair: the fatal ring is itself an example that the lost may be found.—Events which were foredoomed by Heaven must not be lamented.

Dushmn. [*Looking at his ring.*] The fate of this ring, now fallen from a station which it will not easily regain, I may at least deplore.—O gem, thou art removed from the soft finger, beautiful with ruddy tips, on which a place had been assigned thee; and, minute as thou art, thy bad qualities appear from the similarity of thy punishment to mine.

Misr. [*Aside.*] Had it found a way to any other hand its lot would have been truly deplorable.—O Mēnacà, how wouldst thou be delighted with the conversation which gratifies my ears!

Mādh. Let me know, I pray, by what means the ring obtained a place on the finger of Sacontalā.

Dushm. You shall know, my friend.—When I was coming from the holy forest to my capital, my beloved, with tears in her eyes, thus addressed me: 'How long will the son of my lord keep me in his remembrance?'

Mādh. Well; what then?

Dushm. Then, fixing this ring on her lovely finger, I thus answered: 'Repeat each day one of the three syllables engraved on this gem; and before thou hast spelled the word Dushmanta, one of my noblest officers shall attend thee, and conduct my darling to her palace'.—Yet I forgot, I deserted her in my phrensy.

Misr. [*Aside.*] A charming interval of three days was fixed between their separation and their meeting, which the will of Brahmā rendered unhappy.

Mādh. But how came the ring to enter, like a hook, into the mouth of a carp?

Dushm. When my beloved was lifting water to her head in the pool of Sachītirt'ha, the ring must have dropped unseen.

Mādh. It is very probable.

Misr. [*Aside.*] Oh! it was thence that the king, who fears nothing but injustice, doubted the reality of his marriage; but how, I wonder, could his memory be connected with a ring?

Dushm. I am really angry with this gem.

Mādh. [*Laughing.*] So am I with this staff.

Dushm. Why so, Mādhavya?

Mādh. Because it presumes to be so straight when I am so crooked.—Impertinent stick!

Dushm. [*Not attending to him.*] How, O ring, couldst thou leave that hand adorned with soft long fingers, and fall into a pool decked only with water lilies?—The answer is obvious: thou art irrational.—But how could I, who was born with a reasonable soul, desert my only beloved?

Misr. [*Aside.*] He anticipates my remark.

Mādh. [*Aside.*] So; I must wait here during his meditations, and perish with hunger.

Dushm. O my darling, whom I treated with disrespect, and forsook without reason, when will this traitor, whose heart is deeply stung with repentant sorrow, be once more blessed with a sight of thee?

A Damsel enters with a picture.

Dams. Great king, the picture is finished. [*Holding it before him.*]

Dushm. [*Gazing on it.*] Yes; that is her face; those are her beautiful eyes; those her lips embellished with smiles, and surpassing the red lustre of the Catcandhu fruit: her mouth seems, though painted, to speak, and her countenance darts beams of affection blended with a variety of melting tints.

Mādh. Truly, my friend, it is a picture sweet as love itself: my eye glides up and down to feast on every particle of it; and it gives me as much delight as if I were actually conversing with the living Sacontalā.

Misr. [*Aside.*] An exquisite piece of painting!—My beloved friend seems to stand before my eyes.

Dushm. Yet the picture is infinitely below the original; and my warm fancy, by supplying its imperfections, represents, in some degree, the loveliness of my darling.

Misr. [*Aside.*] His ideas are suitable to his excessive love and severe penitence.

Dushm. [*Sighing.*] Alas! I rejected her when she lately approached me, and now I do homage to her picture; like a traveller who negligently passes by a clear and full rivulet, and soon ardently thirsts for a false appearance of water on the sandy desert.

Mādh. There are so many female figures on this canvas, that I cannot well distinguish the lady Sacontalā.

Misr. [*Aside.*] The old man is ignorant of her transcendent beauty; her eyes, which fascinated the soul of his prince, never sparkled, I suppose, on Mādhavya.

Dushm. Which of the figures do you conceive intended for the queen?

Mādh. [*Examining the picture.*] It is she, I imagine, who looks a little fatigued; with the string of her vest rather loose; the slender stalks of her arms falling languidly; a few bright drops on her face, and some flowers dropping from her untied locks. That must be the queen; and the rest, I suppose, are her damsels.

Dushm. You judge well; but my affection requires something more in the piece. Beside, through some defeat in the colouring, a tear seems trickling down her cheek, which ill suits the state in which I desire to see her painted.—[*To the Damsel.*]—The picture, O Chaturicà, is unfinished.—Go back to the painting room and bring the implements of thy art.

Dams. Kind Mādhavya, hold the picture while I obey the king.

Dushm. No; I will hold it. [*He takes the picture; and the Damsel goes out.*]

Mādh. What else is to be painted?

Misr. [*Aside.*] He desires, I presume, to add all those circumstances which became the situation of his beloved in the hermitage.

Dushm. In this landscape, my friend, I wish to see represented the river Mālinì, with some amorous Flamingos on its green margin; farther back must appear some hills near the mountain Himālaya, surrounded with herds of Chamaras; and in the foreground, a dark spreading tree, with some mantles of woven bark suspended on its branches to be dried by the sunbeams; while a pair of black antelopes couch in its shade, and the female gently rubs her beautiful forehead on the horn of the male.

Mādh. Add what you please; but, in my judgement, the vacant places should be filled with old hermits, bent, like me, towards the ground.

Dushm. [*Not attending to him.*] Oh! I had forgotten that my beloved herself must have some new ornaments.

Mādh. What, I pray?

Misr. [*Aside.*] Such, no doubt, as become a damsel bred in a forest.

Dushm. The artist had omitted a Sirīsha flower with its peduncle fixed behind her soft ear, and its filaments waving over part of her cheek; and between her breasts must be placed a knot of delicate fibres, from the stalks of water lilies, like the rays of an autumnal moon.

Mādh. Why does the queen cover part of her face, as if she was afraid of something, with the tips of her fingers,

that glow like the flowers of the Cuvalaya?—Oh! I now perceive an impudent bee, that thief of odours, who seems eager to sip honey from the lotos of her mouth.

Dushm. A bee! drive off the importunate insect.

Mādh. The king has supreme power over all offenders.

Dushm. O male bee, who approachest the lovely inhabitants of a flowery grove, why dost thou expose thyself to the pain of being rejected?—See where thy female sits on a blossom, and, though thirsty, waits for thy return: without thee she will not taste its nectar.

Misr. [*Aside.*] A wild, but apt, address!

Mādh. The perfidy of male bees is proverbial.

Dushm. [*Angrily.*] Shouldst thou touch, O bee, the lip of my darling, ruddy as a fresh leaf on which no wind has yet breathed, a lip from which I drank sweetness in the banquet of love, thou shalt, by my order, be imprisoned in the center of a lotos.—Dost thou still disobey me?

Mādh. How can he fail to obey, since you denounce so severe a punishment?—[*Aside, laughing.*]—He is stark mad with love and affliction; whilst I, by keeping him company, shall be as mad as he without either.

Dushm. After my positive injunction, art thou still unmoved?

Misr. [*Aside.*] How does excess of passion alter even the wise!

Mādh. Why, my friend, it is only a painted bee.

Misr. [*Aside.*] Oh! I perceive his mistake: it shows the perfection of the art. But why does he continue musing?

Dushm. What ill-natured remark was that?—Whilst I am enjoying the rapture of beholding her to whom my soul is attached, thou, cruel remembrancer, tellest me that it is only a picture.—[*Weeping.*]

Misr. [*Aside.*] Such are the woes of a separated lover! He is on all sides entangled in sorrow.

Dushm. Why do I thus indulge unremitted grief? That intercourse with my darling which dreams would give, is prevented by my continued inability to repose; and my tears will not suffer me to view her distinctly even in this picture.

Misr. [*Aside.*] His misery acquits him entirely of having deserted her in his perfect senses.

                              The Damsel *re-enters.*

Dams. As I was advancing, O king, with my box of pencils and colours—

Dushm. [*Hastily.*] What happened?

Dams. It was forcibly seized by the queen Vasumatì, whom her maid Pingalicà had apprised of my errand; and she said: 'I will myself deliver the casket to the son of my lord'.

Mādh. How came you to be released?

Dams. While the queen's maid was disengaging the skirt of her mantle, which had been caught by the branch of a thorny shrub, I stole away.

Dushm. Friend Mādhavya, my great attention to Vasumatì has made her arrogant; and she will soon be here: be it your care to conceal the picture.

Mādh. [*Aside.*] I wish you would conceal it yourself.— [*He takes the picture, and rises.*]—[*Aloud.*]—If, indeed, you will disentangle me from the net of your secret apartments, to which I am confined, and suffer me to dwell on the wall Mēghach'handa, which encircles them, I will hide the picture in a place where none shall see it but pigeons. [*He goes out.*]

Misr. [*Aside.*] How honourably he keeps his former engagements, though his heart be now fixed on another object!

A Warder *enters with a leaf.*

Ward. May the king prosper!

Dushm. Warder, hast thou lately seen the queen Vasumatì?

Ward. I met her, O king; but when she perceived the leaf in my hand, she retired.

Dushm. The queen distinguishes time: she would not impede my publick business.

Ward. The chief minister sends this message: 'I have carefully stated a case which has arisen in the city, and accurately committed it to writing: let the king deign to consider it'.

Dushm. Give me the leaf.—[*Receiving it, and reading.*] —'Be it presented at the foot of the king, that a merchant named Dhanavriddhi, who had extensive commerce at sea, was lost in a late shipwreck: he had child born; and has left a fortune of many millions, which belong, if the king commands, to the royal treasury'.—[*With sorrow.*] —Oh! how great a misfortune it is to die childless![16] Yet with his affluence he must have had many wives:—let an inquiry be made whether any one of them is pregnant.

Ward. I have heard that his wife, the daughter of an excellent man, named Sācētaca, has already performed the cermonies usual on pregnancy.

Dushm. The child, though unborn, has a title to his father's property.—Go: bid the minister make my judgement publick.

Ward. I obey. [*Going.*]

Dushm. Stay a while.—

Ward. [*Returning.*] I am here.

Dushm. Whether he had or had not left offspring, the estate should not have been forfeited.—Let it be proclaimed, that whatever kinsman any one of my subjects may lose, Dushmanta (expecting always the case of forfeiture for crimes) will supply, in tender affection, the place of that kinsman.

Ward. The proclamation shall be made.—[*He goes out.*

[Dushmanta *continues meditating.*]

Re-enter Warder.

O king! the royal decree, which proves that your virtues are awake after a long slumber, was heard with bursts of applause.

Dushm. [*Sighing deeply.*] When an illustrious man dies, alas, without an heir, his estate goes to a stranger; and such will be the fate of all the wealth accumulated by the sons of Puru.[17]

Ward. Heaven avert the calamity! [*Goes out.*]

Dushm. Wo is me! I am stripped of all the felicity which I once enjoyed.

Misr. [*Aside.*] How his heart dwells on the idea of his beloved!

Dushm. My lawful wife, whom I basely deserted, remains fixed in my soul: she would have been the glory of my family, and might have produced a son brilliant as the richest fruit of the teeming earth.

Misr. [*Aside.*] She is not forsaken by all; and soon, I trust, will be thine.

Dams. [*Aside.*] What a change has the minister made in the king by sending him that mischievous leaf! Behold, he is deluged with tears.

Dushm. Ah me! the departed souls of my ancestors, who claim a share in the funeral cake, which I have no son to offer, are apprehensive of losing their due honour, when Dushmanta shall be no more on earth;—when then, alas, will perform in our family those obsequies which the Vēda prescribes?—My forefathers must drink, instead of a pure libation, this flood of tears, the only offering which a man who dies childless can make them. [*Weeping.*]

Misr. [*Aside.*] Such a veil obscures the king's eyes, that he thinks it total darkness, though a lamp be now shining brightly.

Dams. Afflict not yourself immoderately: our lord is young; and when sons illustrious as himself shall be born of other queens, his ancestors will be redeemed from their offences committed here below.

Dushm. [*With agony.*] The race of Puru, which has hither-

---

[16] [Kālidāsa demonstrates his skill as a playwright by using the lawsuit concerning the childless, deceased businessman as a subplot, in order to provide a minor parallel to Dushyanta's own situation.]

[17] [A reference to King Dushyanta's ancestors from the Puru dynasty.]

to been fruitful and unblemished, ends in me; as the river Sereswatì disappears in a region unworthy of her divine stream. [*He faints.*]

Dams. Let the king resume confidence.—[*She supports him.*]

Misr. [*Aside.*] Shall I restore him? No; he will speedily be roused—I heard the nymph Dēvajananì consoling Sacontalā in these words: 'As the gods delight in their portion of sacrifices, thus wilt thou soon be delighted by the love of thy husband'. I go, therefore, to raise her spirits, and please my friend Mēnacà with an account of his virtues and his affection. [*She rises aloft and disappears.*]

Behind the scenes. A Brāhmen must not be slain: save the life of a Brāhmen.

Dushm. [*Reviving and listening.*] Ha! was not that the plaintive voice of Mādhavya?

Dams. He has probably been caught with the picture in his hand by Pingalicà and the other maids.

Dushm. Go, Chaturicà, and reprove the queen in my name for not restraining her servants.

Dams. As the king commands. [*She goes out.*]

Again behind the scenes. I am a Brāhmen, and must not be put to death.

Dushm. It is manifestly some Brāhmen in great danger.— Hola! who is there?

The old Chamberlain *enters.*

Cham. What is the king's pleasure?

Dushm. Inquire why the faint-hearted Mādhavya cries out so piteously.

Cham. I will know in an instant. [*He goes out, and returns trembling.*]

Dushm. Is there any alarm, Pārvatāyana?

Cham. Alarm enough!

Dushm. What causes thy tremour?—Thus do men tremble through age: fear shakes the old man's body, as the breeze agitates the leaves of the Pippala.

Cham. Oh! deliver thy friend.

Dushm. Deliver him! from what?

Cham. From distress and danger.

Dushm. Speak more plainly.

Cham. The wall which looks to all quarters of the heavens, and is named, from the clouds which cover it, Mēghach'handa—

Dushm. What of that?

Cham. From the summit of that wall, the pinnacle of which is hardly attainable even by the blue-necked pigeons, an evil being, invisible to human eyes, has violently carried away the friend of your childhood.

Dushm. [*Starting up hastiliy.*] What! are even my secret apartments infested by supernatural agents?—Royalty is ever subjected to molestation.—A king knows now even the mischiefs which his own negligence daily and hourly occasions:—how then should he know what path his people are treading; and how should he correct their manners when his own are uncorrected?

Behind the scenes. Oh, help! Oh, release me.

Dushm. [*Listening and advancing.*] Fear not, my friend, fear nothing—

Behind the scenes. Not fear, when a monster has caught me by the nape of my neck, and means to snap my backbone as he would snap a sugar-cane!

Dushm. [*Darting his eyes round.*] Hola! my bow—

A Warder *enters with the king's bow and quiver.*

Ward. Here are our great hero's arms. [*Dushmanta takes his bow and an arrow.*]

Behind the scenes. Here I stand; and, thirsting for thy fresh blood, will slay thee struggling as a tyger slays a calf.—Where now is thy protector, Dushmanta, who grasps his bow to defend the oppressed?

Dushm. [*Wrathfully.*] The demon names me with defiance.—Stay, thou basest of monsters.—Here am I, and thou shalt not long exist.—[*Raising his bow.*]—Show the way, Pārvatāyana, to the stairs of the terrace.

Cham. This way, great king!— [*All go out hastily.*]

The Scene changes to a broad Terrace.

Enter Dushmanta.

Dushm. [*Looking round.*] Ah! the place is deserted.

Behind the scenes. Save me, oh! save me.—I see thee, my friend, but thou canst not discern me, who, like a mouse in the claws of a cat, have no hope of life.

Dushm. But this arrow shall distinguish thee from thy foe, in spight of the magick which renders thee invisible.— Mādhavya, stand firm; and thou, blood-thirsty fiend, think not of destroying him whom I love and will protect. —See, I thus fix a shaft which shall pierce thee, who deservest death, and shall save a Brāhmen who deserves long life; as the celestial bird sips the milk, and leaves the water which has been mingled with it. [*He draws the bowstring.*]

Enter Mātali and Mādhavya.

Māt. The god Indra has destined evil demons to fall by way shafts: against them let thy bow be drawn, and cast on thy friends eyes bright with affection.

Dushm. [*Astonished, giving back his arms.*] Oh! Mātali, welcome; I greet the driver of Indra's car.

Mādh. What! this cutthroat was putting me to death, and thou greetest him with a kind welcome!

Māt. [*Smiling.*] O king, live long and conquer! Hear on what errand I am dispatched by the ruler of the firmament.

Dushm. I am humbly attentive.

Māt. There is a race of Dānavas, the children of Cālanēmi, whom it is found hard to subdue—

Dushm. This I have heard already from Nāred.

Māt. The god with an hundred sacrifices, unable to quell that gigantick race, commissions thee, his approved friend, to assail them in the front of battle; as the sun with seven steeds despairs of overcoming the dark legions of night, and gives way to the moon, who easily scatters them. Mount, therefore, with me, the car of Indra, and, grasping thy bow, advance to assured victory.

Dushm. Such a mark of distinction from the prince of good genii honours me highly; but say why you treated so roughly my poor friend Mādhavya.

Māt. Perceiving that, for some reason or another, you were grievously afflicted, I was desirous to rouse your spirits by provoking you to wrath.—The fire blazes when wood is thrown on it; the serpent, when provoked, darts his head against the assailant; and a man capable of acquiring glory, exerts himself when his courage is excited.

Dushm. [*To* Mādhavya.] My friend, the command of Divespetir must instantly be obeyed: go, therefore, and carry the intelligence to my chief minister; saying to him in my name: 'Let thy wisdom secure my people from danger while this braced bow has a different employment'.

Mādh. I obey; but wish it could have been employed without assistance from my terror. [*He goes out.*]

Māt. Ascend, great king. [Dushmanta *ascends, and* Mātali *drives off the car.*]

Act VII

Dushmanta *with* Mātali *in the car of Indra, supposed to be above the clouds.*

Dushmanta.

I am sensible, O Mātali, that, for having executed the commission which Indra gave me, I deserved not such a profusion of honours.

Māt. Neither of you is satisfied. You who have conferred so great a benefit on the god of thunder, consider it as a trifling act of devotion; whilst he reckons not all his kindness equal to the benefit conferred.

Dushm. There is no comparison between the service and the reward.—He surpassed my warmest expectation, when, before he dismissed me, he made me sit on half of his throne, thus exalting me before all the inhabitants of the Empyreum; and smiling to see his son Jayanta, who stood near him, ambitious of the same honour, perfumed my bosom with essence of heavenly sandal wood, throwing over my neck a garland of flowers blown in paradise.

Māt. O king, you deserve all imaginable rewards from the sovereign of good genii; whose empyreal seats have twice been disentangled from the thorns of Danu's race; formerly by the claws of the man-lion, and lately by thy unerring shafts.

Dushm. My victory proceeded wholly from the auspices of the god; as on earth, when servants prosper in great enterprises, they owe their success to the magnificence of their lords.—Could Arun dispel the shades of night if the deity with a thousand beams had not placed him before the car of day?

Māt. That case, indeed, is parallel.—[*Driving slowly.*]— See, O king, the full exaltation of thy glory, which now rides on the back of heaven! The delighted genii have been collecting, among the trees of life, those crimson and azure dyes, with which the celestial damsels tinge their beautiful feet; and they now are writing thy actions in verses worthy of divine melody.

Dushm. [*Modestly.* In my transport, O Mātali, after the rout of the giants, this wonderful place had escaped my notice.—In what path of the winds are we now journeying?

Māt. This is the way which leads along the triple river, heaven's brightest ornament, and causes yon luminaries to roll in a circle with diffused beams: it is the course of a gentle breeze which supports the floating forms of the gods; and this path was the second step of Vishnu, when he confounded the proud Vali.

Dushm. My internal soul, which acts by exterior organs, is filled by the sight with a charming complacency.— [*Looking at the wheels.*]—We are now passing, I guess, through the region of clouds.

Māt. Whence do you form that conjecture?

Dushm. The car itself instructs me that we are moving over clouds pregnant with showers; for the circumference of its wheels disperses pellucid water; the horses of Indra sparkle with lightning; and I now see the warbling Chātacas descend from their nests on the summits of mountains.

Māt. It is even so; and in another moment you will be in the country which you govern.

Dushm. [*Looking down.*] Through the rapid, yet imperceptible, descent of the heavenly steeds, I now perceive the allotted station of men.—Astonishing prospect! It is yet so distant from us, that the low lands appear confounded with the high mountain tops; the trees erect their branchy shoulders, but seem leafless; the rivers look like bright lines, but their waters vanish; and, at this instant, the globe of earth seems thrown upwards by some stupendous power.

Māt. [*Looking with reverence on the earth.*] How delightful is the abode of mankind!—O king, you saw distinctly.

Dushm. Say, Mātali, what mountain is that which, like an evening cloud, pours exhilarating streams, and forms a golden zone between the western and eastern seas?

Māt. That, O king, is the mountain of Gandharvas, named Hēmacūta: the universe contains not a more excellent place for the successful devotion of the pious. There Casyapa, father of the immortals, ruler of men, son of Marīchi, who sprang from the self-existent, resides with his consort Aditi, blessed in holy retirement.

Dushm. [*Devoutly.*] This occasion of attaining good fortune must not be neglected: may I approach the divine pair, and do them complete homage?

Māt. By all means.—It is an excellent idea!—We are now descended on earth.

Dushm. [*With wonder.*] These chariot wheels yield no sound; no dust arises from them; and the descent of the car gave me no shock.

Māt. Such is the difference, O king, between thy car and that of Indra!

Dushm. Where is the holy retreat of Marīchi?

Māt. [*Pointing.*] A little beyond that grove, where you see a pious Yōgì, motionless as a pollard, holding his thick bushy hair, and fixing his eyes on the solar orb.—Mark; his body is half covered with a white ant's edifice made of raised clay; the skin of a snake supplies the place of his sacerdotal thread, and part of it girds his loins; a number of knotty plants encircles and wound his neck; and surrounding birds' nests almost conceal his shoulders.

Dushm. I bow to a man of his austere devotion.

Māt. [*Checking the reins.*] Thus far, and enough.—We now enter the sanctuary of him who rules the world, and the groves which are watered by streams from celestial sources.

Dushm. This asylum is more delightful than paradise itself: I could fancy myself bathing in a pool of nectar.

Māt. [*Stopping the car.*] Let the king descend.

Dushm. [*Joyfully descending.*] How canst thou leave the car?

Māt. On such an occasion it will remain fixed: we may both leave it.—This way, victorious hero, this way.—Behold the retreat of the truly pious.

Dushm. I see with equal amazement both the pious and their awful retreat.—It becomes, indeed, pure spirits to feed on balmy air in a forest blooming with trees of life;

to bathe in rills dyed yellow with the golden dust of the lotos; and to fortify their virtue in the mysterious bath; to meditate in caves, the pebbles of which are unblemished gems; and to restrain their passions, even though nymphs of exquisite beauty frolick around them: in this grove alone is attained the summit of true piety, to which other hermits in vain aspire.

Māt. In exalted minds the desire of perfect excellence continually increases.—[*Turning aside.*]—Tell me, Vriddhasācalya, in what business is the divine son of Marīchi now engaged?—What sayest thou?—Is he conversing with the daughter of Dacsha, who practises all the virtues of a dutiful wife, and is consulting him on moral questions?—Then we must await his leisure.—[*To Dushmanta*] Rest, O king, under the shade of this Asōca tree, whilst I announce thy arrival to the father of Indra.

Dushm. As you judge right—[*Mātali goes out.*—Dushmanta feels his right arms throb.*] Why, O my arm, dost thou flatter me with a vain omen?—My former happiness is lost, and misery only remains.

Behind the scenes. Be not so restless: in every situation thou showest thy bad temper.

Dushm. [*Listening.*] Hah! this is no place, surely, for a malignant disposition.—Who can be thus rebuked?—[*Looking with surprise.*]—I see a child, but with no childish countenance or strength, whom two female anchorites are endeavouring to keep in order; while he forcibly pulls towards him, in rough play, a lion's whelp with a torn mane, who seems just dragged from the half-sucked nipple of the lioness!

A little Boy and two female Attendants are discovered, as described by the king.

Boy. Open thy mouth, lion's whelp, that I may count thy teeth.

First Atten. Intractable child! Why dost thou torment the wild animals of this forest, whom we cherish as if they were our own offspring?—Thou seemest even to sport in anger.—Aptly have the hermits named thee Servademana, since thou tamest all creatures.

Dushm. Ah! what means it that my heart inclines to this boy as if he were my own son?—[*Meditating.*]—Alas! I have no son; and the reflection makes me once more soft-hearted.

Second Atten. The lioness will tear thee to pieces if thou release not her whelp.

Boy. [*Smiling.*] Oh! I am greatly afraid of her to be sure! [*He bites his lip, as in defiance of her.*]

Dushm. [*Aside, amazed.*] The child exhibits the rudiments of heroick valour, and looks like fire which blazes from the addition of dry fuel.

First Atten. My beloved child, set at liberty this young prince of wild beasts; and I will give thee a prettier plaything.

Boy. Give it first.—Where is it? [*Stretching out his hand.*]

Dushm. [*Aside, gazing on the child's palm.*] What! the very palm of his hand bears the marks of empire; and whilst he thus eagerly extends it, shows its lines of exquisite network, and glows like a lotos expanded at early dawn, when the ruddy splendour of its petals hides all other tints in obscurity.

Second Atten. Mere words, my Suvritā, will not pacify him.—Go, I pray, to my cottage, where thou wilt find a plaything made for the hermit's child, Sancara: it is a peacock of earthen-ware painted with rich colours.

First Atten. I will bring it speedily. [*She goes out.*]

Boy. In the mean time I will play with the young lion.

Second Atten. [*Looking at him with a smile.*] Let him go, I entreat thee.

Dushm. [*Aside.*] I feel the tenderest affection for this unmanageable child. [*Sighing.*]—How sweet must be the delight of virtuous fathers, when they soil their bosoms with dust by lifting up their playful children, who charm them with inarticulate prattle, and show the white blossoms of their teeth, while they laugh innocently at every trifling occurrence!

Second Atten. [*Raising her finger.*] What! dost thou show no attention to me?—[*Looking round.*]—Are any of the hermits near?—[*Seeing* Dushmanta.]—Oh! let me request you, gentle stranger, to release the lion's whelp, who cannot disengage himself from the grasp of this robust child.

Dushm. I will endeavour.—[*Approaching the Boy and smiling.*]—O thou, who art the son of a pious anchorite, how canst thou dishonour thy father, whom thy virtues would make happy, by violating the rules of this consecrated forest? It becomes a black serpent only, to infest the boughs of a fragrant sandal tree. [*The Boy releases the lion.*]

Second Atten. I thank you, courteous guest;—but he is not the son of an anchorite.

Dushm. His actions, indeed, which are conformable to his robustness, indicate a different birth: but my opinion arose from the sanctity of the place which he inhabits.— [*Taking the Boy by the hand.—Aside.*]—Oh! since it gives me such delight merely to touch the hand of this child, who is the hopeful scion of a family unconnected with mine, what rapture must be felt by the fortunate man from whom he sprang?

Second Atten. [*Gazing on them alternately.*] Oh wonderful!

Dushm. What has raised your wonder?

Second Atten. The astonishing resemblance between the child and you, gentle stranger, to whom he bears no relation.—It surprised me also to see, that although he has childish humours, and had no former acquaintance with you, yet your words have restored him to his natural good temper.

Dushm. [*Raising the Boy to his bosom.*] Holy matron, if he be not the son of a hermit, what then is the name of his family?

Second Atten. He is descended from Puru.

Dushm. [*Aside.*] Hah! thence, no doubt, springs his disposition, and my affection for him.—[*Setting him down.*] —[*Aloud.*] It is, I know, an established usage among the princes of Puru's race, to dwell at first in rich palaces with stuccoed walls, where they protect and cherish the world, but in the decline of life to seek humbler mansions near the roots of venerable trees, where hermits with subdued passions practice austere devotion.—I wonder, however, that this boy, who moves like a god, could have been born of a mere mortal.

Second Atten. Affable stranger, your wonder will cease when you know that his mother is related to a celestial nymph, and brought him forth in the sacred forest of Casyapa.

Dushm. [*Aside.*] I am transported.—This is a fresh ground of hope.—[*Aloud.*]—What virtuous monarch took his excellent mother by the hand?

Second Atten. Oh! I must not give celebrity to the name of a king who deserted his lawful wife.

Dushm. [*Aside.*] Ah! she means me.—Let me now ask the name of the sweet child's mother.—[*Meditating.*]—But it is against good manners to inquire concerning the wife of another man.

The First Attendant *re-enters with a toy.*

First Atten. Look, Servademana, look at the beauty of this bird, Saconta lāvanyam.

Boy. [*Looking eagerly round.*] Sacontalā! Oh, where is my beloved mother? [*Both attendants laugh.*]

First Atten. He tenderly loves his mother, and was deceived by an equivocal phrase.

Second Atten. My child, she meant only the beautiful shape and colours of this peacock.

Dushm. [*Aside.*] Is my Sacontalā then his mother? Or has that dear name been given to some other woman?—This conversation resembles the fallacious appearance of water in a desert, which ends in bitter disappointment to the stag parched with thirst.

Boy. I shall like the peacock if it can run and fly; not else. [*He takes it.*]

First Atten. [*Looking round in confusion.*] Alas, the child's amulet is not on his wrist!

Dushm. Be not alarmed. It was dropped while he was playing with the lion: I see it, and will put it into your hand.

Both. Oh! beware of touching it.

First Atten. Ah! he has actually taken it up. [*They both gaze with surprise on each other.*]

Dushm. Here it is; but why would you have restrained me from touching this bright gem?

Second Atten. Great monarch, this divine amulet has a wonderful power, and was given to the child by the son of Marīchi, as soon as the sacred rites had been performed after his birth: whenever it fell on the ground, no human being but the father or mother of this boy could have touched it unhurt.

Dushm. What if a stranger had taken it?

First Atten. It would have become a serpent and wounded him.

Dushm. Have you seen that consequence on any similar occasion?

Both. Frequently.

Dushm. [*With transport.*] I may then exult on the completion of my ardent desire. [*He embraces the child.*]

Second Atten. Come, Suvritā, let us carry the delightful intelligence to Sacontalā, whom the harsh duties of a separated wife have so long oppressed. [*The* Attendants *go out.*]

Boy. Farewell; I must go to my mother.

Dushm. My darling son, thou wilt make her happy by going to her with me.

Boy. Dushmanta is my father; and you are not Dushmanta.

Dushm. Even thy denial of me gives me delight.

Sacontalā *enters in mourning apparel, with her long hair twisted in a single braid, and flowing down her back.*

Sac. [*Aside.*] Having heard that my child's amulet has proved its divine power, I must either be strangely diffident of my good fortune, or that event which Misracēsì predicted has actually happened. [*Advancing.*]

Dushm. [*With a mixture of joy and sorrow.*] Ah! do I see the incomparable Sacontalā clad in sordid weeds?—Her face is emaciated by the performance of austere duties; one twisted lock floats over her shoulder; and with a mind perfectly pure, she supports the long absence of her husband, whose unkindness exceeded all bounds.

Sac. [*Seeing him, yet doubting.*] Is that the son of my lord grown pale with penitence and affliction?—If not, who is it, that sullies with his touch the hand of my child,

whose amulet should have preserved him from such indignity?

Boy. [*Going hastily to* Sacontalā.] Mother, here is a stranger who calls me son.

Dushm. Oh! my best beloved, I have treated thee cruelly; but my cruelty is succeeded by the warmest affection; and I implore your remembrance and forgiveness.

Sac. [*Aside.*] Be confident, O my heart!—[*Aloud.*]—I shall be most happy when the king's anger has passed away.—[*Aside.*]—This must be the son of my lord.

Dushm. By the kindness of heaven, O loveliest of thy sex, thou standest again before me, whose memory was obscured by the gloom of fascination; as the star Rōhinī at the end of an eclipse rejoins her beloved moon.[18]

Sac. May the king be—[*She bursts into tears.*]

Dushm. My darling, though the word victorious be suppressed by thy weeping, yet I must have victory, since I see thee again, though with pale lips and a body unadorned.

Boy. What man is this, mother?

Sac. Sweet child, ask the divinity, who presides over the fortunes of us both. [*She weeps.*]

Dushm. O my only beloved, banish from thy mind my cruel desertion of thee.—A violent phrensy overpowered my soul.—Such, when the darkness of illusion prevails, are the actions of the best intentioned; as a blind man, when a friend binds his head with a wreath of flowers, mistakes it for a twining snake, and foolishly rejects it. [*He falls at her feet.*]

Sac. Rise, my husband, oh! rise—My happiness has been long interrupted; but joy now succeeds to affliction, since the son of my lord still loves me.—[*He rises.*]—How was the remembrance of this unfortunate woman restored to the mind of my lord's son?

Dushm. When the dart of misery shall be wholly extracted from my bosom, I will tell you all; but since the anguish of my soul has in part ceased, let me first wipe off that tear which trickles from thy delicate eye-lash; and thus efface the memory of all the tears which my delirium has made thee shed. [*He stretches out his hand.*]

Sac. [*Wiping off her tears and seeing the ring on his finger.*] Ah! is that the fatal ring?

Dushm. Yes; by the surprising recovery of it my memory was restored.

Sac. Its influence, indeed, has been great; since it has brought back the lost confidence of my husband.

---

[18] [The playwright's expertise in mythical astronomy, showing the love-relationship between Rōhinī and the moon as is widely prevalent in Sanskrit literature, is aptly illustrated.]

Dushm. Take it then, as a beautiful plant receives a flower from the returning season of joy.

Sac. I cannot again trust it.—Let it be worn by the son of my lord.

Mātali enters

Māt. By the will of heaven the king has happily met his beloved wife, and seen the countenance of his little son.

Dushm. It was by the company of my friend that my desire attained maturity.—But say, was not this fortunate event previously known to Indra?

Māt. [*Smiling.*] What is unknown to the gods?—But come: the divine Marīcha desires to see thee.

Dushm. Beloved, take our son by the hand; and let me present you both to the father of immortals.

Sac. I really am ashamed, even in thy presence, to approach the deities.

Dushm. It is highly proper on so happy an occasion— Come, I entreat thee. [*They all advance.*]

The scene is withdrawn, and Casyapa *is discovered on a throne conversing with* Aditi.

Cas. [*Pointing to the king.*] That, O daughter of Dacsha, is the hero who led the squadrons of thy son to the front of battle, a sovereign of the earth, Dushmanta; by the means of whose bow the thunder-bolt of Indra (all its work being accomplished) is now a mere ornament of his heavenly palace.

Adi. He bears in his form all the marks of exalted majesty.

Māt. [*To* Dushmanta.] The parents of the twelve Adityas, O king, are gazing on thee, as on their own offspring, with eyes of affection.—Approach them, illustrious prince.

Dushm. Are those, O Mātali, the divine pair, sprung from Marīchi and Dacsha?—Are those the grand-children of Brahmā, to whom the self-existent gave birth in the beginning; whom inspired mortals pronounce the fountain of glory apparent in the form of twelve suns; they who produced my benefactor, the lord of a hundred sacrifices, and ruler of three worlds?

Māt. Even they—[*Prostrating himself with* Dushmanta.] —Great beings, the king Dushmanta, who has executed the commands of your son Vasava, falls humbly before your throne.

Cas. Continue long to rule the world.

Adi. Long be a warriour with a car unshattered in combat.

[Sacontalā *and her son prostrate themselves.*]

Cas. Daughter, may thy husband be like Indra! May thy son resemble Jayanta! And mayst thou (whom no benediction could better suit) be equal in prosperity to the daughter of Pulōman!

Adi. Preserve, my child, a constant unity with thy lord: and may this boy, for a great length of years, be the ornament and joy of you both! Now be seated near us. [*They all sit down.*]

Cas. [*Looking at them by turns.*] Sacontalā is the model of excellent wives; her son is dutiful; and thou, O king, hast three rare advantages, true piety, abundant wealth, and active virtue.

Dushm. O divine being, having obtained the former object of my most ardent wishes, I now have reached the summit of earthly happiness through thy favour, and thy benizon will ensure its permanence.—First appears the flower, then the fruit; first clouds are collected, then the shower falls: such is the regular course of causes and effects; and thus, when thy indulgence preceded, felicity generally followed.

Māt. Great indeed, O king, has been the kindness of the primeval Brāhmens.

Dushm. Bright son of Marīchi, this thy handmaid was married to me by the ceremony of Gandharvas, and, after a time, was conducted to my palace by some of her family; but my memory having failed through delirium, I rejected her, and thus committed a grievous offence against the venerable Canna, who is of thy divine lineage: afterwards, on seeing this fatal ring, I remembered my love and my nuptials; but the whole transaction yet fills me with wonder. My soul was confounded with strange ignorance that obscured my senses; as if a man were to see an elephant marching before him, yet to doubt what animal it could be, till he discovered by the traces of his large feet that it was an elephant.

Cas. Cease, my son, to charge thyself with an offence committed ignorantly, and, therefore, innocently.—Now hear me—

Dushm. I am devoutly attentive.

Cas. When the nymph Mēnacà led Sacontalā from the place where thy desertion of her had afflicted her soul, she brought her to the palace of Aditi; and I knew, by the power of meditation on the Supreme Being, that thy forgetfulness of thy pious and lawful consort had proceeded from the imprecation of Durvāsas, and that the charm would terminate on the sight of thy ring.

Dushm. [*Aside.*] My name then is cleared from infamy.

Sac. Happy am I that the son of my lord, who now recognizes me, denied me through ignorance, and not with real aversion.—The terrible imprecation was heard, I suppose, when my mind was intent on a different object, by my two beloved friends, who, with extreme affection, concealed it from me to spare my feelings, but advised me at parting to show the ring if my husband should have forgotten me.

Cas. [*Turning to* Sacontalā.] Thou art apprised, my

daughter, of the whole truth, and must no longer resent the behaviour of thy lord.—He rejected thee when his memory was impaired by the force of a charm; and when the gloom was dispelled, his conjugal affection revived; as a mirror whose surface has been sullied, reflects no image; but exhibits perfect resemblances when its polish has been restored.

Dushm. Such, indeed, was my situation.

Cas. My son Dushmanta, hast thou embraced thy child by Sacontalā, on whose birth I myself performed the ceremonies prescribed in the Vēda?

Dushm. Holy Marīchi, he is the glory of my house.

Cas. Know too, that his heroick virtue will raise him to a dominion extended from sea to sea: before he has passed the ocean of mortal life, he shall rule, unequalled in combat, this earth with seven peninsulas; and, as he now is called Servademana, because he tames even in childhood the fiercest animals, so, in his riper years, he shall acquire the name of Bherata, because he shall sustain and nourish the world.

Dushm. A boy educated by the son of Marīchi, must attain the summit of greatness.

Adi. Now let Sacontalā, who is restored to happiness, convey intelligence to Canna of all these events: her mother Mēnacà is in my family, and knows all that has passed.

Sac. The goddess proposes what I most ardently wish.

Cas. By the force of true piety the whole scene will be present to the mind of Canna.

Dushm. The devout sage must be still excessively indignant at my frantick behaviour.

Cas. [*Meditating.*] Then let him hear from me the delightful news, that his foster-child has been tenderly received by her husband, and that both are happy with the little warrior who sprang from them.—Hola! who is in waiting?

A Pupil *enters.*

Pup. Great being, I am here.

Cas. Hasten, Gōlava, through the light air, and in my name inform the venerable Canna, that Sacontalā has a charming son by Dushmanta, whose affection for her was restored with his remembrance, on the termination of the spell raised by the angry Durvāsas.

Pup. As the divinity commands. [*He goes out.*]

Cas. My son, reascend the car of Indra with thy consort and child, and return happy to thy imperial seat.

Dushm. Be it as Marīchi ordains.

Cas. Henceforth may the god of the atmosphere with copious rain give abundance to thy affectionate subjects; and mayst thou with frequent sacrifices maintain the Thunderer's friendship! By numberless interchanges of good offices between you both, may benefits reciprocally be conferred on the inhabitants of the two worlds!

Dushm. Powerful being, I will be studious, as far as I am able, to attain that felicity.

Cas. What other favours can I bestow on thee?

Dushm. Can any favours exceed those already bestowed? —Let every king apply himself to the attainment of happiness for his people; let Sereswatì, the goddess of liberal arts, be adored by all readers of the Vēda; and may Siva, with an azure neck and red locks, eternally potent and self-existing, avert from me the pain of another birth in this perishable world, the seat of crimes and of punishment. [*All go out.*]

CRITICISM

Essay on the Arts, Commonly Called Imitative[1]

It is the fate of those maxims, which have been thrown out by very eminent writers, to be received implicitly by most of their followers, and to be repeated a thousand times, for no other reason, than because they once dropped from the pen of a superior genius: one of these is the assertion of *Aristotle*, that *all poetry consists in imitation*, which has been so frequently echoed from author to author, that it would seem a kind of arrogance to controvert it; for almost all the philosophers and criticks, who have written upon the subject of *poetry*, *musick*, and *painting*, how little soever they may agree in some points, seem of one mind in considering them as arts merely *imitative*: yet it must be clear to any one, who examines what passes in his own mind, that he is affected by the finest *poems*, *pieces of musick*, and *pictures*, upon a principle, which, whatever it be, is entirely distinct from *imitation*. M. *le Batteux* has attempted to prove that all the fine arts have a relation to this common principle of *imitating*:[2] but, whatever be said of *painting*, it is prob-

able, that *poetry* and *musick* had a nobler origin; and, if the first language of man was not both *poetical* and *musical*, it is certain, at least, that in countries, where no kind of *imitation* seems to be much admired, there are *poets* and *musicians* both by nature and by art: as in some *Mahometan* nations; where *sculpture* and *painting* are forbidden by the laws, where *dramatick poetry* of every sort is wholly unknown, yet, where the pleasing arts, *of expressing the passions in verse, and of enforcing that expression by melody*, are cultivated to a degree of enthusiasm. It shall be my endeavour in this paper to prove, that, though *poetry* and *musick* have, certainly, a power of *imitating* the manners of men, and several objects in nature, yet, that their greatest effect is not produced by *imitation*, but by a very different principle; which must be sought for in the deepest recesses of the human mind.

To state the question properly, we must have a clear notion of what we mean by *poetry* and *musick*; but we cannot give a precise definition of them, till we have made a few previous remarks on their origin, their relation to each other, and their difference.

It seems probable then that *poetry* was originally no more than a strong and animated expression of the human passions, of *joy* and *grief*, *love* and *hate*, *admiration* and *anger*, sometimes pure and unmixed, sometimes variously modified and combined: for, if we observe the *voice* and *accents* of a person affected by any of the violent passions, we shall perceive something in them very nearly approaching to *cadence* and *measure*; which is remarkably the case in the language of a vehement *Orator*, whose talent is chiefly conversant about *praise* or *censure*; and we may collect from several passages in *Tully*,[3] that the fine speakers of old *Greece* and *Rome* had a sort of rhythm in their sentences, less regular, but not less melodious, than that of the poets.

If this idea be just, one would suppose that the most ancient sort of poetry consisted in *praising the Deity*; for if we conceive a being, created with all his faculties and senses, endued with speech and reason, to open his eyes in a most delightful plain, to view for the first time the serenity of the sky, the splendour of the sun, the verdure of the fields and woods, the glowing colours of the flowers, we can hardly believe it possible, that he should refrain from bursting into an extasy of *joy*, and pouring his praises to the creator of those wonders, and the author of his happiness. This *kind of poetry* is used in all nations; but as it is the sublimest of all, when it is applied to its true object, so it has often been perverted to impious purposes by pagans and idolatres: every one knows that the *dramatick poetry* of the *Europeans* took its rise from the

---

[1] [One of two essays concluding Jones's *Poems* (1772), it articulates an aesthetic theory to undergird the new themes, meters, and subects from the Arabic and Persian literature that he was introducing in *Poems*. It demonstrates his insights into literary criticism, though his contemporaries and posterity have concentrated on a few poems therein, to the neglect of this illuminating essay. Later scholars have incorporated his aesthetic suggestions into better, more advanced criticism. See M.H. Abrams, *The Mirror and the Lamp* (Oxford, 1953), 87-88; René Wellek, *The Rise of English Literary History* (New York, 1966), 51, 71; David Newton-De Molina, 'Sir William Jones' 'Essay on the Arts commonly called Imitative' (1772)', *Anglia* 90 (1972): 147-54; and James S. Malek, 'The Influence of Empirical Psychology on Aesthetic Discourse: Two Eighteenth Century Theories of Art', *Enlightenment Essays* 1 (Spring 1970): 1-16. The essay was reprinted in *Eighteenth-Century Critical Essays*, ed. Scott Elledge (Cornell, 1961) 2: 872-81; and in Carl Gustav Jochmann's *Die Rückschritte der Poesie* (Hamburg, 1982); 79-90.]

[2] [M. le Charles Batteux (1713-80), *Les beaux arts réduits à un méme principe* (Paris, 1747).]

[3] [Marcus Tullius Cicero influenced Jones's ideas and style.]

same spring, and was no more at first than a song in praise of *Bacchus*; so that the only species of poetical composition (if we except the Epick), which can in any sense be called *imitative*, was deduced from a natural emotion of the mind, in which *imitation* could not be at all concerned.

The next source of poetry was, probably, *love*, or the mutual inclination, which naturally subsists between the sexes, and is founded upon personal *beauty*: hence arose the most agreeable *odes*, and love-songs, which we admire in the works of the ancient lyrick poets, not filled, like our *sonnets* and *madrigals*, with the insipid babble of *darts*, and *Cupids*, but simple, tender, natural; and consisting of such unaffected endearments, and mild complaints,

Teneri sdegni, e placide e tranquille
Repulse, e cari vezzi, e liete paci,[4]

as we may suppose to have passed between the first lovers in a state of innocence, before the refinements of society, and the restraints, which they introduced, had made the passion of *love* so fierce, and impetuous, as it is said to have been in *Dido*, and certainly was in *Sappho*, if we may take her own word for it.[5]

The *grief* which the first inhabitants of the earth must have felt at the death of their dearest friends, and relations, gave rise to another species of poetry, which originally, perhaps, consisted of short *dirges*, and was afterwards lengthened into *elegies*.

As soon as vice began to prevail in the world, it was natural for the wise and virtuous to express their *detestation* of it in the strongest manner, and to show their *resentment* against the corrupters of mankind: hence *moral poetry* was derived, which, at first, we find, was severe and passionate; but was gradually melted down into cool precepts of morality, or exhortations to virtue: we may reasonably conjecture that *Epick poetry* had the same origin, and that the examples of heroes and kings were introduced, to illustrate some moral truth, by showing the loveliness and advantages of virtue, or the many misfortunes that flow from vice.

Where there is vice, which is *detestable* in itself, there must be *hate*, since *the strongest antipathy in nature*, as Mr. *Pope* asserted in his writings,[6] and proved by his whole life, *subsists between the good and the bad*: now this passion was the source of that poetry, which we call

Satire, very improperly, and corruptly, since the *Satire* of the *Romans* was no more than a moral piece, which they entitled *Satura* or *Satyra*,[7] intimating, that the poem, like *a dish of fruit and corn offered to Ceres*, contained a variety and plenty of fancies and figures; whereas the true *invectives* of the ancients were called *Iambi*, of which we have several examples in *Catullus*,[8] and in the *Epodes* of *Horace*, who imitated the very measures and manner of *Archilochus*.[9]

These are the principal sources of *poetry*; and of *musick* also, as it shall be my endeavour to show: but it is first necessary to say a few words on *the nature of sound*; a very copious subject, which would require a long dissertation to be accurately discussed. Without entering into a discourse on the *vibrations of chords*, or *the undulations of the air*, it will be sufficient for our purpose to observe that there is a great difference between *a common sound*, and *a musical sound*, which consists chiefly in this, that the former is simple and entire in itself like a *point*, while the latter is always accompanied with other sounds, without ceasing to be *one*; like a *circle*, which is an entire figure, though it is generated by a multitude of points flowing, at equal distances, round a common centre. These accessory sounds, which are caused by the aliquots of a sonorous body vibrating at once, are called *Harmonicks*, and the whole system of modern *Harmony* depends upon them; though it were easy to prove that the system is unnatural, and only made tolerable to the ear by habit: for whenever we strike the perfect accord on a harpsichord or an organ, the harmonicks of the third and fifth have also their own harmonicks, which are dissonant from the principal note: These horrid dissonances are, indeed, almost overpowered by the *natural harmonicks* of the principal chord, but that does not prove them agreeable. Since nature has given us a delightful harmony of her own, why should we destroy it by the additions of art? It is like thinking

---

[4] Two lines of *Tasso*.

[5] See the ode of *Sappho* quoted by *Longinus*, and translated by *Boileau*.

[6] [See *An Essay on Man* 2: iii.]

[7] Some Latin words were spelled either with an *u* or a *y*, as *Sulla* or *Sylla*.

[8] [Gaius Valerius Catullus (84-54 B.C.), a celebrated Roman poet, was known for using galliambic meter in his *Attis* and a new variety of measures in *Lepidus novus libellus*.]

[9] [In his Epodes II, V, and XVI, Horace's elaborate use of iambs, in what modern scholarship has come to call the first, second, third, and fourth Archilochean meters, reflects a medley of lyric experimentations once carried out by Archilochus, the first known Greek lyric poet. For Horace's indebtedness, see his *Epistula* i. 19.22, and for comments on the subject see Eduard Fraenkel, *Horace* (Oxford, 1957).]

—to paint the lily,
And add a perfume to the violet.[10]

Now let us conceive that some vehement passion is expressed in strong words, exactly measured, and pronounced, *in a common voice*, in just cadence, and with proper accents, such an expression of the passion will be *genuine poetry*; and the famous ode of *Sappho* is allowed to be so in the strictest sense: but if the same ode, with all its natural accents, were expressed in a *musical voice* (that is, in sounds accompanied with their *Harmonicks*), if it were sung in due time and measure, in a simple and pleasing tune, that added force to the words without stifling them, it would then be *pure and original musick*; not merely soothing to the ear, but affecting to the heart; not an *imitation* of nature, but the voice of nature herself. But there is another point in which *musick* must resemble *poetry*, or it will lose a considerable part of its effect: we all must have observed, that a speaker, agitated with passion, or an actor, who is, indeed, strictly an *imitator*, are perpetually changing the tone and pitch of their voice, as the sense of their words varies: it may be worth while to examine how this variation is expressed in *musick*. Every body knows that the musical scale consists of seven notes, above which we find a succession of similar sounds repeated in the same order, and above that, other successions, as far as they can be continued by the human voice, or distinguished by the human ear: now each of these seven sounds has no more meaning, when it is heard separately, than a single letter of the alphabet would have; and it is only by thier succession, and their relation to one principal sound, that they take any rank in the scale; or differ from each other, except as they are *graver*, or more *acute*: but in the regular scale each interval assumes a proper character, and every note stands related to the first or principal one by various proportions. Now *a series of sounds relating to one leading note* is called a *mode*, or a *tone*, and, as there are twelve semitones in the scale, each of which may be made in its turn the leader of a mode, it follows that there are twelve modes; and each of them has a peculiar character, arising from the position of the *modal* note, and from some minute difference in the ratio's, as of 81 to 80, or a comma; for there are some intervals, which cannot easily be rendered on our instruments, yet have a surprizing effect in *modulation*, or in the transitions from one mode to another.

The *modes* of the ancients are said to have had a wonderful effect over the mind; and *Plato*, who permits

the *Dorian* in his imaginary republick, on account of its calmness and gravity, excludes the *Lydian*, because of its languid, tender, and effeminate character:[11]not that any series of mere sounds has a power of raising or soothing the passions, but each of these modes was appropriated to a particular kind of poetry, and a particular instrument; and the chief of them, as the *Dorian, Phrygian, Lydian, Ionian, Eolian, Locrian*, belonging originally to the nations, from which they took their names: thus the *Phrygian mode*, which was ardent and impetuous, was usually accompanied with trumpets, and the *Mixolydian*, which, if we believe *Aristoxenus*, was invented by *Sappho*, was probably confined to the pathetick and tragick style: that these modes had a relation to *poetry*, as well as to *musick*, appears from a fragment of *Lasus*, in which he says, '*I sing of Ceres, and her daughter* Meliboea, *the consort of Pluto, in the Eolian mode, full of gravity*'; and *Pindar* calls one of his *Odes* and *Eolian song*.[12] If the *Greeks* surpassed us in the strength of their modulations, we have an advantage over them in our *minor scale*, which supplies us with twelve new modes, where the two semitones are removed from their natural position between the third and fourth, the seventh and eighth notes, and placed between the second and third, the fifth and sixth; this change of the semitones, by giving a minor third to the *modal* note, softens the general expression of the mode, and adapts it admirably to subjects of *grief* and *affliction*: the minor mode of D is tender, that of C, with three flats, plaintive, and that of F, with four, pathetick and mournful to the highest degree, for which reason it was chosen by the excellent *Pergolesi* in his *Stabat Mater*.[13] Now these twenty-four modes, artfully interwoven, and changed as often as the sentiment changes, may, it is evident, express all the variations in the voice of a speaker, and give an additional beauty to the accents of a poet. Consistently with the foregoing principles, we may define *original and native poetry* to be *the language of the violent passions, expressed in exact measure, with strong accents and significant words*; and *true musick* to be no more than *poetry, delivered in a succession of harmonious sounds so disposed as to please the ear*. It is

---

[10] [Shakespeare's *Life and Death of King John* (IV, ii, 11-12, which reads 'To throw a perfume on the violet'.]

[11] [See *The Republic* 3: 389-99.]

[12] [Lasus, one of the major Greek lyric poets celebrated as the founder of the Athenian school of dithyrambic poetry and as Pindar's teacher wrote a hymn to Demeter and her daughter Melibœa.]

[13] [Giovanni Battista Pergolesi (1710-36), Italian composer, wrote *Stabat Mater* (London, 1749) in competition with Alessandro Scarlatti's *Stabat Mater*.]

in this view only that we must consider the musick of the ancient *Greeks*, or attempt to account for its amazing effects, which we find related by the gravest historians, and philosophers; it was wholly passionate or descriptive, and so closely united to poetry, that it never obstructed, but always increased its influence; whereas our boasted harmony, with all its fine accords, and numerous parts, paints nothing, expresses nothing, says nothing to the heart, and consequently can only give more or less pleasure to one of our senses; and no reasonable man will seriously prefer a transitory pleasure, which must soon end in satiety, or even in disgust, to a delight of the soul, arising from sympathy, and founded on the natural passions, always lively, always interesting, always transporting. The old divisions of musick into *celestial* and *earthly*, *divine* and *human*, *active* and *contemplative*, *intellective* and *oratorial*, were founded rather upon metaphors, and chimerical analogies, than upon any real distinctions in nature; but the want of making a distinction between *musick of mere sounds*, and the *musick of the passions*, has been the perpetual source of confusion and contradictions both among the ancients and the moderns: nothing can be more opposite in many points than the systems of *Rameau*[14] and *Tartini*,[15] one of whom asserts that melody springs from harmony, and the other deduces harmony from melody; and both are in the right, if the first speaks only of that musick, which took its rise from *the multiplicity of sounds heard at once in the sonorous body*, and the second, of that, which rose from *the accents and inflexions of the human voice, animated by the passions*: to decide, as *Rousseau* says, whether of these two schools ought to have the preference, we need only ask a plain question, Was the voice made for the instruments, or the instruments for the voice?[16]

In defining what true poetry *ought to be*, according to our principles, we have described what it really *was* among the *Hebrews*, the *Greeks* and *Romans*, the *Arabs* and *Persians*. The lamentation of *David*, and his sacred odes, or psalms, the song of *Solomon*, the prophecies of *Isaiah*, *Jeremiah*, and the other inspired writers, are truly and strictly poetical; but what did *David* or *Solomon* imitate in their divine poems? A man, who is *really* joyful or afflicted, cannot be said to *imitate* joy or affliction. The lyrick verses of *Alcaeus*, *Alcman*, and *Ibycus*, the hymns of *Callimachus*, the elegy of *Moschus* on the death of *Bion*, are all beautiful pieces of poetry;[17] yet *Alcaeus* was no *imitator* of love, *Callimachus* was no *imitator* of religious awe and admiration, *Moschus* was no *imitator* of grief at the loss of an amiable friend. *Aristole* himself wrote a very poetical elegy on the death of a man, whom he had loved; but it would be difficult to say what he imitated in it: '*O virtue, who proposest many labours to the human race, and art still the alluring object of our life; for thy charms, O beautiful goddess, it was always an envied happiness in Greece even to die, and to suffer the most painful, the most afflicting evils: such are the immortal fruits, which thou raisest in our minds; fruits, more precious than gold, more sweet than the love of parents, and soft repose: for thee Hercules the son of Jove, and the twins of Leda, sustained many labours, and by their illustrious actions sought thy favour; for love of thee, Achilles and Ajax descended to the mansion of Pluto; and, through a zeal for thy charms, the prince of Atarnea also was deprived of the sun's light: therefore shall the muses, daughters of memory, render him immortal for his glorious deeds, whenever they sing the god of hospitality, and the honours due to a lasting friendship.*[18]

In the preceding collection of poems, there are some *Eastern* fables, some *odes*, a *panegyrick*, and an *elegy*; yet it does not appear to me, that there is the least *imitation* in either of them: *Petrarch* was, certainly, too deeply affected with real *grief*, and the *Persian* poet was too sincere a lover, to *imitate* the passions of others. As to the rest, a fable in verse is no more an *imitation* than a fable in prose; and if every poetical narrative, which describes the manners, and relates the adventures of men, be called *imitative*, every romance, and even every history, must be called so likewise; since many poems are only *romance*, or parts of *history* told in a regular measure.

14 [Jean Philippe Rameau (1683-1764), French composer and an important innovator in harmonic theory, was known for his *Traite de l' harmonie* (Paris, 1722) and *Nouvean systéme de musique théorique* (Paris, 1726).]

15 [Giuseppe Tartini (1692-1770), Italian composer and violinist, was reputed for his speculative musical thinking as enunciated in his *Trattato di musica* (Padua, 1754). Rousseau and Serre opposed some of his theories.]

16 ['Dissertation sur la musique moderne', and 'Examen de deux principes avancés par M. Rameau', *Oeuvres completes de J.J. Rousseau* 12: (Paris, 1832), 56-7; 339-69, respectively.]

17 [Jones knew these various Greek poets quite well, but modern scholars have proved that Moschus did not write the 'Lament for Bion', even though he considered himself as a pupil of Bion.]

18 [Aristotle, Fragment 5 Diehl; Diogênes Laërtius, 5. 7-8 (ed. H.S. Long, *Vitae philosophorum* 1: (Oxford, 1964) 199-200). He wrote this elegy in memory of his friend, patron, and, later, father-in-law Hermeias, ruler of Atarneus.]

What has been said of *poetry*, may with equal force be applied to *musick*, which is *poetry*, dressed to advantage; and even to *painting*, many sorts of which are poems to the eye, as all poems, merely descriptive, are pictures to the ear: and this way of considering them, will set the refinements of modern artists in their true light; for the *passions*, which were given by nature, never spoke in an unnatural form, and no man, truly affected with *love* or *grief*, ever expressed the one in an *acrostick*, or the other in a *fugue*: these remains, therefore, of the false taste, which prevailed in the dark ages, should be banished from this, which is enlightened with a just one.

It is true, that some kinds of painting are strictly *imitative*, as that which is solely intended to represent the human figure and countenance; but it will be found, that those pictures have always the greatest effect, which represent some *passion*, as the martyrdom of *St. Agnes* by *Domenichino*, and the various representations of the *Crucifixion* by the finest masters of *Italy*; and there can be no doubt, but that the famous *sacrifice of Iphigenia* by *Timanthes* was affecting to the highest degree; which proves, not that painting cannot be said to *imitate*, but that its most powerful influence over the mind arises, like that of the other arts, from *sympathy*.

It is asserted also that *descriptive* poetry, and *descriptive* musick, as they are called, are strict *imitations*; but, not to insist that mere *description* is the meanest part of both arts, if indeed it belongs to them at all, it is clear, that words and sounds have no kind of resemblance to visible objects: and what is an imitation, but a resemblance of some other thing? Besides, no unprejudiced hearer will say that he finds the smallest traces of imitation in the numerous *fugues*, *counterfugues*, and *divisions*, which rather disgrace than adorn the modern musick: even sounds themselves are imperfectly imitated by harmony, and, if we sometimes hear *the murmuring of a brook*, or *the chirping of birds* in a concert, we are generally apprised before-hand of the passages, where we may expect them. Some eminent musicians, indeed, have been absurd enough to think of imitating laughter and other noises, but, if they had succeeded, they could not have made amends for their want of taste in attempting it; for such ridiculous imitations must necessarily destroy the spirit and dignity of the finest poems, which they ought to illustrate by a graceful and natural melody. It seems to me, that, as those parts of *poetry*, *musick*, and *painting*, which relate to the passions, affect by *sympathy*, so those, which are merely descriptive, act by a kind of *substitution*, that is, by raising in our minds, affections, or sentiments, analogous to those, which arise in us, when the respective objects in nature are presented to our senses. Let us suppose that a poet, a musician, and a painter, are striving to give their friend, or patron, a pleasure similar to that, which he feels at the sight of a beautiful prospect. The first will form an agreeable assemblage of lively images, which he will express in smooth and elegant verses of a sprightly measure; he will describe the most delightful objects, and will add to the graces of his description a certain delicacy of sentiment, and a spirit of cheerfulness. The musician, who undertakes to set the words of the poet, will select some mode, which, on his violin, has the character of mirth and gaiety, as the Eolian, or E *flat*, which he will change as the sentiment is varied: he will express the words in a simple and agreeable melody, which will not disguise, but embellish them, without aiming at any fugue, or figured harmony: he will use the bass, to mark the modulation more strongly, especially in the changes; and he will place the *tenour* generally in unison with the bass, to prevent too great a distance between the parts: in the symphony he will, above all things, avoid a *double melody*, and will apply his variations only to some accessory ideas, which the principal part, that is, the voice, could not easily express: he will not make a number of useless repetitions, because the *passions* only repeat the same expressions, and dwell upon the same sentiments, while *description* can only represent a single object by a single sentence. The painter will describe all visible objects more exactly than his rivals, but he will fall short of the other artists in a very material circumstance; namely, that his pencil, which may, indeed, express a simple passion, cannot paint a thought, or draw the shades of sentiment: he will, however, finish his landscape with grace and elegance; his colours will be rich, and glowing; his perspective striking; and his figures will be disposed with an agreeable variety, but not with confusion: above all, he will diffuse over his whole piece such a spirit of liveliness and festivity, that the beholder shall be seized with a kind of rapturous delight, and, for a moment, mistake art for nature.

Thus will each artist gain his end, not by *imitating* the works of nature, but by assuming her power, and causing the same effect upon the imagination, which her charms produce to the senses: this must be the chief object of a poet, a musician, and a painter, who know that *great effects are not produced by minute details, but by the general spirit of the whole piece, and that a gaudy composition may strike the mind for a short time, but that the beauties of simplicity are both more delightful, and more permanent.*

As the *passions* are differently modified in different men, and as even the various objects in nature affect our minds in various degrees, it is obvious, that there must be a great diversity in the pleasure, which we receive from the fine arts, whether that pleasure arises from *sympathy* or *substitution*; and that it were a wild notion in artists to think of pleasing every reader, hearer, or beholder; since every man has a particular set of objects, and a particular inclination, which direct him in the choice of his pleasures, and induce him to consider the productions, both of nature and of art, as more or less elegant, in proportion as they give him a greater or smaller degree of delight: this does not at all contradict the opinion of many able writers, that *there is one uniform standard of taste*; since the *passions*, and, consequently, *sympathy*, are generally the same in all men, till they are weakened by age, infirmity, or other causes.

If the arguments, used in this essay, have any weight, it will appear, that the finest parts of poetry, musick, and painting, are expressive of the *passions*, and operate on our minds by *sympathy*; that the inferior parts of them are *descriptive* of natural *objects*, and affect us chiefly by *substitution*; that the expressions of *love*, *pity*, *desire*, and the *tender* passions, as well as the *descriptions* of objects that delight the senses, produce in the arts what we call the *beautiful*; but that *hate*, *anger*, *fear*, and the *terrible* passions, as well as objects, which are *unpleasing* to the senses, are productive of the *sublime*, when they are aptly expressed, or described.

These subjects might be pursued to infinity; but, if they were amply discussed, it would be necessary to write a series of dissertations, instead of an essay.

An Essay on the Poetry of the Eastern Nations[1]

Arabia, I mean that part of it, which we call the *Happy*, and which the *Asiaticks* know by the name of *Yemen*, seems to be the only country in the world, in which we can properly lay the scene of pastoral poetry; because no nation at this day can vie with the *Arabians* in the delightfulness of their climate, and the simplicity of their manners. There is a valley, indeed, to the north of *Indostan*, called *Cashmîr*, which, according to an account written by a native of it, is a perfect garden, exceedingly fruitful, and watered by a thousand rivulets: but when its inhabitants were subdued by the stratagem of a *Mogul* prince, they lost their happiness with their liberty, and *Arabia* retained its old title without any rival to dispute it. These are not the fancies of a poet: the beauties of *Yemen* are proved by the concurrent testimony of all travellers, by the descriptions of it in all the writings of *Asia*, and by the nature and situation of the country itself, which lies between the eleventh and fifteenth degrees of northern latitude, under a serene sky, and exposed to the most favourable influence of the sun; it is enclosed on one side by vast rocks and deserts, and defended on the other by a tempestuous sea, so that it seems to have been designed by Providence for the most secure, as well as the most beautiful, region of the East.[2]

[1] [This brilliant essay in *Poems* gives Jones's humanistic view regarding Middle Eastern literature and criticizes the contemporary state of European poetry, which had 'subsisted too long' on stale images, fables, symbols, and themes. Arguing philosophically in a rich comparative perspective for the value of Arabic and Persian literature, he urges his readers to learn Oriental languages and the principal Oriental writings. The chief advance in the essay lies in his provocative conclusion that, up to the eighteenth century, Persia had produced more writers (mainly poets) than all of Europe combined, and that new kinds of images and themes might be stimulated in European literature by Arabic, Persian, and perhaps Turkish works. This important study of pastoral literature prompted men of letters to think anew of the pastoral as a vital part of literature. See Harold Mantz, 'Non-Dramatic Pastoral in Europe in the Eighteenth Century', *PMLA* 31 (September 1916): 439-40.]

[2] I am at a loss to conceive, what induced the illustrious Prince *Cantemir* to contend, that *Yemen* is properly a part of *India*; for, not to mention *Ptolemy*, and the other ancients, who considered it as a province of *Arabia*, nor to insist on the language of the country, which is pure *Arabick*, it is described by the *Asiaticks* themselves as a large division of that peninsula

Its principal cities are *Sanaa*, usually considered as its metropolis; *Zebîd*, a commercial town, that lies in a large plain near the sea of *Omman*; and *Aden*, surrounded with pleasant gardens and woods, which is situated eleven degrees from the *Equator*, and seventy-six from the *Fortunate Islands*, or *Canaries*, where the geographers of *Asia* fix their first meridian. It is observable that *Aden*, in the Eastern dialects, is precisely the same word with *Eden*, which we apply to the garden of paradise: it has two senses, according to a slight difference in its pronunciation; its first meaning is *a settled abode*, its second, *delight, softness*, or *tranquillity*: the word *Eden* had, probably, one of these senses in the sacred text, though we use it as a proper name. We may also observe in this place that *Yemen* itself takes its name from a word, which signifies *verdure*, and *felicity*; for in those sultry climates, the freshness of the shade, and the coolness of water, are ideas almost inseparable from that of happiness; and this may be a reason why most of the *Oriental* nations agree in a tradition concerning a delightful spot, where the first inhabitants of the earth were placed before their fall. The ancients, who gave the name of *Eudaimon*, or *Happy*, to this country, either meaned to translate the word *Yemen*, or, more probably, only alluded to the valuable spice-trees, and balsamick plants, that grow in it, and, without speaking poetically, give a real perfume to the air[3]: now it is certain that all poetry receives a very considerable ornament from the beauty of natural images; as the roses of *Sharon*, the verdure of *Carmel*, the vines of *Engaddi*, and the dew of *Hermon*, are the sources of many pleasing metaphors and comparisons in the sacred poetry: thus the odours of *Yemen*, the musk of *Hadramut*, and the pearls of *Omman*, supply the *Arabian* poets with a great variety of allusions; and, if the remark of *Hermogenes* be just, that whatever is *delightful to the senses* produces the *Beautiful* when it is described, where can we find so much beauty as in the *Eastern* poems, which turn chiefly upon the loveliest objects in nature?

To pursue this topick yet farther: it is an observation of *Demetrius* of *Phalera*, in his elegant treatise upon

style, that it is not easy to write on agreeable subjects in a disagreeable manner, and that beautiful *expressions* naturally rise with beautiful images; *for which reason*, says he, *nothing can be more pleasing than Sappho's poetry, which contains the description of gardens, and banquets, flowers and fruits, fountains and meadows, nightingales and turtle-doves, loves and graces*: thus, when she speaks of *a stream softly murmuring among the branches, and the Zephyrs playing through the leaves, with a sound, that brings on a quiet slumber*, her lines flow without labour as smoothly as the rivulet she describes.[4] I may have altered the words of *Demetrius*, as I quote them by memory, but this is the general sense of his remarks, which, if it be not rather specious than just, must induce us to think, that the poets of the *East* may vie with those of *Europe* in *the graces of their diction*, as well as in the liveliness of their images: but we must not believe that the *Arabian* poetry can please only by its descriptions of *beauty*; since the gloomy and terrible objects, which produce the *sublime*, when they are aptly described, are no where more common than in the *Desert* and *Stony Arabia's*; and, indeed, we see nothing so frequently painted by the poets of those countries, as wolves and lions, precipices and forests, rocks and wildernesses.

If we allow the natural objects, with which the *Arabs* are perpetually conversant, to be *sublime* and *beautiful*, our next step must be, to confess that their comparisons, metaphors, and allegories are so likewise; for an allegory is a string of metaphors, a metaphor is a short simile, and the finest similes are drawn from natural objects. It is true that many of the *Eastern* figures are common to other nations, but some of them receive a propriety from the manners of the *Arabians*, who dwell in the plains and woods, which would be lost, if they came from the inhabitants of cities: thus *the dew of liberality*, and the *odour of reputation*, are metaphors used by most people; but they are wonderfully proper in the mouths of those, who have so much need of being refreshed by *the dews*, and who gratify their sense of smelling with the *sweetest odours* in the world. Again; it is very usual in all countries, to make frequent allusions to the brightness of the celestial luminaries, which give their light to all; but the metaphors taken from them have an additional beauty, if we consider them as made by a nation, who pass most of their nights in the open air, or in tents, and consequently see the moon and stars in their greatest splendour. This

---

which they call *Jezeiratul Arab*; and there is no more reason for annexing it to *India*, because the sea, which washes one side of it, is looked upon by some writers as belonging to the great *Indian* ocean, than there would be for annexing it to *Persia*, because it is bounded on another side by the *Persian* gulf.

[3] The writer of an old history of the *Turkish Empire* says, '*The air of Egypt sometimes in summer is like any sweet perfume, and almost suffocates the spirits, caused by the wind that brings the odours of the Arabian spices*'.

[4] [His *De Elocutione* particularly praises Sappho in Sections 132 and 140-67. See W. Rhys Roberts, *Demetrius on Style* (Cambridge, 1902).]

way of considering their poetical figures will give many of them a grace, which they would not have in our languages: so, when they compare *the foreheads of their mistresses to the morning, their locks to the night, their faces to the sun, to the moon, or the blossoms of jasmine, their cheeks to roses or ripe fruit, their teeth to pearls, hail-stones, and snow-drops, their eyes to the flowers of the narcissus, their curled hair to black scorpions, and to hyacinths, their lips to rubies or wine, the form of their breasts to pomegranates, and the colour of them to snow, their shape to that of a pine-tree, and their stature to that of a cypress, a palm-tree, or a javelin, & c.,*[5] these comparisons, many of which would seem forced in our idioms, have undoubtedly a great delicacy in theirs, and affect their minds in a peculiar manner; yet upon the whole their similies are very just and striking, as that of *the blue eyes of a fine woman, bathed in tears, to violets dropping with dew,*[6] and that of *a warriour, advancing at the head of his army, to an eagle sailing through the air, and piercing the clouds with his wings.*

These are not the only advantages, which the natives of *Arabia* enjoy above the inhabitants of most other countries: they preserve to this day the manners and customs of their ancestors, who, by their own account, were settled in the province of *Yemen* above three thousand years ago; they have never been wholly subdued by any nation; and though the admiral of *Selim the First* made a descent on their coast, and exacted a tribute from the people of *Aden*, yet the *Arabians* only keep up a show of allegiance to the Sultan, and act, on every important occasion, in open defiance of his power, relying on the swiftness of their horses, and the vast extent of their forests, in which an invading enemy must soon perish: but here I must be understood to speak of those *Arabians*, who, like the old *Nomades*, dwell constantly in their tents, and remove from place to place according to the seasons; for the inhabitants of the cities, who traffick with the merchants of Europe in spices, perfumes, and coffee, must have lost a great deal of their ancient simplicity: the others have, certainly, retained it; and, except when their tribes are engaged in war, spend their days in watching their flocks and camels, or in repeating their native songs, which they pour out almost extempore, professing a contempt for the stately pillars, and solemn buildings of the cities, compared with the natural charms of the coun-

try, and the coolness of their tents: thus they pass their lives in the highest pleasure, of which they have any conception, in the contemplation of the most delightful objects, and in the enjoyment of perpetual spring; for we may apply to part of *Arabia* that elegant couplet of *Waller* in his poem of the *Summer-island*,

> The gentle spring, that but salutes us here
> Inhabits there, and courts them all year.[7]

Yet the heat of the sun, which must be very intense in a climate so near the Line, is tempered by the shade of the trees, that overhang the valleys, and by a number of fresh streams, that flow down the mountains. Hence it is, that almost all their notions of *felicity* are taken from *freshness* and *verdure*: it is a maxim among them that the three most charming objects in nature are,[8] *a green meadow, a clear rivulet, and a beautiful woman*, and that the view of these objects at the same time affords the greatest delight imaginable. *Mahomed* was so well acquainted with the maxim of his countrymen, that he described the pleasures of heaven to them, under the allegory of *cool fountains, great bowers, and black-eyed girls*, as the word *Houri* literally signifies in *Arabick*;[9] and in the chapter of the *Morning*, towards the end of his *Alcoran*, he mentions a garden, called *Irem*, which is no less celebrated by the *Asiatick* poets than that of the *Hesperides* by the *Greeks*: it was planted, as the commentators say, by a king, named *Shedad*, and was once seen by an *Arabian*, who wandered very far into the deserts in search of a lost camel: it was, probably, a name invented by the impostor, as a type of a future state of happiness. Now it is certain that the genius of every nation is not a little affected by their climate; for, whether it be that the immoderate heat disposes the *Eastern* people to a life of indolence, which gives them full leisure to cultivate their talents, or whether the sun has a real influence on the imagination (as one would suppose that the Ancients believed, by their making *Apollo* the god of poetry); whatever be the cause, it has always been remarked, that the *Asiaticks* excel the inhabitants of our colder regions in the liveliness of their fancy, and the richness of their invention.

To carry this subject one step farther: as the *Arabians*

---

[5] See *Noweiri*, cited by the very learned *Reiske*. [Jones's Arabic quotation is omitted.]

[6] See the *Arabick* Miscellany, entitled *Shecardán*, ch. 14. [Arabic omitted.]

[7] [From Edmund Waller's brief mock-epic, *The Battle of the Summer Islands* (London, 1686), 1:40-41. Waller's polished lines prompted Alexander Pope to call him 'smooth' and 'sweet' (*Epistle to Augustus* 1:267).]

[8] See the life of *Tamerlane*, published by *Golius*, page 299. [Arabic omitted.]

[9] [Sura 37, 11. 41-47.]

are such admirers of *beauty*, and as they enjoy such ease and leisure, they must naturally be susceptible of *that passion*, which is the true spring and source of agreeable poetry; and we find, indeed, that *love* has a greater share in their poems than any other passion: it seems to be always uppermost in their minds, and there is hardly an elegy, a panegyrick, or even a satire, in their language, which does not begin with the complaints of an unfortunate, or the exultations of a successful, lover. It sometimes happens, that the young men of one tribe are in love with the damsels of another; and, as the tents are frequently removed on a sudden, the lovers are often separated in the progress of the courtship: hence almost all the *Arabick* poems open in this manner; the author bewails the sudden departure of his mistress, Hinda, Maia, Zeineb, or Azza, and describes her beauty, comparing her to a wanton fawn, that plays among the aromatick shrubs; his friends endeavour to comfort him, but he refuses consolation; he declares his resolution of visiting his beloved, though the way to her tribe lie through a dreadful wilderness, or even through a den of lions; here he commonly gives a description of the horse or camel, upon which he designs to go, and thence passes, by an easy transition, to the principal subject of his poem, whether it be the praise of his own tribe, or a satire on the timidity of his friends, who refuse to attend him on his expedition; though very frequently the piece turns wholly upon love. But it is not sufficient that a nation have a genius for poetry, unless they have the advantage of a rich and beautiful language, that their expressions may be worthy of their sentiments; the *Arabians* have this advantage also in a high degree: their language is expressive, strong, sonorous, and the most copious, perhaps, in the world; for, as almost every tribe had many words appropriated to itself, the poets, for the convenience of their measure, or sometimes for their singular beauty, made use of them all, and, as the poems became popular, these words were by degrees incorporated with the whole language, like a number of little streams, which meet together in one channel, and, forming a most plentiful river, flow rapidly into the sea.

If this way of arguing *à priori* be admitted in the present case (and no single man has a right to infer the merit of the *Eastern* poetry from the poems themselves, because no single man has a privilege of judging for all the rest), if the foregoing argument have any weight, we must conclude that the *Arabians*, being perpetually conversant with the most beautiful objects, spending a calm and agreeable life in a fine climate, being extremely addicted to the softer passions, and having the advantage of a language singularly adapted to poetry, must be naturally excellent poets, provided that their *manners* and *customs* be favourable to the cultivation of that art; and that they are highly so, it will not be difficult to prove.

The fondness of the *Arabians* for poetry, and the respect which they show to poets, would be scarce believed, if we were not assured of it by writers of great authority: the principal occasions of rejoicing among them, were formerly, and, very probably, are to this day, the birth of a boy, the foaling of a mare, the arrival of a guest, and the rise of a poet in their tribe: when a young *Arabian* has composed a good poem, all the neighbours pay their compliments to his family, and congratulate them upon having a relation capable of recording their actions, and of recommending their virtues to posterity. At the beginning of the seventh century, the *Arabick* language was brought to a high degree of perfection by a sort of poetical Academy, that used to assemble at stated times, in a place called *Ocadh*, where every poet produced his best composition, and was sure to meet with the applause that it deserved: the most excellent of these poems were transcribed in characters of gold upon *Egyptian* paper, and hung up in the temple, whence they were named *Modhahebat*, or *Golden*, and *Moallakat*, or *Suspended*: the poems of this sort were called *Casseida's* or *eclogues*,[10] seven of which are preserved in our libraries, and are considered as the finest that were written before the time of *Mahomed*. The fourth of them, composed by *Lebīd*, is purely pastoral, and extremely like the *Alexis* of *Virgil*, but far more beautiful, because it is more agreeable to nature: the poet begins with praising the charms of the fair *Novâra* (a word, which in *Arabick* signifies a *timorous fawn*) but inveighs against her unkindness; he then interweaves a description of his young camel, which he compares for its swiftness to a stag pursued by the hounds; and takes occasion afterwards to mention his own riches, accomplishments, liberality, and valour, his noble birth, and the glory of his tribe: the diction of this poem is easy and simple, yet elegant, the numbers flowing and musical, and the sentiments wonderfully natural; as the learned reader will see by the following passage

---

[10] These seven poems, clearly transcribed with explanatory notes, are among *Pocock's* manuscripts at *Oxford*, No. 164: the names of the seven poets are *Amralkeis, Tarafa, Zoheir, Lebid, Antara, Amru,* and *Hareth.* In the same collection, No. 174, there is a manuscript, containing above forty other poems, which had the honour of being suspended in the temple at *Mecca*: this volume is an inestimable treasure of ancient *Arabick* literature. [After Jones's attempted stimulation was unsuccessful, he translated the *Mu'allakát* himself.]

which I shall attempt to imitate in verse, that the merit of
the poet may not be wholly lost in a verbal translation:

> *But ah! thou know'st not in what youthful play*
> *Our nights, beguil'd with pleasure, swam away;*
> *Gay songs, and cheerful tales, deceiv'd the time,*
> *And circling goblets made a tuneful chime;*
> *Sweet was the draught, and sweet the blooming*
>   *maid,*
> *Who touch'd her lyre beneath the fragrant shade;*
> *We sip'd till morning purpled ev'ry plain;*
> *The damsels slumber'd, but we sip'd again:*
> *The waking birds, that sung on ev'ry tree*
> *Their early notes were not so blithe as we.*[11]

The *Mahomedan* writers tell a story of this poet, which
deserves to be mentioned here: it was a custom, it seems,
among the old *Arabians*, for the most eminent versifiers
to hang up some chosen couplets on the gate of the
temple, as a publick challenge to their brethren, who
strove to answer them before the next meeting at *Ocadh*,
at which time the whole assembly used to determine the
merit of them all, and gave some mark of distinction to
the author of the finest verses. Now *Lebid*, who, we are
told, had been a violent opposer of *Mahomed*, fixed a
poem on the gate, beginning with the following distich,
in which he apparently meant to reflect upon the new
religion: *Are not all things vain, which come not from
God? and will not all honours decay, but those, which He
confers?*[12] These lines appeared so sublime, that none of
the poets ventured to answer them; till *Mahomed*, who
was himself a poet, having composed a new chapter of
his *Alcoran* (the second, I think), placed the opening of
it by the side of *Lebid's* poem, who no sooner read it, than
he declared it to be something divine, confessed his own
inferiority, tore his verses from the gate, and embraced
the religion of his rival; to whom he was afterwards
extremely useful in replying to the satires of *Amralkeis*,
who was continually attacking the doctrine of *Mahomed*:
the *Asiaticks* add, that their lawgiver acknowledged some
time after, that no heathen poet had ever produced a
nobler distich than that of *Lebid* just quoted.

There are a few other collections of ancient *Arabick*
poetry; but the most famous of them is called *Hamása*,
and contains a number of *epigrams, odes, and elegies,*
composed on various occasions: it was compiled by *Abu
Temam*, who was an excellent poet himself, and used to
say, that *fine sentiments delivered in prose were like gems*

scattered at random, *but that, when they were confined
in a poetical measure, they resembled bracelets and
strings of pearls.*[13] When the religion and language of
Mahomed were spread over the greater part of *Asia*, and
the maritime countries of *Africa*, it became a fashion for
the poets of *Persia, Syria, Egypt, Mauritania*, and even
of *Tartary*, to write in *Arabick*; and the most beautiful
verses in that idiom, composed by the brightest genius's
of those nations, are to be seen in a large miscellany,
entitled *Yateima*; though many of their works are transcribed separately: it will be needless to say much on the
poetry of the *Syrians, Tartarians* and *Africans*, since
most of the arguments, before used in favour of the
Arabs, have equal weight with respect to the other
Mahomedans, who have done little more than imitate
their style, and adopt their expressions; for which reason
also I shall dwell the shorter time on the genius and
manners of the *Persians, Turks*, and *Indians*.

The great empire, which we call PERSIA, is known to
its natives by the name of *Iran*; since the word *Persia*
belongs to a particular province, the ancient *Persis*, and
is very improperly applied by us to the whole kingdom:
but, in compliance with the custom of our geographers, I
shall give the name of *Persia* to that celebrated country,
which lies on one side between the *Caspian* and *Indian*
seas, and extends on the other from the mountains of
Candahar, or *Paropamisus*, to the confluence of the
rivers *Cyrus* and *Araxes*, containing about twenty degrees
from south to north, and rather more from east to west.

In so vast a tract of land there must needs be a great
variety of climates: the southern provinces are no less
unhealthy and sultry, than those of the north are rude and
unpleasant; but in the interior parts of the empire the air
is mild and temperate, and, from the beginning of May to
September, there is scarce a cloud to be seen in the sky:
the remarkable calmness of the summer nights, and the
wonderful splendour of the moon and stars in that country, often tempt the *Persians* to sleep on the tops of their
houses, which are generally flat, where they cannot but
observe the figures of the constellations, and the various
appearances of the heavens; and this may in some
measure account for the perpetual allusions of their poets,
and rhetoricians, to the beauty of the heavenly bodies. We
are apt to censure the oriental style for being so full of
metaphors taken from the sun and moon: this is ascribed
by some to the bad taste of the *Asiaticks; the works of the
Persians, says M. de Voltaire, are like the titles of their*

---

[11] In Arabic [omitted].
[12] In Arabic [omitted].

[13] In Arabic [omitted].

kings, in which the sun and moon are often introduced:[14] but they do not reflect, that every nation has a set of images, and expressions, peculiar to itself, which arise from the difference of its climate, manners, and history. There seems to be another reason for the frequent allusions of the *Persians* to the sun, which may, perhaps, be traced from the old language and popular religion of their country: thus *Mihridâd,* or *Mithridates,* signifies *the gift of the sun,* and answers to the *Theodorus* and *Diodati* of other nations. As to the titles of the *Eastern* monarchs, which seem, indeed, very extravagant to our ears, they are merely formal, and no less void of meaning than those of *European* princes, in which *serenity* and *highness* are often attributed to the most *gloomy,* and *low-minded* of men.

The midland provinces of *Persia* abound in fruits and flowers of almost every kind, and, with proper culture, might be made the garden of *Asia:* they are not watered, indeed, by any considerable river, since the *Tigris* and *Euphrates,* the *Cyrus* and *Araxes,* the *Oxus,* and the five branches of the *Indus,* are at the farthest limits of the kingdom; but the natives, who have a turn for agriculture, supply that defect by artificial canals, which sufficiently temper the dryness of the soil; but in saying they *supply* that defect, I am falling into a common error, and representing the country, not as it *is* at present, but as it *was* a century ago; for a long series of civil wars and massacres have now destroyed the chief beauties of *Persia,* by stripping it of its most industrious inhabitants.

The same difference of climate, that affects the air and soil of this extensive country, gives a variety also to the persons and temper of its natives: in some provinces they have dark complexions, and harsh features; in others they are exquisitely fair, and well made; in some others, nervous and robust: but the general character of the nation is that *softness,* and *love of pleasure,* that *indolence,* and *effeminancy,* which have made them an easy prey to all the western and northern swarms, that have from time to time invaded them. Yet they are not wholly void of martial spirit; and, if they are not naturally brave, they are at least extremely docile, and might, with proper discipline, be made excellent soldiers: but the greater part of them, in the short intervals of peace that they happen

to enjoy, constantly sink into a state of inactivity, and pass their lives in a pleasurable, yet studious, retirement; and this may be one reason, why *Persia* has produced more writers of every kind, and chiefly *poets,* than all *Europe* together, since their way of life gives them leisure to pursue those arts, which cannot be cultivated to advantage, without the greatest calmness and serenity of mind. There is a manuscript at *Oxford,*[15] containing the *lives of an hundred and thirty-five of the finest Persian poets,* most of whom left very ample collections of their poems behind them: but the versifiers, and *moderate poets,* if *Horace* will allow any such men to exist, are without number in *Persia.*

The delicacy of their lives and sentiments has insensibly affected their language, and rendered it the softest, as it is one of the richest, in the world: it is not possible to convince the reader of this truth, by quoting a passage from a *Persian* poet in *European* characters; since the sweetness of sound cannot be determined by the sight, and many words, which are soft and musical in the mouth of a *Persian,* may appear harsh to our eyes, with a number of consonants and gutturals: it may not, however, be absurd to set down in this place, an Ode of the poet *Hafiz,* which, if it be not sufficient to prove the delicacy of his language, will at least show the liveliness of his poetry:

Ai bad nesīmi yârdari,
Zan nefheï mushcbâr dari:
Zinhar mecun diraz-desti!
Ba turreï o che câr dari?
Ai gul, to cujá wa ruyi zeibash.
O taza, wa to kharbâr dari.
Nerkes, to cujâ wa cheshmi mestesh?
O serkhosh, wa to khumâr dari.
Ai seru, to ba kaddi bulendesh,
Der bagh che iytebâr dari?
Ai akl, to ba wujûdi ishkesh
De dest che ikhtiyâr dari?
Rihan to cujâ wa khatti sebzesh?
O mushc, wa to ghubâr dari.
Ruzi bures bewasli Hafiz,
Gher takati yntizâr dari.

That is, word for word, *O sweet gale, thou bearest the fragrant scent of my beloved; thence it is that thou hast this musky odour. Beware! do not steal: what hast thou*

---

[14] [In Chapter 82, 'Sciences de Beaux-Arts aux XIIIe et XIVe siècles, '*Essai sur les moeurs,* in *Oeuvres completes,* ed. Louis Moland (Paris, 1877-85), 12: 62. This is a typical example of the accuracy of Jones's quotations and translations, coming from 'Leurs ouvrages ressemblent aux titres de leurs souverains, dans lesquels il est souvent question du soleil et de la lune.']

[15] In Hyperoo Bodl. 128. There is a prefatory discourse to this curious work, which comprises the lives of ten *Arabian* poets.

to do with her tresses? *O rose, what art thou, to be compared with her bright face? She is fresh, and thou art rough with thorns. O narcissus, what art thou in comparison of her languishing eye? Her eye is only sleepy, but thou art sick and faint. O pine, compared with her graceful stature, what honour hast thou in the garden? O wisdom, what wouldst thou choose, if to choose were in thy power, in preference to her love? O sweet basil, what art thou, to be compared with her fresh cheeks? They are perfect musk, but thou art soon withered. Come, my beloved, and charm Hafez with thy presence, if thou canst but stay with him for a single day.* This little song is not unlike a sonnet ascribed to *Shakespeare*, which deserves to be cited here, as a proof that the Eastern imagery is not so different from the *European* as we are apt to imagine.

The forward violet thus did I chide:
"Sweet thief! whence didst thou steal thy sweet that
*   smells,*
"If not from my love's breath? The purple pride,
"Which on thy soft cheek for complexion dwells,
"In my love's veins thou hast too grossly dyed."
The lily I condemned for thy hand,
And buds of marjoram had stol'n thy hair;
The roses fearfully on thorns did stand,
On blushing shame, another white despair;
A third, nor red nor white had stol'n of both,
And to his robb'ry had annex'd thy breath;
But for his theft, in pride of all his growth,
A vengeful canker eat him up to death.
More flow'rs I noted, yet I none could see,
But scent or colour it had stol'n from thee.[16]

Shakespeare's Poems, p. 207.

The *Persian* style is said to be ridiculously bombast, and this fault is imputed to the slavish spirit of the nation, which is ever apt to magnify the objects that are placed above it: there are bad writers, to be sure, in every country, and as many in *Asia* as elsewhere; but if we take the pains to learn the *Persian* language, we shall find that those authors, who are generally esteemed in *Persia*, are neither slavish in their sentiments, nor ridiculous in their expressions: of which the following passage in a moral work of *Sadi*, entitled *Bostán*, or, *The Garden*, will be a sufficient proof. *I have heard that king Nushirvan, just before his death, spoke thus to his son Hormus: Be a guardian, my son, to the poor and helpless; and be not confined in the chains of thy own idolence. No one can*

be at ease in thy dominion, while thou seekest only thy private rest, and sayest. It is enough. A wise man will not approve the shepherd, who sleeps, while the wolf is in the fold. Go, my son, protect thy weak and indigent people; since through them is a king raised to the diadem. The people are the root, and the king is the tree that grows from it; and the tree, O my son, derives its strength from the root.[17]

Are these mean sentiments, delivered in pompous language? Are they not rather worthy of our most spirited writers? And do they not convey a fine lesson for a young king? Yet *Sadi's* poems are highly esteemed at *Constantinople*, and at *Ispahan*; though, a century or two ago, they would have been suppressed in *Europe*, for spreading with too strong a glare the light of liberty and reason.

As to the great Epick poem of *Ferdusi*, which was composed in the tenth century, it would require a very long treatise, to explain all its beauties with a minute exactness. The whole collection of that poet's works is called *Shahnâma*, and contains the history of *Persia*, from the earliest times to the invasion of the *Arabs*, in a series of very noble poems; the longest and most regular of which is an heroick poem of one great and interesting action, namely, *the delivery of Persia by Cyrus* from the oppressions of *Afrasiab*, king of the *Transoxan Tartary*, who being assisted by the emperors of *India* and *China*, together with all the dæmons, giants, and enchanters of *Asia*, had carried his conquests very far, and become exceedingly formidable to the *Persians*. This poem is longer than the *Iliad*; the characters in it are various and striking; the figures bold and animated; and the diction every where sonorous, yet noble; polished, yet full of fire. A great profusion of learning has been thrown away by some criticks, in comparing *Homer* with the heroick poets, who have suceeded him; but it requires very little judgement to see, that no succeeding poet whatever can with any propriety be compared with *Homer*: that great father of the *Grecian* poetry and literature, had a genius too fruitful and comprehensive to let any of the striking parts of nature escape his observation; and the poets, who have followed him, have done little more than transcribe his images, and give a new dress to his thoughts. Whatever elegance and refinements, therefore, may have been introduced into the works of the moderns, the spirit and invention of *Homer* have ever continued without a rival: for which reasons I am far from pretending to assert that the poet of *Persia* is equal to that of *Greece*; but there is

---

[16] [Sonnet 99.]

[17] [Persian omitted. From opening of Chapter 1, 'On Justice, Management, and Good Judgment'.]

certainly a very great resemblance between the works of those extraordinary men: both drew their images from nature herself, without catching them only by reflection, and painting, in the manner of the modern poets, *the likeness of a likeness*; and both possessed, in an eminent degree, *that rich and creative invention, which is the very soul of poetry*.

As the *Persians* borrowed their poetical measures, and the forms of their poems, from the *Arabians*, so the TURKS, when they had carried their arms into *Mesopotamia* and *Assyria*, took their numbers and their taste for poetry from the *Persians*;

> *Græcia* capta ferum victorem cepit, et artes
> Intulit agresti *Latio*.

In the same manner as the *Greek* compositions were the models of all the *Roman* writers, so were those of *Persia* imitated by the Turks, who considerably polished and enriched their language, naturally barren, by the number of simple and compound words, which they adopted from the *Persian* and *Arabick*. Lady *Wortley Montague* very justly observes, that *we want those compound words, which are very frequent and strong in the Turkish language*; but her interpreters led her into a mistake in explaining one of them, which she translates *stag-eyed*,[18] and thinks *a very lively image of the fire and indifference in the eyes of the royal bride*: now it never entered into the mind of an *Asiatick* to compare his mistress's eyes to those of a stag, or to give an image of their *fire and indifference*; the *Turks* mean to express that *fullness*, and, at the same time, that *soft and languishing lustre*, which is peculiar to the eyes of their beautiful women, and which by no means resembles the unpleasing wildness in those of a stag. The original epithet, I suppose, was[19] *Ahû cheshm*, or, *with the eyes of a young fawn*: now I take the *Ahû* to be the same animal with the *Gazâl* of the *Arabians*, and the *Zabi* of the *Hebrews*, to which their poets allude in almost every page. I have seen one of these animals; it is a kind of antelope, exquisitely beautiful, with eyes uncommonly black and large. This is the same sort of roe, to which *Solomon* alludes in this delicate simile: *Thy two breasts are like two young roes, that are twins, which play among the lilies.*[20]

A very polite scholar, who has lately translated sixteen Odes of *Hafiz*,[21] with learned illustrations, blames the *Turkish* poets for copying the *Persians* too servilely: but, surely, they are not more blameable than *Horace*, who not only imitated the measures and expressions of the *Greeks*, but even translated, almost word for word, the brightest passages of *Alcæus*, *Anacreon*, and others; he took less from *Pindar* than from the rest, because the wildness of his numbers, and the obscurity of his allusions, were by no means suitable to the genius of the *Latin* language; and this may, perhaps, explain his ode to *Julius Antonius*, who might have advised him to use more of *Pindar*'s manner in celebrating the victories of *Augustus*.[22] Whatever we may think of this objection, it is certain that the *Turkish* empire has produced a great number of poets; some of whom had no small merit in their way: the ingenious author just mentioned assured me, that the *Turkish* satires of *Ruhi Bagdadi* were very forcible and striking, and he mentioned the opening of one of them, which seemed not unlike the manner of *Juvenal*. At the beginning of the last century, a work was published at *Constantinople*, containing the finest verses of *five hundred and forty-nine Turkish poets*, which proves at least that they are singularly fond of this art, whatever may be our opinion of their success in it.

The descendants of *Tamerlane* carried into *India* the language and poetry of the *Persians*; and the *Indian* poets to this day compose their verses in imitation of them. The best of their works, that have passed through my hands, are those of *Huzein*, who lived some years ago at *Benāres* with a great reputation for his parts and learning, and was known to the *English*, who resided there, by the time of *the Philosopher*. His poems are elegant and lively, and one of them, *on the departure of his friends*, would suit our language admirably well, but it is too long to be inserted in this essay. The *Indians* are soft and voluptuous, but artful and insincere, at least to the *Europeans*, whom, to say the truth, they have had no great reason to admire for the opposite virtues: but they are fond of poetry, which they learned from the *Persians*, and may, perhaps, before the close of the century, be as fond of a more fomidable art, which they will learn from the *English*.

---

[18] [In letter to Pope of 1 April 1717, *The Complete Letters of Mary Wortley Montagu*, ed. Robert Halsband (Oxford, 1965), 1: 335.]

[19] This epithet seems to answer to the Greek ελιωπις, which our grammarians properly interpret *Quae nigris oculis decora est et venusta*; if it were permitted to make any innovations in a dead language, we might express the *Turkish* adjective by the word δορκωπις, which would, I dare say, have sounded agreeably to the *Greeks* themselves.

[20] [In Canticles 4: 5.]

[21] [Charles Reviczky, *Specimen Poeseos Persicae* (Vienna, 1771).]

[22] [Epistles 2, Ode 1.]

I must request, that, in bestowing these praises on the writings of *Asia*, I may not be thought to derogate from the merit of the *Greek* and *Latin* poems, which have justly been admired in every age; yet I cannot but think that our *European* poetry has subsisted too long on the perpetual repetition of the same images, and incessant allusions to the same fables: and it has been my endeavour for several years to inculcate this truth, that, if the principal writings of the *Asiaticks*, which are reposited in our publick libraries, were printed with the usual advantage of notes and illustrations, and if the languages of the *Eastern* nations were studied in our great seminaries of learning, where every other branch of useful knowledge is taught to perfection, a new and ample field would be opened for speculation; we should have a more extensive insight into the history of the human mind; we should be furnished with a new set of images and similitudes; and a number of excellent compositions would be brought to light, which future scholars might explain, and future poets might imitate.

On the Mystical Poetry of the Persians and Hindus[1]

A figurative mode of expressing the fervour of devotion, or the ardent love of created spirits towards their beneficent Creator, has prevailed from time immemorial in *Asia*; particularly among the *Persian* theists, both ancient *Húshangis* and modern *Súfis*, who seem to have borrowed it from the *Indian* philosophers of the *Vēdānta* school; and their doctrines are also believed to be the source of that sublime, but poetical, theology, which glows and sparkles in the writings of the old *Academicks*. 'PLATO travelled into *Italy* and *Egypt*', says CLAUDE FLEURY,[2] 'to learn the Theology of the Pagans at its fountain head:' its true fountain, however, was neither in *Italy* nor in *Egypt* (though considerable streams of it had been conducted thither by PYTHAGORAS and by the family of MISRA), but in *Persia* or *India*, which the founder of the *Italick* sect had visited with a similar design. What the *Grecian* travellers learned among the sages of the east, may perhaps be fully explained, at a season of leisure, in another dissertation; but we confine this essay to a singular species of poetry, which consists almost wholly of a mystical religious allegory, though it seems on a transient view to contain only the sentiments of a wild and voluptuous libertinism: now, admitting the danger of a poetical style, in which the limits between vice and enthusiasm are so minute as to be hardly distinguishable, we must

---

[1] [The essay was read to the Asiatick Society in December 1791 and later included in *Asiatick Researches* 4 (1794). One of Jones's major essays not directly on an Indian subject, it contains the first English translation of any part of the mystical *Mathnawi*, by Jalál al-Dīn Rumī-see A.J. Arberry, 'Persian Jones', *Asiatic Review* 40 (April 1944): 194. Through quotations from prominent theologians like Isaac Barrow (1630-77) and Jacques Necker (1732-1804), whose metaphysical ideas resemble Vedantic and Sufistic ones, Jones tries to analyze the two philosophical systems, which had stimulated many metaphors and images in Persian and Hindu sacred poetry. He also translates distichs from several of Háfiz's odes to illustrate the religious and secular love poetry of Sufism, which he greatly admired.]

[2] [This French scholar (1640-1725) wrote the twenty-volume *Histoire ecclesiastique* (Paris, 1691-1720), the first comprehensive church-history in Europe.]

beware of censuring it severely, and must allow it to be natural, though a warm imagination may carry it to a culpable excess; for an ardently greateful piety is congenial to the undepraved nature of man, whose mind, sinking under the magnitude of the subject, and struggling to express its emotions, has recourse to metaphors and allegories, which it sometimes extends beyond the bounds of cool reason, and often to the brink of absurdity. BARROW, who would have been the sublimest mathematician, if his religious turn of mind had not made him the deepest theologian of his age,[3] describes Love as 'an affection or inclination of the soul toward an object, proceeding from an apprehension and esteem of some excellence or convenience in it, as its *beauty*, worth, or utility, and producing, if it be absent, a proportionable desire, and consequently an endeavour, to obtain such a property in it, such possession of it, such an *approximation to it, or union with it*, as the thing is capable of; with a regret and displeasure in failing to obtain it, or in the want and loss of it; begetting likewise a complacence, satisfaction, and delight in its presence, possession, or enjoyment, which is moreover attended with a good will toward it, suitable to its nature; that is, with a desire, that it should arrive at, or continue in, its best state; with a delight to perceive it thrive and flourish; with a displeasure to see it suffer or decay; with a consequent endeavour to advance it in all good and preserve it from all evil'.[4] Agreeably to this description, which consists of two parts, and was designed to comprize the tender love of the Creator towards created spirits, the great philosopher bursts forth in another place, with his usual animation and command of language, into the following panegyrick on the pious love of human souls toward the Author of their happiness: 'Love is the sweetest and most delectable of all passions; and, when by the conduct of wisdom it is directed in a rational way toward a worthy, congruous, and attainable object, it cannot otherwise than fill the heart with ravishing delight: such, in all respects superlatively such, is GOD; who, infinitely beyond all other things, deserveth our affection, as most perfectly amiable and desirable: as having obliged us by innumerable and inestimable benefits; all the good, that we have ever enjoyed, or can ever expect, being derived from his pure bounty; all things in the world, in competition with

him being mean and ugly; all things, without him, vain, unprofitable, and hurtful to us. He is the most proper object of our love; for we chiefly were framed, and it is the prime law of our nature, to love him; *our soul, from its original instinct, vergeth toward him as its centre, and can have no rest, till it be fixed on him*: he alone can satisfy the vast capacity of our minds, and fill our boundless desires. He, of all lovely things, most certainly and easily may be attained; for, whereas commonly men are crossed in their affection, and their love is embittered from their affecting things imaginary, which they cannot reach, or coy things, which disdain and reject them, it is with GOD quite otherwise: He is most ready to impart himself; he most earnestly desireth and wooeth our love; he is not only most willing to correspond in affection, but even doth prevent us therein: *He doth cherish and encourage our love by sweetest influences and most consoling embraces*; by kindest expressions of favour, by most beneficial returns; and, whereas all other objects do in the enjoyment much fail our expectation, he doth ever far exceed it. Wherefore in all affectionate motions of our hearts toward GOD; in *desiring* him, or seeking his favour and friendship; in *embracing* him, or setting our esteem, our good will, our confidence on him; in *enjoying* him by devotional meditations and addresses to him; in a reflective sense of our interest and propriety in him; *in that mysterious union of spirit, whereby we do closely adhere to, and are, as it were, inserted in him*; in a hearty complacence in his benignity, a graceful sense of his kindness, and a zealous desire of yielding some requital for it, we cannot but feel very pleasant transports: indeed, that celestial flame, kindled in our hearts by the spirit of love, cannot be void of warmth; we cannot fix our eyes upon *infinite beauty*, we cannot taste infinite sweetness, we cannot cleave to infinite felicity, without also perpetually rejoicing in the first daughter of Love to GOD, Charity toward men; which in complection and careful disposition, doth much resemble her mother; for she doth rid us from all those gloomy, keen, turbulent imaginations and passions, which cloud our mind, which fret our heart, which discompose the frame of our soul; from burning anger, from storming contention, from gnawing envy, from rankling spite, from racking suspicion, from distracting ambition and avarice; and consequently doth settle our mind in an even temper, in a sedate humour, in an harmonious order, in *that plesant state of tranquillity, which naturally doth result from the voidance of irregular passions*.[5] Now this passage from BARROW

---

[3] [This comment on Barrow was quoted in *New Monthly Magazine* 2 (January 1815): 529, alongside comments by Charles II and George III.]

[4] ['Of the Love of God', *The Works of the Learned Isaac Barrow, D.D.* 1 (London, 1687): 311.]

[5] ['Rejoyce evermore', *Works* 3: 119-20.]

(which borders, I admit, on quietism and enthusiastic devotion) differs only from the mystical theology of the Súfi's and Yógis, as the flowers and fruits of *Europe* differ in scent and flavour from those of *Asia*, or as *European* differs from *Asiatick* eloquence: the same strain, in poetical measure, would rise up to the odes of SPENSER on *Divine Love* and *Beauty*, and, in a higher key with richer embellishments, to the songs of HAFIZ and JAYADĒVA the raptures of the *Masnavì*, and the mysteries of the *Bhāgavat*.

Before we come to the *Persians* and *Indians*, let me produce another specimen of *European* theology, collected from a late excellent work of the illustrious M. NECKER. 'Were men animated', says he, 'with sublime thoughts, did they respect the intellectual power, with which they are adorned, and take an interest in the dignity of their nature, they would embrace with transport that sense of religion, which ennobles their faculties, keeps their minds in full strength, and unites them in idea with him, whose immesity overwhelms them with astonishment: *considering themselves as an emanation from that infinite Being,* the source and cause of all things, they would then disdain to be misled by a gloomy and false philosophy, and would cherish the idea of a GOD, who *created,* who *regenerates,* who *preserves* this universe by invariable laws, and by a continued chain of similar causes producing similar effects; who pervades all nature with his divine spirit, as an universal soul, which moves, directs, and restrains the wonderful fabrick of this world. The blissful idea of a GOD sweetens every moment of our time, and embellishes before us the path of life; unites us delightfully to all the beauties of nature, and associates us with every thing that lives or moves. Yes; the whisper of the gales, the murmur of waters, the peaceful agitation of trees and shrubs, would concur to engage our minds and *affect our souls with tenderness,* if our thoughts were elevated to *one universal cause,* if we recognized on all sides the work of *Him, whom we love*; if we marked the traces of his august steps and benignant intentions, if we believed ourselves actually present at the display of his boundless power and the magnificent exertions of his unlimited goodness. Benevolence, among all the virtues, has a character more than human, and a certain amiable simplicity in its nature, which seems analogous to the *first idea,* the original intention of conferring delight, which we necessarily suppose in the Creator, when we presume to seek his motive in bestowing existence: benevolence is that virtue, or, to speak more emphatically, that *primordial beauty,* which preceded all times and all worlds; and, when we reflect on it, there appears an analogy, obscure

indeed at present, and to us imperféctly known, between our moral nature and a time yet very remote, when we shall satisfy our ardent wishes and lively hopes, which constitute perhaps a sixth, and (if the phrase may be used) a distant, sense. It may even be imagined, that love, the brightest ornament of our nature, love, enchanting and sublime, is a mysterious pledge for the assurance of those hopes; since love, by disengaging us from ourselves, by transporting us beyond the limits of our own being, is the first step in our progress to a joyful immortality; and, by affording both the notion and example of a cherished object distinct from our own souls, may be considered as an interpreter to our hearts of something, which our intellects cannot conceive. We may seem even to hear the Supreme Intelligence and Eternal Soul of all nature, give this commission to the spirits, which emaned from him: *Go; admire a small portion of my works, and study them; make your first trial of happiness, and learn to love him, who bestowed it ; but seek not to remove the veil spread over the secret of your existence: your nature is composed of those divine particles, which, at an infinite distance, constitute my own essence; but you would be too near me, were you permitted to penetrate the mystery of our separation and union: wait the moment ordained by my wisdom; and, until that moment come, hope to approach me only by adoration and gratitude'.[6]*

If these two passages were translated into *Sanscrit* and *Persian,* I am confident, that the *Vēdāntis* and *Súfis* would consider them as an epitome of their common system; for they concur in believing, that the souls of men differ infinitely in *degree,* but not at all in *kind,* from the divine spirit, of which they are *particles,* and in which they will ultimately be absorbed; that the spirit of GOD pervades the universe, always immediately present to his work, and consequently always in substance, that he alone is perfect benevolence, perfect truth, perfect beauty; that the love of him alone is *real* and genuine love, while that of all other objects is *absurd* and illusory, that the beauties of nature are faint resemblances, like images in a mirror, of the divine charms; that, from eternity without beginning to eternity without end, the supreme benevolence is occupied in bestowing happiness or the means of attaining it; that men can only attain it by performing their part of

[6] [Jacques Necker, an eminent theologian who was a popular figure with the masses during the French Revolution, was an authority on comparative religion as expressed in his 'Qu'il y a un Dieu', *Sur l' Importance des opinions religieuses* (London, 1788). The original draft of Jones's translation, which Cannon owns, is excessively revised and indicates how he polished his translations before using them in his scholarly essays.]

the *primal covenant* between them and the Creator; that nothing has a pure absolute existence but *mind* or *spirit*; that *material substances*, as the ignorant call them, are no more than gay *pictures* presented continually to our *minds* by the sempiternal Artist; that we must beware of attachment to such *phantoms*, and attach ourselves exclusively to GOD, who truly exists in us, as we exist solely in him; that we retain even in this forlorn state of separation from our beloved, the *idea* of *heavenly beauty*, and the *remembrance* of our *primeval vows*; that sweet musick, gentle breezes, fragrant flowers, perpetually renew the primary *idea*, refresh our fading memory, and melt us with tender affections; that we must cherish those affections, and by abstracting our souls from *vanity*, that is, from all but GOD, approximate to his essence, in our final union with which will consist our supreme beatitude. From these principles flow a thousand metaphors and poetical figures, which abound in the sacred poems of the *Persians* and *Hindus*, who seem to mean the same thing in substance, and differ only in expression as their languages differ in idiom! The modern SÚFIS, who profess a belief in the *Koran*, suppose with great sublimity both of thought and of diction, an *express contract*, on *the day of eternity without beginning*, between the assemblage of created spirits and the supreme soul, from which they were detached, when a celestial voice pronounced these words, addressed to each spirit separately, 'Art thou not with thy Lord?' that is, art thou not bound by a solemn contract with him? and all the spirits answered with one voice, 'Yes:' hence it is, that *alist* or *art thou not*, and *beli*, or *yes*, incessantly occur in the mystical verses of the *Persians*, and of the *Turkish* poets, who imitate them, as the *Romans* imitated the *Greeks*. The *Hindus* describe the same covenant under the figurative notion, so finely expressed by ISAIAH, of a *nuptial contract*; for considering GOD in the three characters of Creator, Regenerator, and Preserver, and supposing the power of *Preservation* and *Benevolence* to have become incarnate in the person of CRISHNA, they represent him as married to RĀDHĀ, a word signifying *atonement*, *pacification*, or *satisfaction*, but applied allegorically to *the soul of man*, or rather to *the whole assemblage of created souls*, between whom and the benevolent Creator they suppose that *reciprocal* love, which BARROW describes with a glow of expression perfectly oriental, and which our most orthodox theologians believe to have been mystically *shadowed* in the song of SOLOMON, while they admit, that, in a *literal* sense, it is an epithalamium on the marriage of the sapient king with the princes of *Egypt*. The very learned author of the prelections on sacred

poetry declared his opinion, that the canticles were founded on historical truth, but involved an allegory of that sort, which he named *mystical*,[7] and the beautiful poem on the loves of LAILI and MAJNUN by the inimitable NIZÁMI (to say nothing of other poems on the same subject) is indisputably built on true history, yet avowedly allegorical and mysterious; for the introduction to it is a continued rapture on *divine love*; and the name of LAILI seems to be used in the *Masnavi* and the odes of HAFIZ for the omnipresent spirit of GOD.

It has been made a question, whether the poems of HAFIZ must be taken in a literal or in a figurative sense; but the question does not admit of a general and direct answer; for even the most enthusiastick of his commentators allow, that some of them are to be taken literally, and his editors ought to have distinguished them, as our SPENSER has distinguished his four Odes on *Love* and *Beauty*, instead of mixing the profane with the divine, by a childish arrangement according to the alphabetical order of the rhymes. HAFIZ never pretended to more than human virtues, and it is known that he had human propensities; for in his youth he was passionately in love with a girl surnamed *Shákhi Nebàt*, or *the Branch of Sugarcane*, and the prince of *Shiraz* was his rival: since there is an agreeable wildness in the story, and since the poet himself alludes to it in one of his odes, I give it you at length from the commentary. There is a place called *Pirisebz*, or *the Green old man*, about four *Persian* leagues from the city; and a popular opinion had long prevailed, that a youth, who should pass forty successive nights in *Pirisebz* without sleep, would infallibly become an excellent poet: young HAFIZ had accordingly made a vow, that he would serve that apprenticeship with the utmost exactness, and for thirty-nine days he rigorously discharged his duty, walking every morning before the house of his coy mistress, taking some refreshment and rest at noon, and passing the night awake at his poetical station; but, on the fortieth morning, he was transported with joy on seeing the girl beckon to him through the lattices, and invite him to enter: she received him with rapture, declared her preference of a bright genius to the son of a king, and would have detained him all night, if he had not recollected his vow, and, resolving to keep it inviolate, returned to his post. The people of *Shiraz* add (and the fiction is grounded on a couplet of HAFIZ), that, early next morning *an old man, in a green mantle*, who was no less a personage than KHIZR himself, approached him at

[7] [Bishop Robert Lowth, *De Sacra Poesi Hebræorum Prælectiones* (Oxford, 1753).]

Pirisebz with a cup brimful of nectar, which the *Greeks* would have called the water of *Aganippe*, and rewarded his perseverance with an inspiring draught of it. After his juvenile passions had subsided, we may suppose that his mind took that religious bent, which appears in most of his compositions; for there can be no doubt that the following distichs, collected from different odes, relate to the mystical theology of the *Súfis*:

'In eternity without beginning, a ray of thy beauty began to gleam; when love sprang into being, and cast flames over all nature;

'On that day thy cheek sparkled even under thy veil, and all this beautiful imagery appeared on the mirror of our fancies.

'Rise, my soul; that I may pour thee forth on the pencil of that supreme Artist, who comprized in a turn of his compass all this wonderful scenery!

'From the moment, when I heard the divine sentence, *I have breathed into man a portion of my spirit*, I was assured, that we were His, and He ours.

'Where are the glad tidings of union with thee, that I may abandon all desire of life? I am a bird of holiness, and would fain escape from the net of this world.

'Shed, O Lord, from the cloud of heavenly guidance one cheering shower, before the moment, when I must rise up like a particle of dry dust!

'The sum of our transactions in this universe, is nothing: bring us the wine of devotion; for the possessions of this world vanish.

'The true object of heart and soul is the glory of union with our beloved: that object really exists, but without it both heart and soul would have no existence.

'O the bliss of that day, when I shall depart from this desolate mansion; shall seek rest for my soul; and shall follow the traces of my beloved:

'Dancing, with love of his beauty, like a mote in a sun-beam, till I reach the spring and fountain of light, whence yon sun derives all his lustre!'

The couplets, which follow, relate as indubitably to human love and sensual gratifications:

'May the hand never shake, which gathered the grapes! May the foot never slip, which pressed them!

'That poignant liquor, which the zealot calls *the mother of sins*, is pleasanter and sweeter to me than the kisses of a maiden.

'Wine two years old and a damsel of fourteen are sufficient society for me, above all companies great or small.

'How delightful is dancing to lively notes and the cheerful melody of the flute, especially when we touch the hand of a beautiful girl!

'*Call for wine, and scatter flowers around*: *what more canst thou ask from fate*? Thus spoke the nightingale this morning: what sayest thou, sweet rose, to his precepts?

'Bring thy couch to the garden of roses, that thou mayest kiss the cheeks and lips of lovely damsels, quaff rich wine, and smell odoriferous blossoms.

'O branch of an exquisite rose-plant, for whose sake dost thou grow? Ah! on whom will that smiling rose-bud confer delight?

'The rose would have discoursed on the beauties of my charmer, but the gale was jealous, and stole her breath, before she spoke.

'In this age, the only friends, who are free from blemish, are a flask of pure wine and a volume of elegant love songs.

'O the joy of that moment, when the self-sufficiency of inebriation rendered me independent of the prince and of his minister!'

Many zealous admirers of HAFIZ insist, that by *wine* he invariably means *devotion*; and they have gone so far as to compose a dictionary of words in the *language*, as they call it, of the *Súfis*: in that vocabulary *sleep* of the divine favour; *gales* are *illapses* of grace; *kisses* and *embraces*, the *raptures* of piety; *idolaters*, *infidels*, and *libertines* are men of the purest *religion*, and their *idol* is the Creator himself; the *tavern* is a retired oratory, and its *keeper*, a sage instructor; *beauty* denotes the *perfection* of the Supreme Being; *tresses* are the *expansion* of his glory; *lips*, the hidden mysteries of his essence; *down* on the cheek, the world of spirits, who encircle his throne; and a *black mole*, the *point* of indivisible unity; lastly, *wantonness*, *mirth*, and *ebriety*, mean religious ardour and abstraction from all terrestrial thoughts. The poet himself gives a colour in many passages to such an interpretation; and without it, we can hardly conceive, that his poems, or those of his numerous imitators, would be tolerated in a *Muselman* country, especially at *Constantinople*, where they are venerated as divine compositions: it must be admitted, that the sublimity of the *mystical allegory*, which, like metaphors and comparisons, should be *general* only, not minutely exact, is diminished, if not destroyed, by an attempt at *particular* and *distinct resemblances*; and that the style itself is open to dangerous misinterpretation, while it supplies real infidels with a pretext for laughing at religion itself.

On this occasion I cannot refrain from producing a most extraordinary ode by a *Súfi of Bokhárà*, who assumed the poetical surname of ISMAT: a more modern poet, by prefixing three lines to each couplet, which rhyme with the first hemistich, has very elegantly and ingeniously converted the *Kasidah* into *Mokhammes*, but I present you only with a literal version of the original distichs:

'Yesterday, half inebriated, I passed by the quarter, where the vintners dwell, to seek the daughter of an infidel who sells wine.

'At the end of the street, there advanced before me a damsel with a fairy's cheeks, who, in the manner of a pagan, wore her tresses dishevelled over her shoulder like the sacerdotal thread. I said: *O thou, to the arch of whose eye-brow the new moon is a slave, what quarter is this and where is thy mansion?*

'She answered: *Cast thy rosary on the ground; bind on thy shoulder the thread of paganism; throw stones at the glass of piety; and quaff wine from a full goblet;*

'*After that come before me, that I may whisper a word in thine ear: thou wilt accomplish thy journey, if thou listen to my discourse.*

'Abandoning my heart and rapt in ecstasy, I ran after her, till I came to a place, in which religion and reason forsook me.

'At a distance I beheld a company, all insane and inebriated, who came boiling and roaring with ardour from the wine of love;

'Without cymbals, or lutes, or viols, yet all full of mirth and melody; without wine, or goblet, or flask, yet all incessantly drinking.

'When the cord of restraint slipped from my hand, I desired to ask her one question, but she said: *Silence!*

'*The is no square temple, to the gate of which thou canst arrive precipitately: this is no mosque to which thou canst come with tumult, but without knowledge. This is the banquet-house of infidels, and within it all are intoxicated; all, from the dawn of eternity to the day of resurrection, lost in astonishment.*

'*Depart then from the cloister, and take the way to the tavern; cast off the cloak of a dervise, and wear the robe of a libertine.*

'I obeyed; and, if thou desirest the same strain and colour with ISMAT, imitate him, and sell this world and the next for one drop of pure wine'.

Such is the strange religion, and stranger language of the *Súfis*; but most of the *Asiatick* poets are of the religion, and, if we think it worth while to read their poems, we must think it worth while to understand them: their great *Maulaví* assures us, 'they profess eager desire, but with no carnal affection, and circulate the cup, but no material goblet; since all things are spiritual in their sect, all is mystery within mystery;' consistently with which decla-

ration he opens his astonishing work, entitled the *Masnaví*, with the following couplets:

Hear, how yon reed in sadly-pleasing tales
Departed bliss and present wo bewails!
'With me, from native banks untimely torn,
Love-warbling youths and soft-ey'd virgins mourn.
O! Let the heart, by fatal absence rent,
Feel what I sing, and bleed when I lament:
Who roams in exile from his parent bow'r,
Pants to return, and chides each ling'ring hour.
My notes, in circles of the grave and gay,
Have hail'd the rising, cheer'd the closing day:
Each in my fond affections claim'd part,
But none discern'd the secret of my heart.
What though my strains and sorrows flow
   combin'd!
Yet ears are slow, and carnal eyes are blind.
Free through each mortal form the spirits roll,
But sight avails not. Can we see the soul?'
Such notes breath'd gently from yon vocal frame:
Breath'd said I? no; 'twas all enliv'ning flame.
'Tis love, that sparkles in the racy wine.
Me, plaintive wand'rer from my peerless maid,
The need has fir'd, and all my soul betray'd.
He gives the bane, and he with balsam cures;
Afflicts, yet sooths; impassions, yet allures.
Delightful pang his am'rous tales prolong;
And LAILI's frantick lover lives in song.
Not he, who reasons best, this wisdom knows:
Ears only drink what rapt'rous tongues disclose.
Nor fruitless deem the reed's heart-piercing pain:
See sweetness dropping from the parted cane.
Alternate hope and fear my days divide:
I courted Grief, and Anguish was my bride.
Flow on, sad stream of life! I smile secure:
THOU livest; THOU, the purest of the pure!
Rise, vig'rous youth! be free; be nobly bold:
Shall chains confine you, though they blaze with
   gold?
Go: to your vase the gather'd main convey:
What were your stores? The pittance of a day!
New plans for wealth your fancies would invent;
Yet shells, to nourish pearls, must lie content.
The man, whose robe love's purple arrows rend
Bids av'rice rest and toils tumultuous end.
Hail, heav'nly love! true source of endless gains!
Thy balm restores me, and thy skill sustains.
Oh, more than GALEN learn'd than PLATO wise!
My guide, my law, my joy supreme arise!

Love warms this frigid clay with mystick fire,
And dancing mountians leap with young desire.
Blest is the soul, that swims in seas of love,
And long the life sustain'd by food above.
With forms imperfect can perfection dwell?
Here pause, my song; and thou, vain world, farewel.

A volume might be filled with similar passages from the *Súfi* poets; from SAÍB, ORFÌ, MÍR KHOSRAU, JÁMI, HAZÍN, and SÁBIK, who are next in beauty of composition to HAFIZ and SADI, but next at a considerable distance; from MESÍHI, the most elegant of their *Turkish* imitators;[8] from a few *Hindi* poets of our own times, and from IBNUL FÁRED, who wrote mystical odes in *Arabick*; but we may close this account of the *Súfis* with a passage from the third book of the BUSTAN, the declared subject of which is *divine love*; referring you for a particular detail of their metaphysicks and theology to the *Dabistan* of MOHSANI FANI, and to the pleasing essay, called the *Junction of two Seas*, by that amiable and unfortunate prince, DÁRÁ SHECÚH:[9]

'The love of a being composed, like thyself, of water and clay, destroys thy patience and peace of mind; it excites thee, in thy waking hours with minute beauties, and engages thee, in thy sleep, with vain imaginations: with such real affection dost thou lay thy head on her foot, that the universe, in comparison of her, vanishes into nothing before thee; and, since thy gold allures not her eye, gold and mere earth appear equal in thine. Not a breath dost thou utter to any one else, for with her thou hast no room for any other; thou declarest that her abode is in thine eye, or, when thou closet it, in thy heart; thou hast no fear of censure from any man; thou hast no power to be at rest for a moment; if she demands thy soul, it runs instantly to thy lip; and if she waves a cimeter over thee, thy head falls immediately under it. Since an absurd love, with its basis on air, affects thee so violently, and commands with a sway so despotic, canst thou wonder, that they, who walk in the true path, are drowned in the sea of mysterious adoration? They disregard life through affection for its giver; they abandon the world through remembrance of its maker; they are inebriated with the melody of amorous

complaints; they remember their beloved, and resign to him both this life and the next. Through remembrance of GOD, they shun all mankind: they are so enamoured of the cup-bearer, that they spill the wine from the cup. No panacea can heal them, for no mortal can be apprized of their malady; so loudly has rung in their ears, from eternity without beginning, the divine word *alest*, with *belì*, the tumultuous exclamation of all spirits. They are a sect fully employed, but sitting in retirement; their feet are of earth, but their breath is a flame: with a single yell they could rend a mountain from its base; with a single cry they could throw a city into confusion: like wind, they are concealed and move nimbly; like stone, they are silent, yet repeat GOD's praises. At early dawn their tears flow so copiously as to wash from their eyes the black powder of sleep: though the courser of their fancy ran so swiftly all night, yet the morning finds them left behind in disorder: night and day are they plunged in an ocean of ardent desire, till they are unable, through astonishment, to distinguish night from day. So enraptured are they with the beauty of Him, who decorated the human form, that with the beauty of the form itself, they have no concern; and, if ever they behold a beautiful shape, they see in it the mystery of GOD's work.

'The wise take not the husk in exchange for the kernel; and he, who makes that choice, has no understanding. He only has drunk the pure wine of unity, who has forgotten, by remembering GOD, all things else in both worlds'.

Let us return to the *Hindus*, among whom we now find the same emblematical theology, which *Pythagoras* admired and adopted. The loves of CRISHNA and RÁDHÁ, or the reciprocal attraction between the divine goodness and the human soul, are told at large in the tenth book of the *Bhágavat*, and are the subject of a little *Pastoral Drama*, entitled *Gítagóvinda*: it was the work of JAYADÉVA, who flourished, it is said, before CÁLIDÁS, and was born, as he tells us himself, in CENDULI, which many believe to be in *Calinga*: but, since there is a town of a similar name in *Berdwan*, the natives of it insist that the finest lyrick poet of *India* was their countryman, and celebrate in honour of him an annual jubilee, passing a whole night in representing his drama, and in singing his beautiful songs. After having translated the *Gítagóvinda* word for word, I reduced my translation to the form, in which it is now exhibited; omitting only those passages, which are too luxuriant and too bold for an *European* taste, and the prefatory ode on the ten incarnations of VISHNU, with which you have been presented on another occasion: the

---

[8] [See Jones's translation, 'A Turkish Ode of Mesíhi', in his *Poems* (1772).]

[9] [Shah Jahán's oldest son, Dára Shikúh was dislodged (and eventually killed in battle) from his legal claim for succession by Aurungzeb because of his liberal religious views and scholarly pursuits. He translated some Upanishads into Persian.]

phrases in *Italicks*, are the *burdens* of the several songs; and you may be assured, that not a single image or idea has been added by the translator.[10]

On the Literature of the Hindus, from the Sanscrit[1]

Communicated by GOVERDHAN CAUL, *translated, with a short Commentary*

THE TEXT

There are eigtheen *Vidyās*, or parts of *true Knowledge*, and some branches of Knowledge *falsely so called*; of both which a short account shall here be exhibited.

The first *four* are the immortal *Vēdas* evidently revealed by GOD; which are entitled, in one compound word, *Rigyajuhsāmāt'harva*, or, in separate words, *Rich, Yajush, Sāman* and *At'harvan*: the *Rigvēda* consists of *five* sections; the *Yajurvēda*, of *eighty-six*; the *Sāmavēda*, of a *thousand*; and the *At'harvavēda*, of *nine*; with eleven hundred *śāc'has*, or Branches, in various divisions and subdivisions. The *Vēdas* in truth are infinite; but were reduced by VYĀSA to this number and order; the principal part of them is that, which explains the Duties of Man in a methodical arrangement; and in the *fourth* is a system of divine ordinances.

From these are deduced the four *Upavēdas*, namely, *Ayush, Gāndharva, Dhanush*, and *St'hapatya*; the first of which, or *Ayurvēda*, was delivered to mankind by BRAHMĀ, INDRA, DHANWANTARI, and *five* other Deities; and comprizes the theory of Disorders and Medicines, with the practical methods of curing Diseases. The second, or Musick, was invented and explained by BHARATA: it is chiefly useful in raising the mind by devotion to the felicity of the Divine nature. The third *Upavēda* was composed by VISWAMITRA on the fabrication and use of arms and implements handled in war by the tribe of *Cshatriyas*. VIŚWACARMAN revealed the fourth in various treatises on *sixty-four* Mechanical Arts, for the improvement of such as exercise them.

[1] [Jones translated and virtually tripled the length of this originally Sanskrit paper on the Vedas by Goverdhana Kaul, by adding critical discussion of the variety and richness of Sanskrit literature. Securing such a description of the sacred literature was a remarkable feat in itself. In Jones's attached commentary he expressed the nationalistic hope that the British would pioneer in giving the West an accurate knowledge of Indian religion and literature. The essay appeared in *Asiatick Researches* 1 (1788).]

[10] [Jones's version of this lyrical drama, which was its first translation, concluded this essay. The translation is included in the Poetry Section earlier in this book.]

Six *Angas*, or *Bodies* of Learning, are also derived from the same source: their names are *Sicshà, Calpa, Vyácarana, Ch'handas, Jyōtish,* and *Niructi.* The *first* was written by PĀNINI, an inspired Saint, on the *pronunciation* of vocal sounds; the *second* contains a detail of religious acts and ceremonies from the first to the last; and from the branches of these works a variety of rules have been framed by ĀSWALĀYANA, and others: the *third,* or the Grammar, entitled *Pān'inīya,* consisting of *eight* lectures or chapters (*Vriddhiradaij,* and so forth), was the production of three *Rishis,* or holy men, and teaches the proper discriminations of words in construction; but other less abstruse Grammars, compiled merely for popular use, are not considered as *Angas:* the *fourth,* or *Prosody,* was taught by a *Muni,* named PINGALA, and treats of charms and incantations in verses aptly framed and variously measured; such as the *Gāyatri,* and a thousand others. *Atronomy* is the *fifth* of the *Vēdāngas,* as it was delivered by SŪRYA, and other divine persons: it is necessary in calculations of time. The *sixth,* or *Niructi,* was composed by YĀSCA (so is the manuscript; but, perhaps, it should be VYĀSA) on the signification of difficult words and phrases in the *Vēdas.*

Lastly, there are four *Upāngas,* called *Purāna, Nyāya, Mīmānsà,* and *Dherma śāstra.* Eighteen *Purānas,* that of BRAHMA, and the rest, were composed by VYĀSA for the instruction and entertainment of mankind in general. *Nyāya* is derived from the root *ni,* to *acquire* or *apprehend;* and, in this sense, the books on *apprehension, reasoning,* and *judgement,* are called *Nyāya:* the principal of these are the work of GAUTAMA in *five* chapters, and that of CANĀDA in *ten;* both teaching the meaning of sacred texts, the difference between just and unjust, right and wrong, and the principles of knowledge, all arranged under *twenty-three* heads. *Mīmānsà* is also *two-fold;* both showing what acts are pure or impure, what objects are to be desired or avoided, and by what means the soul may ascend to the First Principle: the *former,* or *Carma Mīmānsà,* comprized in *twelve* chapters, was written by JAIMINI, and discusses questions of moral Duties and Law; next follows the *Upāsanā Cānda* in four lectures (*Sancarshana* and the rest), containing a survey of Religious Duties; to which part belong the rules of SĀNDILYA, and others, on devotion and duty to GOD. Such are the contents of the *Pūrva,* or *former, Mīmānsà.* The *Uttara,* or *latter,* abounding in questions on the Divine Nature and other sublime speculations, was composed by VYĀSA, in *four* chapters and *sixteen* sections: it may be considered as the brain and spring of all the *Angas;* it exposes the heretical opinions of RĀMĀNUJA MĀDHWA,

VALLABHA, and other Sophists; and, in a manner suited to the comprehension of adepts, it treats on the true nature of GANĒSA, BHĀSCARA, or the Sun, NĪLACANTA, LAC'SHMĪ, and other *forms* of One Divine Being. A similar work was written by ŚRI ŚANCARA, demonstrating the Supreme Power, Goodness, and Eternity of GOD.

The Body of *Law,* called *Smriti,* consists of *eighteen* books, each divided under three general heads, the duties of *religion,* the administration of *justice,* and the punishment of *expiation* of crimes: they were delivered, for the instruction of the human species, by MENU, and other sacred personages.

As to *Ethicks,* the *Vēdas* contain all that relates to the duties of Kings; the *Purānas,* what belongs to the relation of husband and wife; and the duties of friendship and society (which complete the triple division) are taught succinctly in both: this double division of *Angas* and *Upāngas* may be considered as denoting the double benefit arising from them in *theory* and *practice.*

The *Bhārata* [*Mahābhārata*] and *Rāmāyana,* which are both Epick Poems, comprize the most valuable part of ancient History.

For the information of the lower classes in religious knowledge, the *Pāsūpata,* the *Pancharātra,* and other works, fit for nightly meditation, were composed by ŚIVA, and others in an hundred and ninety-two parts on different subjects.

What follow are not really divine, but contain infinite contradictions. *Sānc'hya* is twofold, that with ĪSWARA and that without ĪSWARA: the *former* is entitled in one chapter of four sections, and is useful in removing doubts by pious comtemplation; the *second,* or *Cāpila,* is in six chapters on the production of all things by the union of PRACRITI, or *Nature,* and PURUSHA, or the *First Male:* it comprizes also, in eight parts, rules for devotion, thoughts on the invisible power, and other topicks. Both these works contain a studied and accurate *enumeration* of natural bodies and their principles; whence this philosophy is named *Sanc'hya.* Others hold, that it was so called from its *reckoning three sorts of pain.*

The *Mīmānsà,* therefore, is in *two* parts; the *Nyāya,* in *two;* and the *Sānc'hya* in *two;* and these *six* Schools comprehend all the doctrine of the Theists.

Last of all appears a work written by BUDDHA; and there are also *six* Atheistical systems of Philosophy, entitled *Yōgāchāra, Saudhānta, Vaibhāshica, Mādhyamica, Digambara,* and *Chārvāc;* all full of indeterminate phrases, errors in sense, confusion between distinct qualities, incomprehensible notions, opinions not duly weighed, tenets destructive of natural equality, containing a jumble of Atheism and Ethicks; distributed, like our Orthododx books, into a number of sections, which omit what ought to be expressed, and express what ought to be omitted; abounding in false propositions, idle propositions, impertinent propositions: some assert, that the heterodox Schools have no *Upāngas*; others, that they have six *Angas*, and as many *Sāngas*, or *Bodies* and other *Appendices*.

Such is the analysis of universal knowledge, *Practical* and *Speculative*.

THE COMMENTARY

This first chapter of a rare *Sanscrit* Book, entitled *Vidyādersa*, or a *View of Learning*, is written in so close and concise a style, that some parts of its are very obscure, and the whole requires an explanation. From the beginning of it we learn, that the *Vēdas* are considered by the *Hindus* as the fountain of all knowledge human and divine; whence the verses of them are said in the *Gītā* to be *leaves* of that holy tree, to which the Almighty Himself is compared:

ūrdhwa mūlam adhah śāc'ham aś watt'ham
prāhuravyayam
ch'handānsi yasya pernāni yastam vēda sa vēdavit.

'The wise have called the Incorruptible One as *Aśwatt'ha* with its roots above and its branches below; the leaves of which are the sacred measures: he, who knows this tree, knows the *Vēdas*'.

All the *Pandits* insist, that *Aśwatt'ha* means the *Pippala*, or *Religious Fig-tree* with heart-shaped pointed and tremulous leaves; but the comparison of heavenly knowledge, descending and taking root on earth, to the *Vat'a* or great *Indian* Fig-tree, which has most conspicuously its roots on high, or at least has radicating branches, would have been far more exact and striking.

The *Vēdas* consist of three *Cānḍas* or *General Heads*: namely, *Carma, Jnyāna, Upāsana*, or *Works, Faith*, and *Worship*; to the first of which the Author of the *Vidyādersa* wisely gives the preference, as MENU himself prefers *universal benevolence* to the *ceremonies* of religion:

Japyēnaiva tu sansiddhyèdbrāhmanō nātra
sansayah:
Curyādanyatravā curyānmaitrō brāhmana uchyatè.

that is: 'By silent adoration undoubtedly a *Brāhman* attains holiness; but every *benevolent man*, whether he perform or omit that ceremony, is justly styled a *Brāhman*'. This triple division of the *Vēdas* may seem at first to throw light on a very obscure line in the*Gītā*:

Traigunyavishayah vēdà nistraigunya bhavārjuna

or, 'The *Vēdas* are attended with *three* qualities: be not thou a man of *three* qualities, O ARJUNA'.[2]

But several *Pandits* are of opinion, that the phrase must relate to the three *gunas, qualities* of the mind, that of *excellence*, that of *passion*, and that of *darkness*; from the last of which a Hero should be wholly exempt, though examples of it occur in the *Vēdas*, where animals are ordered to be *sacrificed*, and where horrid incantations are inserted for the *destruction* of enemies.

It is extremely singular, as Mr. WILKINS has already observed, that, notwithstanding the fable of BRAHMĀ's *four* mouths, each of which uttered of *Vēda*, yet most ancient writers mention only *three Vēdas*, in order as they occur in the compound word *Rigyajuhsāma*; whence it is inferred, that the *At'harvan* was written or collected after the three first; and the two following arguments, which are entirely new, will strongly confirm this inference. In the eleventh book of MENU, a work ascribed to the *first* age of mankind, and certainly of high antiquity, the *At'harvan* is mentioned by name, and styled the *Vēda* of *Vēdas*; a phrase, which countenances the notion of DĀRĀ SHECŪH, who asserts, in the preface to his *Upanishat*, that 'the *three* first *Vēdas* are named separately, because the *At'harvan* is a corollary from them all, and contains the quintessence of them'. But this verse of MENU, which occurs in a modern copy of the work brought from *Bānāras*, and which would support the antiquity and excellence of the *fourth Vēda*, is entirely omitted in the best copies, and particularly in a very fine one written at *Gayā*, where it was accurately collated by a learned *Brāhman*; so that, as MENU himself in other places names only three *Vēdas*, we must believe this line to be an interpolation by some admirer of the *At'harvan*; and such an artifice overthrows the very doctrine, which it was intended to sustain.

The next argument is yet stronger, since it arises from *internal* evidence; and of this we are now enabled to judge by the noble zeal of Colonel POLIER in collecting *Indian* curiosities; which has been so judiciously applied and so happily exerted, that he now possesses a complete copy of the *four Vēdas* in eleven large volumes.[3]

On a cursory inspection of those books it appears, that even a learner of *Sanscrit* may read a considerable part

---

[2] [Passion, goodness, and ignorance, in Chapter 4 of the *Bhagavad-Gita*.]

[3] [Anthony Louis Henry Polier (1741-95), a Swiss engineer in the British Army, lent Jones this copy for the preparation of Jones's commentary and other research.]

of the *At'harvavēda* without a dictionary; but that the style of the other three is so obsolete, as to seem almost a different dialect: when we are informed, therefore, that few *Brāhmans* at *Bānāras* can understand any part of the *Vēdas*, we must presume, that none are meant, but the *Rich, Yajush,* and *Sāman,* with an exception of the *At'harvan,* the language of which is comparatively modern; as the learned will perceive from the following specimen:

Yatra brahmavidò yānti dicshayà tapasà saha ag-nirmāntatra nayatwagnirmēdhān dedhātumē. agnayē swāhà. vāyurmān tatra nayatu vāyuh prānān dedhātu mè, vāyuwè swāhà. suryò mān tatra nayatu chacshuh suryò dedhātu me, sùryāya swāhà; chandrò mān tatra nayatu manaschandrò dedhātu mē, chandrāya swāhà. sōmò mān tatra nayatu payah sōmò dedhātu mē, sōmāya swāhà. Indrò mān tatra nayatu balamindrò dedhātu mē, indrāya swāhà. āp ò mān tatra nayatwāmritammōpatishtatu, adbhyah swāhà. yatra brahmavidò yānti dicshayà tapasà saha, brāhmà mān tatra nayatu brahma brahmà dedhātu mē brahmanè swāhà.

that is, 'Where they, who know the Great One, go, through holy rites and through piety, thither may *fire* raise me! May fire receive my sacrifices! Mysterious praise to fire! May *air* waft me thither! May air increase my spirits! Mysterious praise to air! May the *Sun* draw me thither! May the sun enlighten my eye! Mysterious praise to the sun! May the Moon bear me thither! May the moon receive my mind! Mysterious praise to the moon! May the plant *Sōma* lead me thither! May *Sōma* bestow on me its hallowed milk! Mysterious praise to *Sōma*! May INDRA, or the *firmament,* carry me thither! May INDRA give me strength! Mysterious praise to INDRA! May *water* bear me thither! May water bring me the stream of immortality! Mysterious praise to the waters! Where they, who know the Great One, go, through holy rites and through piety, tither may BRAHMĀ, conduct me! May BRAHMĀ lead me to the Great One! Mysterious praise to BRAHMĀ!'

Several other passages might have been cited from the first book of the *At'harvan* particularly a tremendous *incantation* with consecrated *grass,* called *Darbha,* and a sublime Hymn to *Cāla,* or time; but a single passage will suffice to show the style and language of this extraordinary work. It would not be so easy to produce extract from the other *Vēdas*: indeed, in a book, entitled *Sivavēdānta,* written in *Sanscrit,* but in *Cāshmirian* letters, a stanza from the *Yajurvēda* is introduced; which

deserves for its sublimity to be quoted here; though the regular cadence of the verses, and the polished elegance of the language, cannot but induce a suspicion, that it is a more modern paraphrase of some text in the ancient Scripture:

natatra sūryò bhāti nach chandra tāracau, nēmā vidyutō bhānti cuta ēva vahnih: tamēva bhāntam anubhāti servam, tasya bhāsā servamidam vibhāti.

that is, 'There the sun shines not, nor the moon and stars: these lightnings flash not *in that place*; how should even fire blaze *there*? GOD irradiates all this bright substance; and by its effulgence the universe is enlightened'.

After all, the books on divine *Knowledge,* called *Vēda,* or what is *known,* and *Sruti,* or what has *heard,* from revelation, are still supposed to be very numerous; and the *four* here mentioned are thought to have been selected, as containing all the information necessary for man. MOHSANI FÁNÍ, the very candid and ingenious author of the *Dabistàn,* describes in his first chapter a race of old *Persian* sages, who appear from the whole of his account to have been *Hindus*; and we cannot doubt, that the book of MAHĀBĀD, or MENU, which was written, he *says,* in a *celestial dialect,* means the *Vēda*; so that, as ZERĀTUSHT was only a reformer, we find in *India* the true source of the ancient *Persian* religion. To this head belong the numerous *Tantra, Mantra, Agama,* and *Nigama, Sāstras,* which consist of *incantations* and other texts of the *Vēdas,* with remarks on the occasions, on which they may be successfully applied. It must not be omitted, that the *Commentaries* on the *Hindu* Scriptures, among which that of VASISHTHA seems to be reputed the most excellent, are innumerable; but, while we have access to the fountains, we need not waste our time in tracing the rivulets.

From the *Vēdas* are immediately deduced the practical arts of *Chirurgery* and *Medicine, Musick* and *Dancing, Archery,* which comprizes the whole art of war, and *Architecture,* under which the system of *Mechanical* arts is included. According to the *Pandits,* who instructed ABU'LFAZL, each of the *four* Scriptures gave rise to one of the *Upavēdas,* or *Sub-scriptures,* in the order in which they have been mentioned; but this exactness of analogy seems to savour of refinement.

Infinite advantage may be derived by *Europeans* from the various *Medical* books in *Sanscrit,* which contain the names and descriptions of *Indian* plants and minerals, with their uses, discovered by experience, in curing disorders: there is a vast collection of them from the

Cheraca, which is considered as a work of SIVA, to the *Rōganirūpana* and the *Nidāna*, which are comparatively modern. A number of books, in prose and verse, have been written on *Musick*, with specimens of *Hindu* airs in a very elegant notation; but the *Silpa śastra*, or Body of Treatises on *Mechanical arts*, is believed to be lost.

Next in order to these are the six *Vēdāngas*, three of which belong to *Grammar*; one relates to religious ceremonies; a fifth to the whole compass of Mathematicks, in which the author of *Līlāwatī* was esteemed the most skilful man of his time; and the *sixth*, to the explanation of obscure words or phrases in the *Vēdas*. The grammatical work of PĀNINI, a writer supposed to have been inspired, is entitled *Siddhānta Caumudi*, and is so abstruse, as to require the lucubrations of many years, before it can be perfectly understood. When *Cāśīnāt'ha Serman*, who attended Mr. WILKINS, was asked what he thought of the *Pānināya*, he answered very expressively, that 'it was a forest;' but, since Grammar is only an instrument, not the end, of true knowledge, there can be little occasion to travel over so rough and gloomy a path; which contains, however, probably some acute speculations in *Metaphysicks*.[4] The *Sanscrit* Prosody is easy and beautiful: the learned will find in it almost all the measures of the *Greeks*; and it is remarkable, that the language of the Brahmans runs very naturally into *Sapphicks, Alcaicks,* and *Iambicks*. Astronomical works in this language are exceedingly numerous: seventy-nine of them are specified in one list; and, if they contain the names of the principal stars visible in *India*, with observations on their positions in different ages, what discoveries may be made in Science, and what certainty attained in ancient Chronology?

Subordinate to these *Angas* (though the reason of the arrangement is not obvious) are the series of *Sacred Poems*, the Body of *Law*, and the *six* Philosophical *sāstras*; which the author of our text reduces to *two*, each consisting of *two* parts, and rejects a *third*, in *two* parts also, as not perfectly *orthodox*, that is, not strictly conformable to his own principles.

The first *Indian* Poet was VĀLMĪCI, author of the *Rāmāyana*, a complete Epick Poem on one continued, interesting, and heroick, action; and the next in celebrity, if it be not superior in reputation for holiness, was the *Mahābhārata* of VYĀSA: to him are ascribed the sacred Puranas, which are called, for their excellence, *the Eighteen*, and which have the following titles: BRAHME, or the *Great One*, PEDMA, or the *Lotos*, BRĀHMĀND'A, or the *Mundane Egg*, and AGNI, or *Fire* (these *four* relate to the *Creation*), VISHNU, or the *Pervader*, GARUDA, or his *Eagle*, the Transformations of BRAHMĀ, SIVA, LINGA, NĀREDA, son of BRAHMĀ, SCANDA son of SIVA, MARCANDĒYA, or the Immortal Man, and BHAWISHYA, or the *Prediction* of *Futurity* (these nine belong to the *attributes* and *powers* of the Deity), and *four* others, MATSYA, VARAHA, CŪRMA, VĀMENA, or as many incarnations of the Great One in his character of *Preserver*: all containing ancient traditions embellished by poetry or disguised by fable: the *eighteenth* is the BHĀGAWATA, or Life of CRISHNA, with which the same Poet is by some imagined to have crowned the whole series; though others, with more reason, assign them different composers.

The system of *Hindu* Law, besides the fine work, called MENUSMRITI, or 'what is *remembered* from MENU', that of YĀJNYAWALCYA, and those of *sixteen* other *Munis*, with *Commentaries* on them all, consists of many tracts in high estimation, among which those current in *Bengal* are, an excellent treatise on *Inheritances* by JĪMŪTA VĀHANA, and a complete *Digest*, in *twenty-seven* volumes, compiled a few centuries ago by RAGHUNANDAN, the TRIBONIAN of *India*, whose work is the grand repository of all that can be known on a subject so curious in itself, and so interesting to the *British* Government.[5]

Of the Philosophical Schools it will be sufficient here to remark, that the first *Nyāya* seems analogous to the *Peripatetick*, the *second*, sometimes called *Vaiśēshica*, to the *Ionick*, the two *Mimānsas*, of which the *second* is often distinguished by the name of *Vēdānta*, to the *Platonick*, the first *Sānch'ya* to the *Italick*, and the second, or *Pātanjala*, to the *Stoick*, Philosophy; so that GAUTAMA corresponds with ARISTOTLE; CANĀDA, with THALES; JAIMINI with SOCRATES; VYĀSA with PLATO; CAPILA with PYTHAGORAS; and PATANJALI with ZENO: but an accurate comparison between the *Grecian* and *Indian* Schools would require a considerable volume. The original works of those Philosophers are very succinct; but, like all the other *Sāstras*, they are explained, or obscured, by the *Upadersana* or *Commentaries* without end: one of the finest compositions on the Philosophy of the *Vēdānta* is entitled *Yōga Vāsīsht'ha*, and contains the instructions of the great VASISHTHA to his pupil, RĀMA, king of Ayōdhyà.

---

[4] [This early conception of applied linguistics dominated Jones's view of language, the findings from which should be made useful to humanity. Later he worked extensively with the formulations of Pāṇini, the great Sanskrit grammarian who was so to influence modern Western linguistics.]

[5] [An English translation of Raghu Nandana Bhaṭṭacharya's *Dayatattwa* was finally published in Madras in 1911.]

It results from this analysis of *Hindu* Literature, that the *Vēda, Upavēda, Vēdānga, Purāna, Dherma,* and *Dersana* are the *Six* great *Sāstras,* in which all knowledge, divine and human, is supposed to be comprehended; and here we must not forget, that the word *Sāstra,* derived from a root signifying *to ordain,* means generally an *Ordinance,* and particularly a *Sacred Ordinance* delivered by inspiration: properly, therefore, this word is applied only to *sacred literature,* of which the text exhibits an accurate sketch.

The *Sūdras,* of *fourth* class of *Hindus,* are not permitted to study the *six* proper *Sāstras* before-enumerated; but an ample field remains for them in the study of *profane literature,* comprized in a multitude of *popular* books, which correspond with the several *Sāstras,* and abound with beauties of every kind. All the tracts on *Medicine* must, indeed, be studied by the *Vaidyas,* or those, who are born Physicians; and they have often more learning. with far less pride, than any of the *Brāhmanas:* they are usually Poets, Grammarians, Rhetoricians, Moralists; and may be esteemed in general the most virtuous and amiable of the *Hindus.* Instead of the *Vēdas* they study the *Rājanīti,* or *Instruction of Princes,* and instead of *Law,* the *Nītisāstra,* or general system of *Ethicks:* their *Sahitia,* or *Cāvya Sāstra,* consists of innumerable poems, written chiefly by the *Medical* tribe, and supplying the place of the *Purānas,* since they contain all the stories of the *Rāmāyana, Bhārata,* and *Bhāgawata:* they have access to many treatises of *Alāncara,* or Rhetorick, with a variety of works in modulated prose; to *Upāc'hyana,* or Civil History, called also *Rājatarangini;* to the *Nātaca,* which answers to the *Gāndharvavēda,* consisting of regular *Dramatick* pieces in *Sanscrit* and *Prācrit:* besides which they commonly get by heart some entire Dictionary and Grammar. The best Lexicon or Vocabulary was composed in verse, for the assistance of the memory, by the illustrious AMARASINHA; but there are *seventeen* others in great repute: the best Grammar is the *Mugdhabōdha,* or the *Beauty of Knowledge,* written by *Gōswami,* named VŌPADĒVA, and comprehending, in two hundred short pages, all that a learner of the language can have occasion to know. To the *Cōshas,* or dictionaries, are usually annexed very ample *Tīcas,* or *Etymological* Commentaries.

We need say no more of the heterodox writings, than that those on the religion and philosophy of BUDDHA seem to be connected with some of the most curious parts of *Asiatick* History, and contain, perhaps, all that could be found in the *Pāli,* or *sacred language* of the Eastern *Indian* peninsula. It is asserted in *Bengal,* that AMARA-

SINHA himself was a *Bauddha;* but he seems to have been a theist of tolerant principles, and, like ABUL'FAZL, desirous of reconciling the different religions of *India.*

Wherever we direct our attention to *Hindu* Literature, the notion of *infinity* presents itself; and the longest life would not be sufficient for the perusal of near five hundred thousand stanzas in the *Purānas,* with a million more perhaps in the other works before mentioned: we may, however, select the best from each *Sāstra,* and gather the fruits of science, without loading ourselves with the leaves and branches; while we have the pleasure to find, that the learned *Hindus,* encouraged by the mildness of our government and manners, are at least as eager to communicate their knowledge of all kinds, as we can be to receive it. Since *Europeans* are indebted to the *Dutch* for almost all they know of *Arabick,* and to the *French* for all they know of *Chinese,* let them now receive from our nation the first accurate knowledge of *Sanscrit,* and of the valuable works composed in it; but, if they wish to form a correct idea of *Indian* religion and literature, let them begin with forgetting all that has been written on the subject, by ancients or moderns, before the publications of the *Gītà.*

Jones's English Advertisement to His Printing of the *Ṛitusaṃhāra*[1]

This book is the first ever printed in *Sanscrit*: and it is by the press alone, that the ancient literature of *India* can long be preserved: a learner of that most interesting language who had carefully perused one of the popular grammars, could hardly begin his course of study with an easier or more elegant work, than the *Ṛitusaṃhāra*, or *Assemblage of Seasons*. Every line composed by CĀLIDĀS is exquisitely polished; and every couplet in the poem exhibits an *Indian* landscape, always beautiful, sometimes highly coloured, but never beyond nature: four copies of it have been diligently collated; and where they differed, the clearest and most natural reading has constantly had the preference.

W.J.

II. LANGUAGE AND LINGUISTICS

Because Jones was such a varied, pioneering scholar of language and linguistics, it is difficult to select representative works that are of manageable length for inclusion in a one-volume collection. By far his best-known and perhaps best philosophical work is *A Grammar of the Persian Language*, which is much too long to be reprinted here, not to mention that it was reprinted in a convenient facsimile edition by the Scolar Press in 1969. Arbitrarily, his richly humanistic Preface has been selected. His comparatively known proposal to make a much-needed revision of Meninski's monumental *Thesaurus* is included to show his attitude toward lexical and linguistic scholarship, as well as his methodology and goals in his work with Persian. The phonological importance of the Jonesian System prompted the inclusion of his orthographic dissertation. The philological basis of his programmatic, first essay to the charter members of his Asiatic Society of Bengal is a central reason for the inclusion of that essay, where language is innovationally shown to be often central to research. By far the most frequently quoted material by Jones in linguistics books is his short, lucid formulation about language derivation, which is buried in a fascinating essay about India, the entirely of which is reprinted here so as to reveal both the context and his generalizations about India.

Some eventful, philological letters to fellow-scholars like Thomas Maurice and Charles Wilkins and letters to Jones by Lord Monboddo and Edmund Burke are not included in this collection, for they can be perused in Cannon's *Letters*. This section, thus, contains two extracts, a proposal, and two essays to exemplify Jones's major contributions to the development of modern linguistics.

---

[1] [Fearing that the long-preserved Sanskrit manuscripts might still be lost, Jones wanted to make available to the Indian people an inexpensive first publication of a complete literary work from their cultural heritage, while giving Western scholars a convenient text for study. Thus he anticipated the College of Fort William and its major vernacular publications. The only English in the 144-quatrain book, which Jones published at his own expense in Calcutta in 1792, is this Advertisement. Herman Krevenborg did a facsimile edition of Jones's text (Hannover, 1924.)]

A Grammar of the Persian Language[1]
[Extract]

The Preface

The Persian language is rich, melodious, and elegant; it has been spoken for many ages by the greatest princes in the politest courts of Asia; and a number of admirable works have been written in it by historians, philosophers, and poets, who found it capable of expressing with equal advantage the most beautiful and the most elevated sentiments.

It must seem strange, therefore, that the study of this language should be so little cultivated at a time when a taste for general and diffusive learning seems universally to prevail; and that the fine productions of a celebrated nation should remain in manuscript upon the shelves of our publick libraries, without a single admirer who might open their treasures to his countrymen, and display their beauties to the light; but if we consider the subject with a proper attention, we shall discover a variety of causes which have concurred to obstruct the progress of Eastern literature.

Some men never heard of the Asiatick writings, and others will not be convinced that there is any thing valuable in them; some pretend to be busy, and others are really idle; some detest the Persians, because they believe in Mahomed, and others despise their language, because they do not understand it: we all love to excuse, or to conceal, our ignorance, and are seldom willing to allow

any excellence beyond the limits of our own attainments: like the savages, who thought that the sun rose and set for them alone, and could not imagine that the waves, which surrounded their island, left coral and pearls upon any other shore.

Another obvious reason for the neglect of the Persian language is the great scarcity of books, which are necessary to be read before it can be perfectly learned: the greater part of them are preserved in the different museums and libraries of Europe, where they are shewn more as objects of curiosity than as sources of information; and are admired, like the characters on a Chinese screen, more for their gay colours than for their meaning.

Thus, while the excellent writings of Greece and Rome are studied by every man of a liberal education, and diffuse a general refinement through our part of the world, the works of the Persians, a nation equally distinguished in ancient history, are either wholly unknown to us, or considered as entirely destitute of taste and invention.

But if this branch of literature has met with so many obstructions from the ignorant, it has, certainly, been checked in its progress by the learned themselves; most of whom have confined their study to the minute researches of verbal criticism; like men who discover a precious mine, but instead of searching for the rich ore, or for gems, amuse themselves with collecting smooth pebbles and pieces of crystal. Others mistook reading for learning, which ought to be carefully distinguished by every man of sense, and were satisfied with running over a great number of manuscripts in a superficial manner, without condescending to be stopped by their difficulty, or to dwell upon their beauty and elegance. The rest have left nothing more behind them than grammars and dictionaries; and though, they deserve the praises due to unwearied pains and industry, yet they would, perhaps, have gained a more shining reputation, if they had contributed to beautify and enlighten the vast temple of learning, instead of spending their lives in adorning only its porticos and avenues.

There is nothing which has tended more to bring polite letters into discredit, than the total insensibility of commentators and criticks to the beauties of the authors whom they profess to illustrate: few of them seem to have received the smallest pleasure from the most elegant compositions, unless they found some mistake of a transcriber to be corrected, or some established reading to be changed, some obscure expression to the explained, or some clear passage to be made obscure by their notes.

It is a circumstance equally unfortunate, than men of

---

[1] [This immediately famous partly pedagogical grammar (London, 1771), is a good eighteenth-century analysis of written Persian, chiefly poetry dating back to Middle Persian. Intended to help East India Company officials learn Persian, it became a model for Oriental grammars like those by Francis Gladwin, George Hadley, Robert Jones, and John Richardson. It went through nine editions by 1828 and was translated into German (1773) and French (1772), with a second French edition in 1845. See Cannon, 'Sir William Jones's Persian Linguistics', *Journal of the American Oriental Society*, 78 (October 1958): 263-73. Jones's Preface, in which he persuasively paints the beauties and values of Persian literature and invites other scholars to join him in Persian studies, is extracted here.]

the most refined taste and the brightest parts are apt to look upon a close application to the study of languages as inconsistent with their spirit and genius: so that the state of letters seems to be divided into two classes, men of learning who have no taste, and men of taste who have no learning.

M. de Voltaire, who excels all writers of his age and country in the elegance of his style, and the wonderful variety of his talents, acknowledges the beauty of the Persian images and sentiments, and has versified a very fine passage from Sadi, whom he compares to Petrarch: if that extraordinary man had added a knowledge of the Asiatick languages to his other acquisitions, we should by this time have seen the poems and histories of Persia in an European dress, and any other recommendation of them would have been unnecessary.

But there is yet another cause which has operated more strongly than any before mentioned towards preventing the rise of oriental literature; I mean the small encouragement which the princes and nobles of Europe have given to men of letters. It is an indisputable truth, that learning will always flourish most where the amplest rewards are proposed to the industry of the learned; and that the most shining periods in the annals of literature are the reigns of wise and liberal princes, who know that fine writers are the oracles of the world, from whose testimony every king, statesman, and hero must expect the censure or approbation of posterity. In the old states of Greece the highest honours were given to poets, philosophers, and orators; and a single city (as an eminent writer[2] observes) in the memory of one man, produced more numerous and splendid monuments of human genius than most other nations have afforded in a course of ages.

The liberality of the Ptolemies in Egypt drew a number of learned men and poets to their court, whose works remain to the present age the models of taste and elegance; and the writers, whom Augustus protected, brought their composition to a degree of perfection, which the language of mortals cannot surpass. Whilst all the nations of Europe were covered with the deepest shade of ignorance, the Califs in Asia encouraged the Mahomedans to improve their talents, and cultivate the fine arts; and even the Turkish Sultan, who drove the Greeks from Constantinople, was a patron of literary merit, and was himself an elegant poet. The illustrious family of Medici invited to Florence the learned men whom the Turks had driven from their country, and a

general light succeeded the gloom which ignorance and superstition had spread through the western world.[3] But that light has not continued to shine with equal splendour; and though some slight efforts have been made to restore it, yet it seems to have been gradually decaying for the last century: it grows very faint in Italy; it seems wholly extinguished in France; and whatever sparks of it remain in other countries are confined to the closets of humble and modest men, and are not general enough to have their proper influence.

The nobles of our days consider learning as a subordinate acquisition, which would not be consistent with the dignity of their fortunes, and should be left to those who toil in a lower sphere of life: but they do not reflect on the many advantages which the study of polite letters would give, peculiarly to persons of eminent rank and high employments; who, instead of relieving their fatigues by a series of unmanly pleasures, or useless diversions, might spend their leisure in improving their knowledge, and in conversing with the great statesmen, orators, and philosophers of antiquity.

If learning in general has met with so little encouragement, still less can be expected for that branch of it, which lies so far removed from the common path, and which the greater part of mankind have hitherto considered as incapable of yielding either entertainment or instruction: if pains and want be the lot of a scholar, the life of an orientalist must certainly be attended with peculiar hardships. Gentius, who published a beautiful Persian work called *The Bed of Roses*, with an useful but inelegant translation, lived obscurely in Holland, and died in misery. Hyde, who might have contributed greatly towards the progress of eastern learning, formed a number of expensive projects with that view, but had not the support and assistance which they deserved and required. The labours of Meninski immortalized and ruined him: his dictionary of the Asiatick languages is, perhaps, the most laborious compilation that was ever undertaken by any single man; but he complains in his preface that his patrimony was exhausted by the great expense of employing and supporting a number of writers and printers, and of raising a new press for the oriental characters. M. d'Herbelot, indeed, received the most splendid reward of his industry: he was invited to Italy by Ferdinand II, duke of Tuscany, who entertained him with that striking munificence which always distinguished the race of the Medici: after the death of Ferdinand, the illustrious Colbert recalled

---

[2] Ascham. [Athens, in 'The First Booke for the Youth', *The Scholemaster* (London, 1570), 17.]

[3] [Here Jones anticipated the Oriental Renaissance in the West which he caused.]

him to Paris, where he enjoyed the fruits of his labour, and spent the remainder of his days in an honourable and easy retirement. But this is a rare example: the other princes of Europe have not imitated the duke of Tuscany; and Christian VII was reserved to be the protector of the eastern muses in the present age.

Since the literature of Asia was so much neglected, and the causes of that neglect were so various, we could not have expected that any slight power would rouze the nations of Europe from their inattention to it; and they would, perhaps, have persisted in despising it, if they had not been animated by the most powerful incentive that can influence the mind of man: interest was the magick wand which brought them all within one circle; interest was the charm which gave the languages of the East a real and solid importance. By one of those revolutions, which no human prudence could have foreseen, the Persian language found its way into India; that rich and celebrated empire, which, by the flourishing state of our commerce, has been the source of incredible wealth to the merchants of Europe. A variety of causes, which need not be mentioned here, gave the English nation a most extensive power in that kingdom: our India company began to take under their protection the princes of the country, by whose protection they gained their first settlement; a number of important affairs were to be transacted in peace and war between nations equally jealous of one another, who did not the common instrument of conveying their sentiments; the servants of the company received letters which they could not read, and were ambitious of gaining titles of which they could not comprehend the meaning; it was found highly dangerous to employ the natives as interpreters, upon whose fidelity they could not depend; and it was at last discovered, that they must apply themselves to the study of the Persian language, in which all the letters from the Indian princes were written. A few men of parts and taste, who resided in Bengal, have since amused themselves with the literature of the East, and have spent their leisure in reading the poems and histories of Persia; but they found a reason in every page to regret their ignorance of the Arabick language, without which their knowledge must be very circumscribed and imperfect. The languages of Asia will now, perhaps, be studied with uncommon ardour; they are known to be useful, and will soon be found instructive and entertaining; the valuable manuscripts that enrich our publick libraries will be in a few years elegantly printed; the manners and sentiments of the eastern nations will be perfectly known; and the limits of our knowledge will be no less extended than the bounds of our empire.

It was with a view to facilitate, the progress of this branch of literature, that I reduced to order the following instructions for the Persian language, which I had collected several years ago; but I would not present my grammar to the publick till I had considerably enlarged and improved it: I have, therefore, endeavoured to lay down the clearest and most accurate rules, which I have illustrated by select examples from the most elegant writers; I have carefully compared my work with every composition of the same nature that has fallen into my hands; and though on so general a subject I must have made several observations which are common to all, yet I flatter myself that my own remarks, the disposition of the whole book, and the passages quoted in it, will sufficiently distinguish it as an original production. Though I am not conscious that there are any essential mistakes or omissions in it, yet I am sensible that it falls very short of perfection, which seems to withdraw itself from the pursuit of mortals, in proportion to their endeavours of attaining it; like the talisman in the Arabian tales, which a bird carried from tree to tree as often as its pursuer approached it. But it has been my chief care to avoid all the harsh and affected terms of art which render most didactick works so tedious and unpleasant, and which only perplex the learner, without giving him any real knowledge: I have even refrained from making any enquiries into general grammar, or from entering into those subjects which have already been so elegantly discussed by the most judicious philosopher,[4] the most learned divine,[5] and the most laborious scholar of the present age.[6]

It was my first design to prefix to the grammar a history of the Persian language from the time of Xenophon to our days, and to have added a copious praxis of tales and poems extracted from the classical writers of Persia; but as those additions would have delayed the publication of the grammar, which was principally wanted, I thought it advisable to reserve them for a separate volume, which the publick may expect in the course of the ensuing winter. I have made a large collection of materials for a general history of Asia, and for an account of the geography, philosophy, and literature of the eastern nations, all which I propose to arrange in

---

[4] See Hermes. [By James Harris (London, 1751).]

[5] A short Introduction to English Grammar. [By Bishop Robert Lowth.]

[6] The grammar prefixed to the Dictionary of the English Language. [By Johnson.]

order, if my more solid and more important studies will allow me any intervals of leisure.[7]

I cannot forbear acknowledging in this place the signal marks of kindness and attention, which I have received from many learned and noble persons; but General Carnac has obliged me the most sensibly of them, by supplying me with a valuable collection of Persian manuscripts on every branch of eastern learning, from which many of the best examples in the following grammar are extracted. A very learned Professor[8] at Oxford has promoted my studies with that candour and benevolence which so eminently distinguish him; and many excellent men that are the principal ornaments of that university have conferred the highest favours on me, of which I shall ever retain a grateful sense: but I take a singular pleasure in confessing that I indebted to a foreign nobleman[9] for the little knowledge which I have happened to acquire of the Persian language; and that my zeal for the poetry and philology of the Asiaticks was owing to his conversation, and to the agreeable correspondence with which he still honours me.

Before I conclude this Preface it will be proper to add a few remarks upon the method of learning the Persian language, and upon the advantages which the learner may expect from it. When the student can read the characters with fluency, and has learned the true pronunciation of every letter from the mouth of a native, let him peruse the grammar with attention, and commit to memory the regular inflexions of the nouns and verbs: he needs not burden his mind with those that deviate from the common form, as they will be insensibly learned in a short course of reading. By this time he will find a dictionary necessary, and I hope he will believe me, when I assert from a long experience, that, whoever possesses the admirable work of Meninski, will have no occasion for any other dictionary of the Persian tongue. He may proceed by the help of this work to analyse the passages quoted in the grammar, and to examine in what manner they illustrate the rules; in the meantime he must not neglect to converse with his living instructor, and to learn from him the phrases of common discourse, and the names of visible objects, which he will soon imprint on his memory, if he will take the trouble to look for them in the dictionary: and here I must caution him against condemning a work as defec-

tive, because he cannot find in it every word which he hears; for sounds in general are caught imperfectly by the ear, and many words are spelled and pronounced very differently.

The first book that I would recommend to him is the Gulistan or *Bed of Roses*, a work which is highly esteemed in the East, and of which there are several translations in the languages of Europe: the manuscripts of this book are very common; and by comparing them with the printed edition of Gentius, he will soon learn the beautiful flowing hand used in Persia, which consists of bold strokes and flourishes, and cannot be imitated by our types. It will then be a proper time for him to read some short and easy chapter in this work, and to translate it into his native language with the utmost exactness; let him then lay aside the original, and after a proper interval let him turn the same chapter back into Persian by the assistance of the grammar and dictionary; let him afterwards compare his second translation with the original, and correct its faults according to that model. This is the exercise so often recommended by the old rhetoricians, by which a student will gradually acquire the style and manner of any author, whom he desires to imitate, and by which almost any language may be learned in six months with ease and pleasure. When he can express his sentiments in Persian with tolerable facility, I would advise him to read some elegant history or poem with an intelligent native, who will explain to him in common words the refined expressions that occur in reading, and will point out the beauties of learned allusions and local images. The most excellent book in the language is, in my opinion, the collection of tales and fables called *Anvah Soheili* by Aussein Vaéz, surnamed Cashefi, who took the celebrated work of Bidpai or Pilpay for his text, and has comprised all the wisdom of the eastern nations in fourteen beautiful chapters. At some leisure hour he may desire his Munshi or writer to transcribe a section from the Gulistan, or a fable of Cashefi, in the common broken hand used in India, which he will learn perfectly in a few days by comparing all its turns and contractions with the more regular hands of the Arabs and Persians: he must not be discouraged by the difficulty of reading the Indian letters, for the characters are in reality the same with those in which our books are printed, and are only rendered difficult by the frequent omission of the diacritical points, and the want of regularity in the position of the words: but we all know that we are often at a loss to read letters which we receive in our native tongue; and it has been proved that a man who has a perfect knowledge of any language, may, with a proper attention, decypher

---

[7] See the *History of the Persian Language*, a *Description of Asia*, and a *Short History of Persia*, published with my *Life of Nader Shah* in the year 1773.

[8] Dr. [Thomas] HUNT.

[9] Baron REVISKI.

a letter in that idiom, though it be written in characters which he has never seen before, and of which he has no alphabet.

In short, I am persuaded, that whoever will study the Persian language according to my plan, will in less than a year be able to translate and to answer any letter from an Indian prince, and to converse with the natives of India, not only with fluency, but with elegance.[10] But if he desires to distinguish himself as an eminent translator, and to understand not only the general purport of a composition, but even the graces and ornaments of it, he must necessarily learn the Arabick tongue, which is blended with the Persian in so singular a manner, that one period often contains both languages, wholly distinct from each other in expression and idiom, but perfectly united in sense and construction. This must appear strange to an European reader; but he may form some idea of this uncommon mixture, when he is told that the two Asiatick languages are not always mixed like the words of Roman and Saxon origin in this period, 'The true law is right reason, conformable to the nature of things; which calls us to duty by commanding, deters us from sin by forbidding;'[11] but as we may suppose the Latin and English to be connected in the following sentence. '*The true* lex *is* recta ratio, *conformable* naturæ, *which by commanding* vocet ad officium, *by forbidding* à fraude deterreat'.

A knowledge of these two languages will be attended with a variety of advantages to those who acquire it: the Hebrew, Chaldaick, Syriack, and Ethiopean tongues are dialects of the Arabick, and bear as near a resemblance to it as the Ionick to the Attick Greek; the jargon of Indostan, very improperly called the language of the Moors, contains so great a number of Persian words, that I was able with very little difficulty to read the fables of Pilpay which are translated into that idiom: the Turkish contains ten Arabick or Persian words for one originally Scythian, by which it has been so refined, that the modern kings of Persia were fond of speaking it in their courts: in short, there is scarce a country in Asia or Africa, from the source of the Nile to the wall of China, in which a man who understands Arabick, Persian, and Turkish, may not travel with satisfaction, or transact the most important affairs with advantage and security.

As to the literature of Asia, it will not, perhaps, be essentially useful to the greater part of mankind, who have neither leisure nor inclination to cultivate so extensive a branch of learning; but the civil and natural history of such mighty empires as India, Persia, Arabia, and Tartary, cannot fail of delighting those who love to view the great picture of the universe, or to learn by what degrees the most obscure states have risen to glory, and the most flourishing kingdoms have sunk to dacay; the philosopher will consider those works as highly valuable, by which he may trace the human mind in all its various appearances, from the rudest to the most cultivated state: and the man of taste will undoubtedly be pleased to unlock the stores of native genius, and to gather the flowers of unrestrained and luxuriant fancy.[12]

---

[10] [Jones used this naive method and initially could not be understood by Persian speakers.]

[11] See Middleton's *Life of Cicero* 3: 351.

[12] My professional studies having wholly engaged my attention, and induced me not only to abandon oriental literature, but even to efface, as far as possible, the very traces of it from my memory, I committed the conduct and revisal of this edition of my Grammar, and the composition of the Index to Mr. [John] Richardson, in whose skill I have a perfect confidence, and from whose application to the eastern languages, I have hopes that the learned world will reap no small advantage. [Jones's encouragement bore rich fruit, as Richardson published an Arabic grammar (1776) and a Persian-Arabic-English dictionary (1777).]

London, October 1, 1770

Proposals for Re-printing by Subscription[1]
A Dictionary of the Arabick, Persian, and Tur̓kish Languages,

Compiled and first Published by MENINSKI,
in Four Volumes Folio:
Revised and Corrected by
WILLIAM JONES, ESQ.
Fellow of University-College, Oxford,
Translator of the Life of Nader Chah from the
Persian,With an English Translation and Index.

It is unnecessary in these Proposals to point out minutely the particular Excellence of the Dictionary which we propose to reprint, since it has been for near a Century admired by all Europe, and its extreme Scarcity universally regretted. A Work which for so long a Period has met with general Approbation, and has been sought with such Avidity by Men of the every Rank and Profession, must, certainly, have a real and Intrinsick Merit, to which no Commendation can add, and from which no Censure can detract.

It may not however be improper to subjoin a few Arguments in Favour of this Work, and demonstrate its peculiar Usefulness in facilitating the Attainment of the Persian (a Language at present of such Importance to the Honourable East-India Company) more especially as those who are unacquainted with the Subject might possibly imagine that a Dictionary including the Arabick and Turkish, as well as the Persian, was not necessary for those who wish to acquire a Knowledge of the Persian *only*.

The Dialects of Arabia and Persia were extremely different in their Origin; but they were entirely united and blended together when the Arabs spread over Asia their Arms, their Religion, and their Learning. The conquered Nations received with Eagerness the Literature of their Conquerors. They considered it as a Mark of Erudition and of Elegance to use Arabick Words and Expressions, and to fill their Writings with Allusions to the Arabian Poem and Proverbs: Besides, from a Principle of Piety, they adopted the Language and even the Letters of their Koran. The Ottoman Conquests in Aftertimes diffused their Dialect over half the East; it became even the Language of the Court of Persia, where it is still spoken, and made its Way into many Parts of India; thus, by Degrees, the Idioms of those Tongues were almost changed, and became Dialects of one vast Language, which mutually enrich and illustrate each other; like a Number of little States, resigning their private Laws, and concurring to form one extensive Empire.

Several very learned Turks, it may also be observed, have translated and explained in their Dialect the most celebrated Poems of Persia, which would be absolutely

---

[1] [When Jones first began studying Persian in 1765, he used Franciscus A. Mesgnien Meninski's famous *Thesaurus linguarum orientalium Turcicæ, Arabicæ, Persicæ* (Vienna, 1680-87) as his lexicon. Realizing the need for an updated, expanded version that incorporated words from the fine Bodleian manuscripts, he formally approached the East India Company for assistance, which was encouraged but not aided financially. By October 1770 he was frequently advertising his proposals in the *London Chronicle* and the *Gazette*. These richly reveal contemporary views of the three languages and attempt to solicit subscribers and to convince the Company of the need to subsidize the huge project. But even the announced addition of an English translation and a comprehensive index was insufficient to attract many subscribers, though both Oxford and Cambridge lent their names (but no money) to the edition, and the Turkey Company similarly encouraged Jones.

As his books were still being published partly at his own expense, he decided to add a Portuguese translation in order to capitalize on Portugal's Indian possessions, where, as in British India, none of the languages represented would be of real advantage. Upon finally realizing in 1774 that the East India Company was not going to provide funds, Jones sadly returned the subscriptions and abandoned this major linguistic project, which would have bankrupted him, like Meninski earlier. Though he encouraged John Richardson to do an abridged edition, which became *A Dictionary, Persian, Arabic, and English* (Oxford, 1777), it was not until the Company subsidized Francis Johnson that a full-scale revision appeared in 1,420 pages (London, 1852). Jones's ill-fated project was important because it required a deepening of his knowledge of Persian, an intimate use of the Bodleian manuscripts, and close association with philologists like John Uri, who combed the Oxford manuscript dictionaries for additional Persian words. Jones's Persian knowledge was crucial to his famous comparative formulation in 1786. Newly discovered letters throw light on his unsuccessful effort to secure help from the East India Company with the Meninski revision. See Abu Taher Mojumder. 'Three New Letters by Sir William Jones', *India Office Library and Records Report 1981*: 24-35.]

unintelligible to us without such Explanation; and even their Military Terms in some of the modern Persian Histories, are borrowed from the Turks. As the proper understanding of one Word or Phrase, therefore, determines often the Sense of a whole Passage, where such Word or Phrase in the Writings of a Persian happens to be Turkish or Arabick, it is obvious that a Dictionary where these three Languages are displayed at one View can only solve the Difficulty; and that a Dictionary of the Persian alone could in such Case afford no Manner of Assistance. Exclusive also of the Advantage which the Persic Student must receive from an Acquaintance with the Arabick and the Turkish, these Languages may be of Importance to the Honourable East-India Company's Servants in their commercial or political Concerns at Mocha, Bassora, and other Places under the Turkish Dominion and the Red-Sea or Persian Gulph, the Arabick being in many of those Countries the vulgar Tongue, and Turkish the Language of the Great Officers of State.

To conclude: These three Languages, or, more properly, these three Dialects of one Language, are by no Means so difficult as we are taught to believe: A moderate Application for a few Months will enable the Learner to proceed with Ease and Pleasure, and in few Years he may read and write them with Facility; in which Study the Dictionary of Meninski will afford him all the Assistance he can require; as the best Orientalists have derived their Knowledge from it, and its Utility been so peculiarly experienced by Gentlemen in India, that they have sent Commissions to England to purchase it at any Price; whilst Dictionaries of the Persian only, have been considered of no Value, as inadequate to teach the Language.

The Oriental Words and Phrases are explained by Meninski in Five European Languages, Latin, French, Italian, German, and Polish: But for the Use of our Countrymen, to whom those Languages may not be familiar, we shall add an English Translation, and an Index of every Word.

The Types designed for this Work are entirely new, and are copied from the most elegant Arabian Characters, in which all the Books of polite Literature in the East are written; and which are intelligible to every Mahometan, as the Koran is expressed in no other: But that nothing may be omitted, we shall insert several Plates of Persian Odes, engraved from the most beautiful Manuscripts in the common flowing Hand used in India.

From Fifty to Sixty Guineas is the present Price of a Copy of Meninski: But in order to bring this great and useful Work within the Compass of general Purchase, the proposed improved Edition will be delivered to Sub-scribers at Twelve Guineas; It is hoped therefore the Subscription will be proportionably liberal; and that those who consider this Publication as an Object of Encouragement, will send their Address as early as possible.—A Copy will not be sold to Non-Subscribers under Sixteen Guineas.—This Work it is expected will be published in 1773.

A List of the Subscribers will be prefixed to the Work.

Subscriptions are received and Receipts given by W. and J. Richardson, N° 12 and 13 Salisbury-Court, and N° 76 in Fleet-Street, where may be had Proposals at large.

Six Guineas to be paid at Subscribing, and Six Guineas on the Delivery of the Book in Sheets.

Printed: *London Gazette*, 2 February 1771.

A Dissertation on the Orthography of Asiatick Words in Roman Letters [1]

[Extract]

Every man, who has occasion to compose tracts on *Asiatick* Literature, or to translate from the *Asiatick* Languages, must always find it convenient, and sometimes necessary, to express *Arabian*, *Indian*, and *Persian* words, or sentences, in the characters generally used among *Europeans*; and almost every writer in those circumstances has a method of notation peculiar to himself; but none has yet appeared in the form of a complete system; so that each original sound may be rendered invariably by one appropriated symbol, conformably to the natural order of articulation, and with a due regard to the primitive power of the *Roman* alphabet, which modern *Europe* has in general adopted. A want of attention to this object has occasioned great confusion in History and Geography. The ancient *Greeks*, who made a voluntary sacrifice of truth to the delicacy of their ears, appear to have altered by design almost all the oriental names, which they introduced into their elegant, but romantick, Histories; and even their more modern Geographers, who were too vain, perhaps, of their own language to learn any other, have so strangely disguised the

proper appellations of countries, cities, and rivers in *Asia*, that, without the guidance of the sagacious and indefatigable M. D'ANVILLE,[2] it would have been as troublesome to follow ALEXANDER through the *Panjàb* on the Ptolemaick map of AGATHODÆMON, as actually to travel over the same country in its present state of rudeness and disorder. They had an unwarrantable habit of moulding foreign names to a *Grecian* form, and giving them a resemblance to some derivative word in their own tongue: thus, they changed the *Gogra* into *Agoranis*, or *a river of the assembly*, *Uchah* into *Oxydracæ*, or *sharp-sighted*, and *Renas* into *Aornos*, or *a rock inaccessible to birds*; whence their poets, who delighted in wonders, embellished their works with new images, distinguishing regions and fortresses by properties, which existed only in imagination. If we have less liveliness of fancy than the Ancients, we have more accuracy, more love of truth, and, perhaps, more solidity of judgement; and, if our works shall afford less delight to those, in respect of whom we shall be Ancients, it may be said without presumption, that we shall give them more correct information on the History and Geography of this eastern world; since no man can perfectly describe a country, who is unacquainted with the language of it. The learned and entertaining work of M. D'HERBELOT, which professes to interpret and elucidate the names of persons and places, and the titles of books, abounds also in citations from the best writers of *Arabia* and *Persia*; yet, though his orthography will be found less defective than that of other writers on similar subjects, without excepting the illustrious Prince KANTEMIR, still it requires more than a moderate knowledge of *Persian*, *Arabick*, and *Turkish*, to comprehend all the passages quoted by him in *European* characters;[3] one instance of which I cannot forbear giving. In the account of *Ibnu Zaidùn*, a celebrated *Andalusian* poet, the first couplet of an elegy in *Arabick* is praised for its elegance, and expressed thus in *Roman* letters:

Iekad heïn tenagikom dhamairna;
Iacdha âlaïna alassa laula tassina.

'The time', adds the translator', 'will soon come, when you will deliver us from all our cares: the remedy is assured, provided we have a little patience'. When

---

[1] [The two transliterative methods advanced by the Orientalists Major William Davy and Charles Wilkins had not resolved the confusion about proper nouns in certain Oriental languages as caused by inconsistent representation in English spelling. So Jones devised his innovational Jonesian System for transliterating Persian, Arbic, and Indic languages into Roman orthography. First published in *Asiatick Researches* 1 (1788), his essay took a long step toward the International Phoenetic Alphabet by following the great principle of getting closer to the original sounds and their rapid-speech arrangement through transliteration rather than through translation, while not introducing strange new diacritics. Recognizing that Roman orthography poorly represented English vowels, Jones added French diacritics to English letters so as to gain the needed vowels, so that 'each original sound may be rendered invariably by one appropriated symbol, conformably to the natural order of articulation'.]

[2] [Jean-Baptiste d'Anville (1697-1782), *Antiquité géographique de l'Inde* (Paris, 1775).]

[3] [Barthélemy d'Herbelot (1625-1695), *Bibliothèque orientale, ou dictionnaire universal* (Paris, 1697).]

Dr. HUNT of *Oxford*,[4] whom I am bound to name with gratitude and veneration, together with two or three others, attempted at my request to write the same distich in *Arabian* characters, they all wrote it differently, and all, in my present opinion, erroneously. I was then a very young student, and could not easily have procured *Ibnu Zaidún*'s works, which are, no doubt, preserved in the *Bodley* library, but which have not since fallen in my way. This admired couplet, therefore, I have never seen in the original characters, and confess myself at a loss to render them with certainty. Both verses are written by *D'Herbelot* without attention to the grammatical points, that is, in a form which no learned *Arb* would give them in recitation; but, although the *French* version be palpably erroneous, it is by no means easy to correct the errour. If *álásà* or a *remedy* be the true reading, the negative particle must be absurd, since *taássainā* signifies *we are patient*, and not *we despair*, but, if *álásay* or *affliction* be the proper word, some obscurity must arise from the verb, with which it agrees. On the whole I guess, that the distich should thus be written:

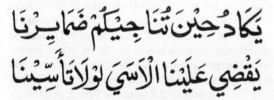

Yecádu bhīna tunájīcum d'emáirunā
Yakdì àlainà 'lāsay là taássinà .

'When our bosoms impart their secrets to you, anguish would almost fix our doom, if we are not mutually to console ourselves'.

The principal verbs may have a future tense, and the last word may admit of a different interpretation. Dr. HUNT, I remember, had found in GIGGEIUS the word *dhemáyer*, which he conceived to be in the original. After all, the rhyme seems imperfect, and the measure irregular. Now I ask, whether such perplexities could have arisen, if *D'Herbelot* or his Editor had formed a regular system of expressing *Arabick* in *Roman* characters, and had apprized his readers of it in his introductory dissertation?

If a further proof be required, that such a system will be useful to the learned and essential to the student, let me remark, that a learner of *Persian*, who should read in our best histories the life of Sultan AZIM, and wish to write his name in *Arabick* letters, might express it *thirty-nine* different ways, and be wrong at last: the word should be written *Aâzem* with three points on the first consonant.

There are two general modes of exhibiting *Asiatick* words in our own letters: they are founded on principles nearly opposite, but each of them has its advantages, and each has been recommended by respectable authorities. The first professes to regard chiefly the *pronunciation* of the words intended to be expressed; and this method, as far as it can be pursued, is unquestionably useful: but new sounds are very inadequately presented to a sense not formed to receive them; and the reader must in the end be left to pronounce many letters and syllables precariously; besides, that by this mode of orthography all grammatical analogy is destroyed, simple sounds are represented by double characters, vowels of one denomination stand for those of another; and possibly with all our labour we perpetuate a provincial or inelegant pronunciation: all these objections may be made to the usual way of writing *Kummerbund*, in which neither the letters nor the true sound of them are preserved, while *Kemerbend*, or *Cemerbend*, as an ancient *Briton* would write it, clearly exhibits both the original characters and the *Persian* pronunciation of them. To set this point in a strong light, we need only suppose, that the *French* had adopted a system of letters wholly different from ours, and of which we had no types in our printing-houses: let us conceive an *Englishman* acquainted with their language to be pleased with MALHERBE's well-known imitation of *Horace*, and desirous of quoting it in some piece of criticism. He would read thus:

'La mort a des rigueurs à nulle autre pareilles;
    On a beau la prier:
La cruelle qu'elle est se bouche les oreilles,
    Et nous laisse crier.

Le pauvre en sa cabane, ou le chaume le couvre,
    Est sujet à ses loix,
Et la garde, qui veille aux barrieres du *Louvre*,
    N'en défend pas nos rois!'[5]

Would he then express these eight verses, in *Roman* characters, exactly as the *French* themselves in fact express them, or would he decorate his composition with a

---

[4] [The Oxford Orientalist Thomas Hunt (1696-1774), who assisted Jones in the controversy over Anquetil-Duperron's pionerring translation of the *Zend-Avesta*.]

[5] [François de Malherbe (1555-1628), 'Consolation à Monsieur Du Périer, gentilhomme d'Aix-en-Provence, sur la mort de sa fille', 11. 73-80.]

passage more resembling the dialect of savages, than that of a polished nation? His pronunciation, good or bad, would, perhaps, be thus represented:

'Law more aw day reegyewrs aw nool otruh
parellyuh,
Onne aw bo law preeay:
Law crooellyuh kellay suh booshuh lays orellyuh,
Ay noo laysuh creeay.

Luh povre ong saw cawbawn oo luh chomuh
luh coovruh,
Ay soozyet aw say lwaw,
Ay law gawrduh kee velly ò bawryayruh dyoo
Loovruh,
Nong day-long paw no rwaw!'

The second system of *Asiatick* Orthography consists in scrupulously rendering letter for letter, without any particular care to preserve the pronunciation; and, as long as this mode proceeds by unvaried rules, it seems clearly entitled to preference.

For the first method of writing *Persian* words the warmest advocate, among my acquaintance, was the late Major DAVY, a Member of our Society, and a man of parts, whom the world lost prematurely at a time, when he was meditating a literary retirement, and hoping to pass the remainder of his life in domestick happiness, and in the cultivation of his very useful talents. He valued himself particularly on his pronunciation of the *Persian* language, and on his new way of exhibiting it in our characters, which he instructed the learned and amiable Editor of his *Institutes of Timour* at *Oxford* to retain with minute attention throughout his work.[6] Where he had acquired his refined articulation of the *Persion*, I never was informed; but it is evident, that he spells most proper names in a manner, which a native of *Persia*, who could read our letters, would be unable to comprehend. For instance: that the capital of *Azarbáïján* is now called *Tabríz*, I know from the mouth of a person born in that city, as well as from other *Iranians*; and that it was so called sixteen hundred years ago, we all know from the Geography of *Ptolemy*; yet Major DAVY always wrote it *Tubburaze*, and insisted that it should thus be pronounced. Whether the natives of *Semerkand*, or *Samarkand*, who probably speak the dialect of *Soghd* with a *Turanian* pronunciation, call their birthplace, as DAVY spelled it, *Summarkund*, I have yet to learn; but I cannot

believe it, and am convinced, that the former mode of writing the word expresses both the letters and the sound of them better than any other combination of characters. His method, therefore, has every defect; since it renders neither the original elements of words, nor the sounds represented by them in *Persia*, where alone we must seek for genuine *Persian*, an for *French* in *France*, and for *Italian* in *Italy*.

The second method has found two able supporters in Mr. HALHED and Mr. WILKINS; to the first of whom the publick is indebted for a perspicuous and ample grammar of the Bengal language, and to the second for more advantages in *Indian* literature than *Europe*, or *India*, can ever sufficiently acknowledge.[7]

Mr. HALHED, having justly remarked, 'that the two greatest defects in the orthography of any language are the application of the same letter to several different sounds, and of different letters to the same sound', truly pronounces them both to be 'so common in *English*, that he was exceedingly embarrassed in the choice of letters to express the sound of the *Bengal* vowels, and was at last by no means satisfied with his own selection'. If anything dissatisfies me, in his clear and accurate system, it is the use of *double* letters for the long vowels (which might however be justified) and the frequent intermixture of *Italick* with *Roman* letters in the same word; which both in writing and printing must be very inconvenient: perhaps it may be added, that his diphthongs are not expressed analogously to the sounds, of which they are composed.

The system of Mr. WILKINS has been equally well considered, and Mr. HALHED himself has indeed adopted it in his preface to the *Compilation of Hindu Laws*: it principally consists of double letters to signify our third and fifth vowels, and of the common prosodial marks to ascertain their brevity or their length; but those marks are so generally appropriated to books of prosody, that they never fail to convey an idea of metre; nor, if *either* prosodial sign were adopted, would *both* be necessary; since the omission of a long mark would evidently denote the shortness of the unmarked vowel, or conversely. On the whole, I cannot but approve this notation for *Sanscrit* words, yet require something more universally expressive of *Asiatick* letters: as it is perfect, however, in its

---

[6] [William Davy (d. 1784), Warren Hastings's Persian secretary in Calcutta, authored *Institutes Political and Military, Written Originally in the Mogul Language* (Oxford, 1783).]

[7] [See Nathaniel Brassey Halhed (1751-1830), *A Grammar of the Bengal Language* (Hooghly, 1778); and Wilkins, *The Bhăgvăt-Gēētā* (London, 1785), where Wilkins says, 'a regular mode hath been followed in the orthography of the proper names, and other original words' (p. 26).]

kind, and will appear in the works of its learned inventor, I shall annex, among the examples, four distichs from the *Bhāgawat* expressed both in his method and mine: a translation of them will be produced on another occasion; but, in order to render this tract as complete as possible, a fuller specimen of Sanscrit will be subjoined with the original printed in the characters of *Bengal*, into which the *Brāhmans* of that province transpose all their books, few of them being able to read the *Dēvanāgari* letters: so far has their indolence prevailed over their piety

Let me now proceed, not prescribing rules for others, but explaining those which I have prescribed for myself, to unfold my own system, the convenience of which has been proved by careful observation and long experience.

It would be superfluous to discourse on the organs of speech, which have been a thousand times dissected, and as often described by musicians or anatomists; and the several powers of which every man may perceive either by the touch or by sight, if he will attentively observe another person pronouncing the different classes of letters, or pronounce them himself distinctly before a mirror: but a short analysis of articulate sounds may be proper to introduce an examination of every separate symbol.

All things abound with errour, as the old searchers for truth remarked with despondence; but it is really deplorable, that our first step from total ignorance should be into gross inaccuracy, and that we should begin our education in *England* with learning to read *the five vowels, two* of which, as we are taught to pronounce them, are clearly diphthongs. There are, indeed, five simple vocal sounds in our language, as in that of *Rome*: which occur in the words *an innocent bull*, though not precisely in their natural order, for we have retained the true arrangement of the letters, while we capriciously disarrange them in pronunciation; so that our eyes are satisfied, and our ears disappointed. The primary elements of articulation are the *soft* and *hard breathings*, the *spiritus lenis* and *spiritus asper* of the *Latin* Grammarians. If the lips be opened ever so little, the breath suffered gently to pass through them, and the feeblest utterance attempted, a sound is formed of so simple a nature, that, when lengthened, it continues nearly the same, except that, by the least acuteness in the voice it becomes a cry, and is probably the first sound uttered by infants; but if, while this element is articulated, the breath be forced with an effort through the lips, we form an *aspirate* more or less harsh in proportion to the force exerted. When, in pronouncing the simple vowel, we open our lips wider, we express a sound completely articulated, which most na-

tions have agreed to place the *first* in their symbolical systems: by opening them wider still with the corners of them a little drawn back, we give birth to the *second* of the *Roman* vowels, and by a large aperture, with a farther inflexion of the lips and a higher elevation of the tongue, we utter the *third* of them. By pursing up our lips in the least degree, we convert the simple element into another sound of the same nature with the *first* vowel, and easily confounded with it in a broad pronunciation: when this new sound is lengthened, it approaches very nearly to the *fourth* vowel, which we form by a bolder and stronger rotundity of the mouth; a farther contraction of it produces the *fifth* vowel, which in its elongation almost closes the lips, a small passage only being left for the breath. These are all short vowels; and, if an *Italian* were to read the words *an innocent bull*, he would give the sound of each corresponding long vowel, as in the monosyllables of his own language, *sà, si, sò, se, sù*. Between these ten vowels are numberless gradations, and nice inflexions, which use only can teach; and, by the composition of them all, might be formed an hundred diphthongs, and a thousand triphthongs; many of which are found in *Italian*, and were probably articulated by the *Greeks*; but we have only occasion, in this tract, for two diphthongs, which are compounded of the *first* vowel with the *third*, and with the *fifth*, and should be expressed by their constituent letters: as to whose vocal compounds which begin with the *third* and *fifth* short vowels, they are generally and not inconveniently rendered by distinct characters, which are improperly ranged among the consonants. The tongue, which assists in forming some of the vowels, is the principal instrument in articulating two liquid sounds, which have something of a vocal nature; one, by striking the roots of the upper teeth, while the breath passes gently through the lips, another, by an inflexion upwards with a tremulous motion; and these two liquids coalesce with such ease, that a mixed letter, used in some languages, may be formed by the first of them followed by the second: when the breath is obstructed by the pressure of the tongue, and forced between the teeth on each side of it, a liquid is formed peculiar to the *British* dialect of the *Celtick*.

We may now consider in the same order, beginning with the root of the tongue and ending with the perfect close of the lips, those less musical sounds, which require the aid of a *vowel*, or atleast of the *simple breathing*, to be fully articulated; and it may here be premised, that the *harsh breathing* distinctly pronounced after each of these *consonants*, as they are named by grammarians, constitutes its proper *aspirate*.

By the assistance of the tongue and the palate are produced two congenial sounds, differing only as *hard* and *soft*; and these two may be formed still deeper in the throat, so as to imitate, with a long vowel after them, the voice of a raven; but if, while they are uttered, the breath he harshly protruded, two analogous articulations are heard, the second of which seems to characterize the pronunciation of the *Arabs*; while the nasal sound, very common among the *Persians* and *Indians*, may be considered as the *soft palatine* with part of the breath passing through the nose; which organ would by itself rather produce a *vocal* sound, common also in *Arabia*, and not unlike the cry of a young antelope and some other quadrupeds.

Next come different classes of *dentals*, and among the first of them should be placed the *sibilants*, which most nations express by an *indented* figure: each of the *dental* sounds is hard or soft, sharp or obtuse, and, by thrusting the tip of the tongue between the teeth, we form two sounds exceedingly common in *Arabick* and *English*, but changed into lisping sibilants by the *Persians* and *French*, while they on the other hand have a sound unknown to the *Arabs*, and uncommon in our language, though it occurs in some words by the composition of the hard sibilant with our last vowel pronounced as a diphthong. The liquid *nasal* follows these, being formed by the tongue and roots of the teeth, with a little assistance from the other organ; and we must particularly remember, when we attend to the pronunciation of *Indian* dialects, that most sounds of this class are varied in a singular manner by turning the tongue upwards, and almost bending it back towards the palate, so as to exclude them nearly from the order, but not from the analogy, of dentals.

The *labials* form the last series, most of which are pronounced by the appulse of the lips on each other or on the teeth, and one of them by their perfect close: the letters, by which they are denoted, represent in most alphabets the curvature of one lip or of both; and *a natural character* for all articulate sounds might easily be agreed on, if nations would agree on any thing generally beneficial, by delineating the several organs of speech in the act of articulation, and selecting from each a distinct and elegant outline. A perfect language would be that, in which every idea, capable of entering the human mind, might be neatly and emphatically expressed by one specifick word, simple if the idea were simple, complex, if complex; and on the same principle a perfect system of letters ought to contain one specifick symbol for every sound used in pronouncing the language to which they belonged:[8] in this respect the old *Persian* or *Zend* approaches to perfection; but the *Arabian* alphabet, which all *Mohammedan* nations have inconsiderately adopted, appears to me so complete for the purpose of writing *Arabick*, that not a letter could be added or taken away without manifest inconvenience, and the same may indubitably be said of the *Dēvanāgarī* system; which, as it is more naturally arranged then any other, shall here be the standard of my particular observations on *Asiatick* letters. Our *English* alphabet and orthography are disgracefully and almost ridiculously imperfect; and it would be impossible to express either *Indian, Persian,* or *Arabian* words in *Roman* characters, as we are absurdly taught to pronounce them; but a mixture of new characters would be inconvenient, and by the help of the diacritical marks used by the *French*, with a few of those adopted in our own treatises on *fluxions*, we may apply our present alphabet so happily to the notation of all *Asiatick* languages, as to equal the *Dēvanāgarī* itself in precision and clearness, and so regularly that any one, who knew the original letters, might rapidly and unerringly transpose into them all the proper names, appellatives, or cited passages, occurring in tracts of *Asiatick* literature . . . .

[8] [While using the primarily auditory (rather than the modern physiological or acoustic) description employed by eighteenth-century scholars, Jones innovationally hints at a universal notational system and description of the particular articulatory processes.]

A Discourse on the Institution
of a Society for Inquiring into
the History, Civil and Natural,
the Antiquities, Arts, Sciences,
and Literature of Asia[1]

Gentlemen,

When I was at sea last August, on my voyage to this country, which I had long and ardently desired to visit, I found one evening, on inspecting the observations of the day, that *India* lay before us, and *Persia* on our left, whilst a breeze from *Arabia* blew nearly on our stern. A situation so pleasing in itself, and to me so new, could not fail to awaken a train of reflections in a mind, which had early been accustomed to contemplate with delight the eventful histories and agreeable fictions of this eastern world. It gave me inexpressible pleasure to find myself in the midst of so noble an amphitheatre, almost encircled by the vast regions of *Asia*, which has ever been esteemed the nurse of sciences, the inventress of delightful and useful arts, the scene of glorious actions, fertile in the productions of human genius, abounding in natural wonders, and infinitely diversified in the forms of religion and government, in the laws, customs, and languages, as well as in the features and complexions, of men. I could not help remarking, how important and extensive a field was yet unexplored, and how many solid advantages unimproved; and when I considered, with pain, that, in this fluctuating, imperfect, and limited condition of life, such

---

[1] [First published (with two other pieces) as a pamphlet (London, 1784), this essay dramatically describes Jones's almost mystical vision in the Indian Ocean. It innovationally recognizes the necessity of intellectual and natural intercourse between the West and Asia, which was atypically lauded in that colonial day as being fertile in human genius, abounding in natural wonders, and infinitely diverse. The vision led to the founding of the Asiatic Society of Bengal in 1784, which was to become the inspiration for other Asiatic Societies all over the world and served as a springboard for Jones's and others' introduction of Sanskrit literature and other riches to the West. The essay programmatically sets vast, inspired guidelines for the Asiatic Society that still underpin it.]

inquiries and improvements could only be made by the united efforts of many, who are not easily brought, without some pressing inducement or strong impulse, to converge in a common point, I consoled myself with a hope, founded on opinions which it might have the appearance of flattery to mention, that, if in any country or community, such a union could be effected, it was among my countrymen in *Bengal*, with some of whom I already had, and with most was desirous of having, the pleasure of being intimately acquainted.

You have realized that hope, gentlemen, and even anticipated a declaration of my wishes, by your alacrity in laying the foundation of a society for inquiring into the history and antiquities, the natural productions, arts, sciences, and literature of *Asia*. I may confidently foretel, that an institution so likely to afford entertainment, and convey knowledge, to mankind, will advance to maturity by slow, yet certain, degrees; as the Royal Society, which at first was only a meeting of a few literary friends at *Oxford*, rose gradually to that splendid zenith, at which a *Halley* was their secretary, and a *Newton* their president.

Although it is my humble opinion, that, in order to ensure our success and permanence, we must keep a middle course between a languid remissness, and an over zealous activity, and that the tree, which you have auspiciously planted, will produce fairer blossoms, and more exquisite fruit, if it be not at first exposed to too great a glare of sunshine, yet I take the liberty of submitting to your consideration a few general ideas on the plan of our society; assuring you, that, whether you reject or approve them, your correction will give me both pleasure and instruction, as your flattering attentions have already conferred on me the highest honour.

It is your design, I conceive, to take an ample space for your learned investigations, bounding them only by the geographical limits of *Asia*; so that, considering *Hindustan* as a centre, and turning your eyes in idea to the North, you have on your right, many important kingdoms in the Eastern peninsula, the ancient and wonderful empire of *China* with all her *Tartarian* dependencies, and that of *Japan*, with the cluster of precious islands, in which many singular curiosities have too long been concealed: before you lies that prodigious chain of mountains, which formerly perhaps were a barrier against the violence of the sea, and beyond them the very interesting country of *Tibet*, and the vast regions of *Tartary*, from which, as from the *Trojan* horse of the poets, have issued so many consummate warriors, whose domain has extended at least from the banks of the *Ilissus* to the mouths of the *Ganges*: on your left are the beautiful and celebrated provinces of *Iran* or *Persia*, the unmeasured, and perhaps unmeasurable deserts of *Arabia*, and the once flourishing kingdom of *Yemen*, with the pleasant isles that the *Arabs* have subdued or colonized; and farther westward, the *Asiatick* dominions of the *Turkish* sultans, whose moon seems approaching rapidly to its wane.—By this great circumference, the field of your useful researches will be inclosed; but, since *Egypt* had unquestionably an old connexion with this country, if not with *China*, since the language and literature of the *Abyssinians* bear a manifest affinity to those of *Asia*, since the *Arabian* arms prevailed along the *African* coast of the *Mediterranean*, and even erected a powerful dynasty on the continent of *Europe*, you may not be displeased occasionally to follow the streams of *Asiatick* learning a little beyond its natural boundary; and, if it be necessary or convenient, that a short name or epithet be given to our society, in order to distinguish it in the world, that of *Asiatick* appears both classical and proper, whether we consider the place or the object of the institution, and preferable to *Oriental*, which is in truth a word merely relative, and, though commonly used in *Europe*, conveys no very distinct idea.

If now it be asked, what are the intended objects of our inquiries within these spacious limits, we answer, MAN and NATURE; whatever is performed by the one, or produced by the other. Human knowledge has been elegantly analysed according to the three great faculties of the mind, *memory*, *reason*, and *imagination*, which we constantly find employed in arranging and retaining, comparing and distinguishing, combining and diversifying, the ideas, which we receive through our senses, or acquire by reflection; hence the three main branches of learning are *history*, *science*, and *art*: the first comprehends either an account of natural productions, or the genuine records of empires and states; the second embraces the whole circle of pure and mixed mathematicks, together with ethicks and law, as far as they depend on the reasoning faculty; and the third includes all the beauties of imagery and the charms of invention, displayed in modulated language, or represented by colour, figure, or sound.

Agreeably to this analysis, you will investigate whatever is rare in the stupendous fabrick of nature, will correct the geography of *Asia* by new observations and discoveries; will trace the annals, and even traditions, of those nations, who from time to time have peopled or desolated it; and will bring to light their various forms of government, with their institutions civil and religious; you will examine their improvements and methods in arithmetick and geometry, in trigonometry, mensuration,

mechanicks, opticks, astronomy, and general physicks; their systems of morality, grammar, rhetorick, and dialectick; their skill in chirurgery and medicine, and their advancement, whatever it may be, in anatomy and chymistry. To this you will add reserches into their agriculture, manufactures, trade; and, whilst you inquire with pleasure into their musick, architecture, painting, and poetry, will not neglect those inferior arts, by which the comforts and even elegances of social life are supplied or improved. You may observe, that I have omitted their languages, the diversity and difficulty of which are a sad obstacle to the progress of useful knowledge; but I have ever considered languages as the mere instruments of real learning, and think them improperly confounded with learning itself: the attainment of them is, however, indispensably necessary; and if to the *Persian*, *Armenian*, *Turkish*, and *Arabick*, could be added not only the *Sanscrit*, the treasures of which we may now hope to see unlocked, but even the *Chinese*, *Tartarian*, *Japanese*, and the various insular dialects, an immense mine would then be open, in which we might labour with equal delight and advantage.

Having submitted to you these imperfect thoughts on the *limits* and *objects* of our future society, I request your permission to add a few hints on the *conduct* of it in its present immature state.

Lucian begins one of his satirical pieces against historians, with declaring that the only true proposition in his work was, that it should contain nothing true;[2] and perhaps it may be advisable at first, in order to prevent any difference of sentiment on particular points not immediately before us, to establish but one rule, namely, to have no rules at all. This only I mean, that, in the infancy of any society, there ought to be no confinement, no trouble, no expense, no unnecessary formality. Let us, if you please, for the present, have weekly evening meetings in this hall, for the purpose of hearing original papers read on such subjects, as fall within the circle of our inquiries. Let all curious and learned men be invited to send their tracts to our secretary, for which they ought immediately to receive our thanks; and if, towards the end of each year, we should be supplied with a sufficiency of valuable materials to fill a volume, let us present our *Asiatick* miscellany to the literary world, who have derived so much pleasure and information from the agreeable work of *Kœmpfer*,[3] than which we can scarce pro-

---

[2] [Stated in Part I of the Greek satirist's *A True Story*.]

[3] [Engelbert Kæmpfer (1651-1716), the German 'Humboldt of the 17th century' and a rich source on contemporary Persia

pose a better model, that they will accept with eagerness any fresh entertainment of the same kind. You will not perhaps be disposed to admit mere translations of considerable length, except of such unpublished essays or treatises as may be transmitted to us by native authors; but, whether you will enrol as members any number of learned natives, you will hereafter decide, with many other questions as they happen to arise; and you will think, I presume, that all questions should be decided on a ballot, by a majority of two thirds, and that nine members should be requisite to constitute a board for such decisions. These points, however, and all others I submit entirely, gentlemen, to your determination, having neither wish nor pretension to claim any more than my single right of suffrage. One thing only, as essential to your dignity, I recommend with earnestness, on no account to admit a new member, who has not expressed a voluntary desire to become so; and in that case, you will not require, I suppose, any other qualification than a love of knowledge, and a zeal for the promotion of it.

Your institution, I am persuaded, will ripen of itself, and your meetings will be amply supplied with interesting and amusing papers, as soon as the object of your inquiries shall be generally known. There are, it may not be delicate to name them, but there are many, from whose important studies I cannot but conceive high expectations; and, as far as mere labour will avail, I sincerely promise, that, if in my allotted sphere of jurisprudence, or in any intellectual excursion, that I may have leisure to make, I should be so fortunate as to collect, by accident, either fruits or flowers, which may seem valuable or pleasing, I shall offer my humble *Nezr*[4] to your society with as much respectful zeal as to the greatest potentate on earth.

The Third Anniversary Discourse[1]
Delivered 2 February, 1786

In the former discourses, which I had the honour of addressing to you, Gentlemen, on the *institution* and *objects* of our Society, I confined myself purposely to general topics; giving in the first a distant prospect of the vast career, on which we were entering, and, in the second, exhibiting a more diffuse, but still superficial, sketch of the various discoveries in History, Science, and Art, which we might justly expect from our inquiries into the literature of *Asia*. I now propose to fill up that outline so comprehensively as to omit nothing essential, yet so concisely as to avoid being tedious; and, if the state of my health shall suffer me to continue long enough in this climate, it is my design, with your permission, to prepare for our annual meetings a series of short dissertations, unconnected in their titles and subjects, but all tending to a common point of no small importance in the pursuit of interesting truths.

Of all the works, which have been published in our own age, or, perhaps, in any other, on the History of the Ancient World, and *the first population of this habitable globe*, that of Mr. JACOB BRYANT, whom I name with reverence and affection, has the best claim to the praise

[1] [This celebrated essay, a direct result of his study of Sanskrit, was read to the Asiatic Society in February 1786 and published in *Asiatick Researches* (1). It posited the modern theory that languages are related because of common descent from an earlier language (rather than because of borrowings or coincidence to explain similar inflections and/or etymologies). In this brilliant speculative essay about India, Jones provided the clear formulation of relationships among Indo-European languages, grouping them as a family and articulating them in a wider and clearer net than had Halhead, Monboddo, Coeurdoux, etc. (See Rosane Rocher, 'Nathaniel Brassey Halhed, Sir William Jones, and Comparative Indo-European linguistics', *Recherches de Linguistique*, 1980: 173-80). This made Jones the founder of comparative linguistics and stimulated the German philologists' striking phonological reconstructions a few decades later. This dramatically lucid linguistic formulation and his founding of the Asiatic Society were probably his greatest contributions to posterity. See P.J. Marshall's scholarly edition of the Discourse, in the *British Discovery of Hinduism in the Eighteenth Century* (Cambridge, 1970): 246-61.]

and Japan, wrote *Amœnitatum exoticarum* (Lemgo, 1712).]
[4] [Hindustani 'gift'.]

of deep erudition ingeniously applied, and new theories happily illustrated by an assemblage of numberless converging rays from a most extensive circumference: it falls, nevertheless, as every human work must fall, short of perfection; and the least satisfactory part of it seems to be that, which relates to the derivation of words from *Asiatick* languages. Etymology has, no doubt, some use in historical reserches; but it is a medium of proof so very fallacious, that, where it elucidates one fact, it obscures a thousand, and more frequently borders on the ridiculous, than leads to any solid conclusion: it rarely carries with it any *internal* power of conviction from a resemblance of sounds or similarity of letters; yet often, where it is wholly unassisted by those advantages, it may be indisputably proved by *extrinsick* evidence.[2] We know *à posteriori*, that both *fitz* and *hijo*, by the nature of two several dialects, are derived from *filius*; that *uncle* comes from *avus*, and *stranger* from *extra*; that *jour* is deducible, through the *Italian*, from *dies*; and *rossignol* from *luscinia*, or the *singer in groves*; that *sciuro*, *écureuil*, and *squirrel* are compounded of two *Greek* words descriptive of the animal; which etymologies, though they could not have been demonstrated *à priori*, might serve to confirm, if any such confirmation were necessary, the proofs of a connection between the members of one great Empire; but, when we derive our *hanger*, or *short pendent sword*, from the *Persian*, because ignorant travellers thus misspell the work *khanjar*, which in truth means a different weapon, or *sandal-wood* from the *Greek*, because we suppose, that *sandals* were sometimes made of it, we gain no ground in proving the affinity of nations, and only weaken arguments, which might otherwise be firmly supported. That CÚS then, or, as it certainly is written in one ancient dialect, CÚT, and in others, probably, CÁS, enters into the composition of many proper names, we may very reasonably believe; and that *Algeziras* takes its name from the *Arabick* word for an *island*, cannot be doubted; but, when we are told from *Europe*, that places and provinces in *India* were clearly denominated from those words, we cannot but observe, in the first instance, that the town, in which we now are assembled, is properly written and pronounced *Calicātà*; that both *Cātā* and *Cūt* unquestionably mean *places of strength*, or, in general, any *inclosures*; and that

---

[2] [Jones's longtime friend Jacob Bryant (1715-1804) wrote the famous *A New System, or an Analysis of Ancient Mythology*, 3 vols. (London, 1774-76). It was among Jones's 'books of entertainment', for he found Bryant's mythological system quite speculative but nonetheless admired the theorist's dramatic sweep and larger purposes.]

Gujràt is at least as remote from *Jezirah* in sound, as it is in situation.

Another exception (and a third could hardly be discovered by any candid criticism) to the *Analysis of Ancient Mythology*, is, that the *method* of reasoning and arrangement of topicks adopted in that learned work are not quite agreeable to the title, but almost wholly *synthetical*; and, though *synthesis* may be the better mode in pure *science*, where the principles are undeniable, yet it seems less calculated to give complete satisfaction in *historical* disquisitions, where every postulatum will perhaps be refused, and every definition controverted: this may seem a slight objection, but the subject is in itself so interesting, and the full conviction of all reasonable men so desirable, that it may not be lost labour to discuss the same or a similar theory in a method purely analytical, and, after beginning with facts of general notoriety or undisputed evidence, to investigate such truths, as are at first unknown or very imperfectly discerned.

The *five* principal nations, who have in different ages divided among themselves, as a kind of inheritance, the vast continent of *Asia*, with the many islands depending on it, are the *Indians*, the *Chinese*, the *Tartars*, the *Arabs*, and the *Persians*: *who* they severally were, *whence*, and *when* they came, *where* they now are settled, and *what advantage* a more perfect knowledge of them all may bring to our *European* world, will be shown, I trust, in *five* distinct essays; the last of which will demonstrate the connexion or diversity between them, and solve the great problem, whether they had *any* common origin, and whether that origin was *the same*, which we generally ascribe to them.

I begin with *India*, not because I find reason to believe it the true centre of population or of knowledge, but, because it is the country, which we now inhabit, and from which we may best survey the regions around us; as, in popular language, we speak of the *rising* sun, and of his *progress through the Zodiack*, although it had long ago been imagined, and is now demonstrated, that he is himself the centre of our planetary system. Let me here premise, that, in all these inquiries concerning the history of *India*, I shall confine my researches downwards to the *Mohammedan* conquests at the beginning of the *eleventh* century, but extend them upwards, as high as possible, to the earliest authentick records of the human species.

India then, on its most enlarged scale, in which the ancients appear to have understood it, comprises an area of near *forty* degrees on each side, including a space almost as large as all *Europe*; being divided on the west from *Persia* by the *Arachosian* mountains, limited on the east by the *Chinese* part of the farther peninsula, confined on the north by the wilds of *Tartary*, and extending to the south as far as the isles of *Java*. This trapezium, therefore, comprehends the stupendous hills of *Potyid* or *Tibet*, the beautiful valley of *Cashmīr*, and all the domains of the old *Indoscythians*, the countries of *Nēpāl* and *Butānt*, *Cāmrùp* or *Asàm*, together with *Siam*, *Ava*, *Racan*, and the bordering kingdoms, as far as the *Chīna* of the *Hindus* or *Sīn* of the *Arabian* Geographers; not to mention the whole western peninsula with the celebrated island of *Sinhala*, or *Lion-like men*, at its southern extremity. By *India*, in short, I mean that whole extent of country, in which the primitive religion and languages of the *Hindus* prevail at this day with more or less of their ancient purity, and in which the *Nāgarì* letters are still used with more or less deviation from their original form.

The *Hindus* themselves believe their own country, to which they give the vain epithets of *Medhyama* or *Central*, *Punyabhūmi*, or the *Land of Virtues*, to have been the portion of BHARAT, one of *nine* brothers, whose father had the dominion of the whole earth; and they represent the mountains of *Himālaya* as lying to the north, and, to the west, those of *Vindhya*, called also *Vindian* by the *Greeks*; beyond which the *Sindhu* runs in several branches to the sea, and meets it nearly opposite to the point of *Dwāracà*, the celebrated seat of their Shepherd God: in the *south-east* they place the great river *Saravatya*; by which they probably mean that of *Ava*, called also *Airāvati* in part of its course, and giving perhaps its ancient name to the gulf of *Sabara*. This domain of *Bharat* they consider as the middle of the *Jambudwīpa*, which the *Tibetians* also call the Land of *Zambu*; and the appellation is extremely remarkable; for *Jambu* is the Sanscrit name of a delicate fruit called *Jàman* by the *Muselmàns*, and by us *rose-apple*; but the largest and richest sort is named *Amrita*, or *Immortal*; and the Mythologists of *Tibet* apply the same word to a celestial tree bearing *ambrosial fruit*, and adjoining to *four* vast rocks, from which as many sacred rivers derive their several streams.

The inhabitants of this extensive tract are described by Mr. LORD with great exactness, and with a picturesque elegance peculiar to our ancient language: 'A people', says he, 'presented themselves to mine eyes, clothed in linen garments somewhat low descending, of a gesture and garb, as I may say, maidenly and well nigh effeminate, of a countenance shy and somewhat estranged, yet smiling out a glozed and bashful familiarity'.[3]

[3] [From Henry Lord, 'The Sect of the Banians', *A Display of Two Forraigne Sects in the East Indies* (London, 1630).]

Mr. ORME,[4] the *Historian* of *India*, who uniters an exquisite taste for every fine art with an accurate knowledge of *Asiatick* manners, observes, in his elegant preliminary Dissertation, that this 'country has been inhabited from the earliest antiquity by a people, who have no resemblance, either in their figure or manners, with any of the nations contiguous to them', and that, 'although conquerors have established themselves at different times in different parts of *India*, yet the original inhabitants have lost very little of their original character'. The ancients, in fact, give a description of them, which our early travellers confirmed, and our own personal knowledge of them nearly verifies; as you will perceive from a passage in the Geographical Poem of DIONYSIUS, which the Analyst of Ancient Mythology has translated with great spirit:

'To th' east a lovely country wide extends,
INDIA, whose borders the wide ocean bounds;
On this the sun, new rising from the main,
Smiles pleas'd, and sheds his early orient beam.
Th' inhabitants are swart, and in their locks
Betray the tints of the dark hyacinth.
Various their functions; some the rock explore,
And from the mine extract the latent gold;
Some labour at the woof with cunning skill,
And manufacture linen; others shape
And polish iv'ry with the nicest care:
Many retire to rivers shoal, and plunge
To seek the beryl flaming in its bed,
Or glitt'ring diamond. Oft the jasper's found
Green, but diaphanous; the topaz too
Of ray serene and pleasing; last of all
The lovely amethyst, in which combine
All the mild shades of purple. The rich soil,
Wash'd by a thousand rivers, from all sides
Pours on the natives wealth without control'.[5]

Their sources of wealth are still abundant even after so many revolutions and conquests; in their manufactures of cotton they still surpass all the world; and their features have, most probably, remained unaltered since the time of DIONYSIUS; nor can we reasonably doubt, how degenerate and abased so ever the *Hindus* may now appear, that in some early age they were splendid in arts and arms, happy in government, wise in legislation, and eminent in various knowledge: but, since their civil history beyond

[4] [Jones's friend Robert Orme (1728-1801), who wrote *A History of the Military Transactions of the British Nation in Indostan from the Year 1745* (London, 1763-78), corresponded with him about Indian judicature.]

[5] [Bryant, *New System*, 3: 227-28.]

the middle of the *nineteenth* century from the present time, is involved in a cloud of fables, we seem to posses only *four* general media of satisfying our curiosity concerning it; namely, first, their *Languages* and *Letters*; secondly, their *Philosophy* and *Religion*; thirdly, the actual remains of their old *Sculpture* and *Architecture*; and fourthly, the written memorials of their *Sciences* and *Arts*. I. It is much to be lamented, that neither the *Greeks*, who attended ALEXANDER into *India*, nor those who were long connected with it under the *Bactrian* Princes, have left us any means of knowing with accuracy, what vernacular languages they found on their arrival in this Empire. The *Mohammedans*, we know, heard the people of proper *Hindustan*, or *India* on a limited scale, speaking a *Bháshá*, or living tongue of a very singular construction, the purest dialect of which was current in the districts round *Agrà*, and chiefly on the poetical ground of *Mat'hurà*; and this is commonly called the idiom of *Vraja*. Five words in six, perhaps, of this language were derived from the *Sanscrit*, in which books of religion and science were composed, and which appears to have been formed by an exquisite grammatical *arrangement*, as the name itself implies, from some unpolished idiom; but the basis of the *Hindustàni*, particularly the inflexions and regimen of verbs, differed as widely from both those tongues, as *Arabick* differs from *Persian*, or *German* from *Greek*. Now the general effect of conquest is to leave the current language of the conquered people unchanged, or very little altered, in its groundwork, but to blend with it a considerable number of exotick names both for things and for actions; as it has happened in every country, that I can recollect, where the conquerors have not preserved their own tongue unmixed with that of the natives, like the *Turks* in *Greece*, and the *Saxons* in *Britain*; and this analogy might induce us to believe, that the pure *Hindì*, whether of *Tartarian* or *Chaldean* origin, was primeval in Upper *India*, into which the *Sanscrit* was introduced by conquerors from other kingdoms in some very remote age; for we cannot doubt that the language of the *Vēdas* was used in the great extent of country, which has before been delineated, as long as the religion of *Brahmá* has prevailed in it.

The *Sanscrit* language, whatever be its antiquity, is of a wonderful structure; more perfect than the *Greek*, more copious than the *Latin*, and more exquisitely refined than either, yet bearing to both of them a stronger affinity, both in the roots of verbs and in the forms of grammar, than could possibly have been produced by accident; so strong indeed, that no philologer could examine them all three, without believing them to have sprung from some common source, which, perhaps, no longer exists: there is a similar reason, though not quite so forcible, for supposing that both the *Gothick* and the *Celtick*, though blended with a very different idiom, had the same origin with the *Sanscrit*; and the old *Persian* might be added to the same family, if this were the place for discussing any question concerning the antiquities of *Persia*.[6]

The *characters*, in which the languages of *India* were originally written, are called *Nāgarī*, from *Nagara*, a City, with the word *Dēva* sometimes prefixed, because they are believed to have been taught by the Divinity himself, who prescribed the artificial order of them in a voice from heaven. These letters, with no greater variation in their form by the change of straight lines to curves, or conversely, than the *Cusick* alphabet has received in its way to *India*, are still adopted in more than twenty kingdoms and states, from the borders of *Cashgar* and *Khoten*, to *Rāma*'s bridge, and from the *Sindhu* to the river of *Siam*; nor can I help believing, although the polished and elegant *Dēvanāgarī* may not be so ancient as the monumental characters in the caverns of *Jarasandha*, that the square *Chaldaick* letters, in which most *Hebrew* books are copied, were originally the same, or derived from the same prototype, both with the *Indian* and *Arabian* characters: that the *Phenician*, from which the *Greek* and *Roman* alphabets were formed by various changes and inversions, had a similar origin, there can be little doubt; and the inscriptions at *Canārah*, of which you now possess a most accurate copy, seem to be compounded of *Nāgarī* and *Ethiopick* letters, which bear a close relation to each other, both in the mode of writing from the left hand, and in the singular manner of connecting the vowels with the consonants. These remarks may favour an opinion entertained by many, that all the symbols of *sound*, which at first, probably, were only rude outlines of the different organs of speech, had a common origin: the symbols of *ideas*, now used in *China* and *Japan*, and formerly, perhaps, in *Egypt* and *Mexico*, are quite of distinct nature; but it is very remarkable, that the order of *sounds* in the *Chinese* grammars corresponds nearly with that observed in *Tibet*, and hardly differs from that, which the *Hindus* consider as the invention of their Gods.

II. Of the *Indian* Religion and Philosophy, I shall here say but little; because a full account of each would require a separate volume: it will be sufficient in this dissertation to assume, what might be proved beyond controversy,

---

[6] [This paragraph is the philologer's formulation, which has been reprinted more than anything else from Jones's works.]

that we now live among the adorers of those very deities, who were worshipped under different names in old *Greece* and *Italy*, and among the professors of those philosophical tenets, which the *Ionick* and *Attick* writers illustrated with all the beauties of their melodious language. On one hand we see the trident of NEPTUNE, the eagle of JUPITER, the satyrs of BACCHUS, the bow of CUPID, and the chariot of the *Sun*; on another we hear the cymbals of RHEA, the songs of the *Muses*, and the pastoral tales of APOLLO NOMIUS. In more retired scenes, in groves, and in seminaries of learning, we may perceive the *Brāhmans* and the *Sarmanes* mentioned by CLEMENS,[7] disputing in the forms of *logick*, or discoursing on the vanity of human enjoyments, on the immortality of the soul, her emanation from the eternal mind, her debasement, wanderings, and final union with her source. The *six* philosophical schools, whose principles are explained in the *Dersana Sāstra*, comprise all the metaphysicks of the old *Academy*, the *Stoa*, the *Lyceum*; nor is it possible to read to *Vēdānta*, or the many fine compositions in illustration of it, without believing, that PYTHAGORAS and PLATO derived their sublime theories from the same fountain with the sages of *India*. The *Scythian* and *Hyperborean* doctrines and mythology may also be traced in every part of these eastern regions; nor can we doubt, that WOD or ODEN, whose religion, as the northern historians admit, was introduced into *Scandinavia* by a foreign race, was the same with BUDDH, whose rites were probably imported into *India* nearly at the same time, though received much later by the Chinese, who soften his name into FÓ.

This may be a proper place to ascertain an important point in the Chronology of the *Hindus*; for the priests of BUDDHA left in *Tibet* and *China* the precise epoch of his appearance, real or imagined, in this Empire; and their information, which had been preserved in writing, was compared by the *Christian* Missionaries and scholars with our own era. COUPLET, DE GUIGNES, GIORGI, and BAILLY, differ a little in their accounts of this epoch, but that of *Couplet* seems the most correct:[8] on taking, how-

ever, the medium of the four several dates, we may fix the time of BUDDHA, or the *ninth* great incarnation of VISHNU, in the year *one thousand* and *fourteen* before the birth of CHRIST, or *two thousand seven hundred and ninety-nine* years ago. Now the *Cāshmirians*, who boast of his descent in their kingdom, assert that he appeared on earth about *two* centuries after CHRISHNA the *Indian* APOLLO, who took so decided a part in the war of the *Mahābhārat*; and, if an Etymologist were to suppose, that the *Athenians* had embellished their poetical history of PANDION'S expulsion and the restoration of ÆGEUS with the *Asiatick* tale of the PĀNDUS and YUDHISHTIR, neither of which words they could have articulated, I should not hastily deride his conjecture: certain it is, that *Pāndumandel* is called by the Greeks the country of PANDION. We have, therefore, determined another interesting epoch, by fixing the age of CRISHNA near the *three thousandth* year from the present time; and, as the three first *Avatàrs*, or descents of VISHNU, relate no less clearly to an Universal Deluge, in which eight persons only were saved, than the *fourth* and *fifth* do to the *punishment of impiety* and the *humiliation* of the *proud*, we may for the present assume, that the *second*, or *silver*, age of the *Hindus* was subsequent to the dispersion from *Babel*; so that we have only a dark interval of about a *thousand* years, which were employed in the settlement of nations, the foundation of states or empires, and the cultivation of civil society. The great incarnate Gods of this intermediate age are both name RĀMA but with different epithets; one of whom bears a wonderful resemblance to the *Indian* BACCHUS, and his wars are the subject of several heroick poems. He is represented as a descendent from SŪRYA, or the SUN, as the husband of SĪTĀ, and the son of a princess named CAŪSELYĀ: it is very remarkable, that the *Peruvians*, whose *Incas* boasted of the same descent, styled their greatest festival *Ramasitoa*; whence we may suppose, that South *America* was peopled by the same race, who imported into the farthest parts of *Asia* the rites and fabulous history of RĀMA. These rites and this history are extremely curious; and, although I cannot believe with NEWTON, that ancient mythology was nothing but historical truth in a poetical dress, nor, with BACON, that it consisted solely of moral and metaphysical allegories,[9] nor with BRYANT, that all the heathen divinities are only different attributes and representations of the Sun of deceased progenitors, but conceive that the

---

[7] [Clemens Alexandrinus, third century A.D., in *Stromata*.]

[8] [Jones used Philippe Couplet's *Confucius, Sinarum Philosophus sive Scientia Sinensis Latine Exposita* (Paris, 1687), which contains translations of major Chinese classical works. Jones's other three sources were Joseph de Guignes, 'Recherches historiques sur la religion Indienne', *Mémoires de l' Académie Royale des Inscriptions et Belles Lettres*, 40 (1773-76): 195, 210; Augustino Antonio Giorgi, *Alphabetum Tibetanum* (Rome, 1762): 14-15; and Jean-Sylvain Bailly, *Histoire de l'astronomie ancienne* (Paris, 1775): 76-77, 342.

[9] [Newton, *The Chronology of Ancient Kingdoms Amended* (London, 1728); and Bacon, *De Sapientia Veterum* (London, 1609).]

whole system of religious fables rose, like the *Nile*, from several distinct sources, yet I cannot but agree, that one great spring and fountain of all idolatry in the four quarters of the globe was the veneration paid by men to the vast body of fire, which 'looks from his sole dominion like the God of this world;' and another, the immoderate respect shown to the memory of powerful or virtuous ancestors, especially the founders of kingdoms, legislators, and warriors, of whom the *Sun* or the *Moon* were wildly supposed to be the parents.

III.   The remains of *architecture* and *sculpture* in *India*, which I mention here as mere monuments of antiquity, not as specimens of ancient art, seem to prove an early connection between this country and *Africa*: the pyramids of *Egypt*, the colossal statues described by PAUSANIAS[10] and others, the sphinx, and the HERMES *Canis*, which last bears a great resemblance to the *Varāh āvatār*, or the incarnation of VISHNU in the form of a *Boar*, indicate the style and mythology of the same indefatigable workmen, who formed the vast excavations of *Cānārah*, the various temples and images of BUDDHA, and the idols, which are continually dug up at *Gayā*, or in its vicinity. The letters on many of those monuments appear, as I have before intimated, partly of *Indian*, and partly of *Abyssinian* or *Ethiopick*, origin; and all these indubitable facts my induce no ill-grounded opinion, that *Ethiopia* and *Hindustàn* were peopled or colonized by the same extraordinary race; in confirmation of which, it may be added, that the mountaineers of *Bengal* and *Bahàr* can hardly be distinguished in some of their features, particularly their lips and noses, from the modern *Abyssinians*, whom the *Arabs* call the children of CŪSH: and the ancient *Hindus*, according to STRABO, differed in nothing from the *Africans*, but in the straitness and smoothness of their hair, while that of the others was crisp or woolly;[11] a difference proceeding chiefly, if not entirely, from the respective humidity or dryness of their atmospheres: hence the people who *received the first light of the rising sun*, according to the limited knowledge of the ancients, are said by APULEIUS to be the *Arü* and *Ethiopians*, by which he clearly meant certain nations of *India*;[12] where we frequently see figures of BUDDHA with *curled hair* apparently designed for a representation of it in its natural state.

---

[10] [In the second century A.D. in 'Description of Greece', 2: xli]

[11] [See J.W. McCrindle, *Ancient India as Described in Classical Literature* (London, 1901): 29-30.]

[12] [Lucius Apuleius, *Florida*, vi.]

IV.   It is unfortunate, that the *Silpi Sāstra*, or *collection of treatises on Arts and Manufactures*, which must have contained a treasure of useful information on *dying*, *painting*, and *metallurgy*, has been so long neglected, that few, if any, traces of it are to be found; but the labours of the *Indian* loom and needle have been universally celebrated; and *fine linen* is not improbably supposed to have been called *Sindon*, from the name of the river near which it was wrought in the highest perfection: the people of *Colchis* were also famed for this manufacture, and the *Egyptians* yet more, as we learn from several passages in scripture, and particularly from a beautiful chapter in EZEKIAL containing the most authentick delineation of ancient commerce, of which *Tyre* had been the principal mart. Silk was fabricated immemorially by the *Indians*, though commonly ascribed to the people of *Serica* of *Tancùt*, among whom probably the word *Sèr*, which the *Greeks* applied to the *silk-worm*, signified *gold*; a sense, which it now bears in *Tibet*. That the *Hindus* were in early ages a *commercial* people, we have many reasons to believe; and in the first of their sacred law-tracts, which they suppose to have been revealed by MENU many *millions* of years ago, we find a curious passage on the legal *interest* of money, and the limited rate of it in different cases, with an exception in regard to *adventures* at *sea*; and exception, which the sense of mankind approves, and which commerce absolutely requires, though it was not before the reign of CHARLES I that our own jurisprudence fully admitted it in respect of maritime contracts.

We are told by the *Grecian* writers, that the *Indians* were the wisest of nations; and in moral wisdom, they were certainly eminent: their *Nīti Sāstra*, or *System or Ethicks*, is yet preserved, and the Fables of VISHNUSERMAN, whom we ridiculously call *Pilpay*, are the most beautiful, if not the most ancient, collection of apologues in the world: they were first translated from the *Sanscrit*, in the *sixth* century, by the order of BUZERCHUMIHR, or *Bright as the Sun*, the chief physician and afterwards *Vēzīr* of the great ANŪSHIREVĀN, and are extant under various names in more than twenty languages; but their original title is *Hitōpadēsa*, or *Amicable Instruction*; and, as the very existence of ESOP, whom the *Arabs* believe to have been an *Abyssinian*, appears rather doubtful, I am not disinclined to suppose, that the first *moral fables*, which appeared in *Europe*, were of *Indian* or *Ethiopian* origin.

The *Hindus* are said to have boasted of *three* inventions, all of which, indeed, are admirable, the method of instructing by *apologues*, the *decimal scale* adopted now by all civilized nations, and the game of *Chess*, on which they have some curious treatises; but, if their numerous works on Grammar, Logick, Rhetorick, Musick, all which are extant and accessible, were explained in some language generally known, it would be found, that they had yet higher pretensions to the praise of a fertile and inventive genius. Their lighter Poems are lively and elegant; their Epick, magnificent and sublime in the highest degree; their *Purānas* comprise a series of mythological Histories in blank verse from the *Creation* to the supposed incarnation of BUDDHA; and their *Vēdas*, as far as we can judge from that compendium of them, which is called *Upanishat*, abound with noble speculations in metaphysicks, and fine discourses on the being and attributes of GOD. Their most ancient medical book, entitled *Chereca*, is believed to be the work of SIVA; for each of the divinities in their *Triad* has at least one *sacred* composition ascribed to him; but, as to mere human works on *History* and *Geography*, though they are said to be extant in *Cashmīr*, it has not been yet in my power to procure them. What their *astronomical* and *mathematical* writings contain, will not, I trust, remain long a secret: they are easily procured, and their importance cannot be doubted. The Philosopher, whose works are said to include a system of the universe founded on the principle of *Attraction* and the *Central* position of the sun, is named YAVAN ACHĀRYA, because he had travelled, we are told, into *Ionia*: if this be true, he might have been one of those, who conversed with PYTHAGORAS; this at least is undeniable, that a book on astronomy in *Sanscrit* bears the title of *Yavana Jātica*, which may signify the *Ionic Sect*; nor is it improbable, that the names of the planets and *Zodiacal* stars, which the *Arabs* borrowed from the *Greeks*, but which we find in the oldest *Indian* records, were originally devised by the same ingenious and enterprising race, from whom both *Greece* and *India* were peopled; the race, who, as DIONYSIUS describes them,

> first assayed the deep,
> And wafted merchandize to coasts unknown,
> Those, who digested first the starry choir,
> Their motions mark'd, and call'd them by their
>   names.[13]

Of these cursory observations on the *Hindus*, which it would require volumes to expand and illustrate, this is the result: that they had an immemorial affinity with the old *Persians*, *Ethiopians*, and *Egyptians*, the *Phenicians*, *Greeks*, and *Tuscans*, the *Scythians* or *Goths*, and *Celts*, the *Chinese*, *Japanese*, and *Peruvians*; whence, as no

reason appears for believing, that they were a colony from any one of those nations, or any of those nations from them, we may fairly conclude that they all proceeded from some *central* country, to investigate which will be the object of my future Discourses; and I have a sanguine hope, that your collections during the present year will bring to light many useful discoveries; although the departure for *Europe* of a very ingenious member, who first opened the inestimable mine of *Sanscrit* literature, will often deprive us of accurate and solid information concerning the languages and antiquities of *India*.[14]

13 [In Bryant's version, *New System*, 3: 230.]

14 [Wilkins was the first to translate the *Bhagavad-Gita* (1785). The vast universalism and comparative anthropology permeating Jones's 'Third Anniversary Discourse' literally revolutionized European thought about India and the interconnections of mankind.]

III. RELIGION, MYTHOLOGY, AND METAPHYSICS

This is the only purely Indian section in the book. It is difficult today to appreciate the intense emotion and scholarly reaction to several of Jones's essays on Indian topics, written in India, and presented to the Asiatic Society for suggestions before appearance in the first four volumes of *Asiatick Researches*, which caught the imagination of Western scholars. Though there were errors in his pioneering 'On the Gods of Greece, Italy, and India' and 'On the Antiquity of the Indian Zodiac', the comparative sweep and universalism in those essays prompted scholars to seek out much of the truth about those important topics. The length of these two essays dictates that only extracts can be included here. His last anniversary discourse to the Asiatic Society is significant in a different way, by reflecting the state of his great Indian knowledge just before his death, as well as the intellectual and practical values which he perceived therefrom. The length of his monumental *Institutes of Hindu Law* dictates that only two sections from that distinguished translation can be reprinted, though Jones's universalism and humanistic appreciation so richly endow his Preface that it too is included. Impressive passages from the Vedas which he collected and translated in sometimes unpolished drafts are also included, because of their intellectual and spiritual beauty, the insight which they provide into Jones's attitude toward Hinduism vis-a-vis Christianity, and the needed correction of Teignmouth's first printing of them. The same reasons prompt the inclusion of his translation of the *Mohamudgara*.

Overall, these works richly show his cosmopolitanism, his capacity to appreciate an antique non-European culture and to present that appreciation influentially to a colonial West. Indeed, his scholarly, comparative bent let him make valuable, objective comparisons with Christianity that few other scholars of the day could have made and that, overall, led Europe to reject completely the century-old Christian bias and to recognize Hinduism as one of the great, contributing religions and philosophies of the world.

On the Gods of Greece, Italy, and India[1]
Written in 1784, and since Revised
[Extract]

We cannot justly conclude, by arguments preceding the proof of facts, that one idolatrous people must have borrowed their deities, rites, and tenets from another; since Gods of all shapes and dimensions may be framed by the boundless powers of imagination, or by the frauds and follies of men, in countries never connected; but, when features of resemblance, too strong to have been accidental, are observable in different systems of polytheism, without fancy or prejudice to colour them and improve the likeness, we can scarce help believing, that some connection has immemorially subsisted between the several nations, who have adopted them: it is my design in this essay, to point out such a resemblance between the popular worship of the old *Greeks* and *Italians* and that of the *Hindus*; nor can there be room to doubt of a great similarity between their strange religions and that of *Egypt, China, Persia, Phrygia, Phœnice, Syria*; to which, perhaps, we may safely add some of the southern kingdoms and even islands of *America*; while the *Gothick* system, which prevailed in the northern regions of *Europe*, was not merely similar to those of *Greece* and *Italy*, but almost the same in another dress with an embroidery of images apparently *Asiatick*. From all this, if it be satisfactorily proved, we may infer a general union or affinity between the most distinguished

---

[1] [Published in *Asiatick Researches* (I), this second longest Indian essay by Jones makes innovational comparisons between Hindu divinities and the chief classical ones. Unfortunately, his yearning for universal connections among languages, cultures, and religions, based on the few Sanskrit sources then available, leads him into dubious conclusions partly based on accidental phonetic similarity of divinities' names. This once-celebrated essay, of which almost a fourth is extracted here, gave a tremendous impetus toward the development of the modern discipline of comparative religion, but today is chiefly of historical interest apart from its imaginative, zealous approach. See P.J. Marhall's fine critical edition of this essay, in *The British Discovery of Hinduism in the Eighteenth Century* (Cambridge, 1970): 196-245, whose notes have been used in the following annotations.]

inhabitants of the primitive world, at the time when they deviated, as they did too early deviate, from the rational adoration of the only true GOD.

There seem to have been four principal sources of all mythology.

I. Historical, or natural, truth has been perverted into fable by ignorance, imagination, flattery, or stupidity; as a king of *Crete*, whose tomb had been discovered in that island, was conceived to have been the God of *Olympus*,[2] and MINOS, a legislator of that country, to have been his son, and to hold a supreme appellate jurisdiction over departed souls; hence too probably flowed the tale of CADMUS, as BOCHART[3] learnedly traces it; hence beacons or volcanos became one-eyed giants and monsters vomiting flames; and two rocks, from their appearance to mariners in certain positions, were supposed to crush all vessels attempting to pass between them; of which idle fictions many other instances might be collected from the *Odyssey* and the various *Argonautick* poems. The less we say of *Julian* stars, deifications of princes or warriours, altars raised, with those of APOLLO, to the basest of men, and divine titles bestowed on such wretches as CAJUS OCTAVIANUS, the less we shall expose the infamy of grave senators and fine poets, or the brutal folly of the low multitude: but we may be assured, that the mad apotheosis of truly great men, or of little men falsely called great, has been the origin of gross idolatrous errors in every part of the pagan world. II. The next source of them appears to have been a wild admiration of the heavenly bodies, and, after a time, the systems and calculations of Astronomers: hence came a considerable portion of *Egyptian* and *Grecian* fable; the *Sabian* worship in *Arabia*; the *Persian* types and emblems of *Mihr* or the sun, and the far extended adoration of the elements and the powers of nature; and hence perhaps, all the artificial Chronology of the *Chinese* and *Indians*, with the invention of demigods and heroes to fill the vacant niches in their extravagant and imaginary periods. III. Numberless divinities have been created solely by the magick of poetry; whose essential business it is, to personify the most abstract notions, and to place a nymph or a genius in every grove and almost in every flower: hence *Hygieia* and *Jaso*, health and remedy, are the poetical daughters of ÆSCULAPIUS, who was either a distinguished physician, or medical skill personified; and hence *Chloris*, or

verdure, is married to the *Zephyr*. IV. The metaphors and allegories of moralists and metaphysicians have been also very fertile in Deities; of which a thousand examples might be adduced from PLATO, CICERO, and the inventive commentators on HOMER in their pedigrees of the Gods, and their fabulous lessons of morality: the richest and noblest stream from this abundant fountain is the charming philosophical tale of PSYCHE, or the *Progress of the Soul*;[4] than which, to my taste, a more beautiful, sublime, and well supported allegory was never produced by the wisdom and ingenuity of man. Hence also, the *Indian* MĀYĀ, or, as the word is explained by some *Hindu* scholars, 'the first inclination of the Godhead to diversify himself (such is their phrase) by creating worlds', is feigned to be the mother of universal nature, and of all the inferiour Gods; as a *Cashmirian* informed me, when I asked him, why CĀMA, or *Love*, was represented as her son; but the word MĀYĀ, or *delusion*, has a more subtile and recondite sense in the *Vēdānta* philosophy, where it signifies the system of *perceptions*, whether of secondary or of primary qualities, which the Deity was believed by EPICHARMUS, PLATO, and many truly pious men, to raise by his omnipresent spirit in the minds of his creatures, but which had not, in their opinion, any existence independent of mind.

In drawing a parallel between the Gods of the *Indian* and *European* heathens, from whatever source they were derived, I shall remember, that nothing is less favourable to enquiries after truth than a systematical spirit, and shall call to mind the saying of a *Hindu* writer, 'that whoever obstinately adheres to any set of opinions, may bring himself to believe that the freshest sandal-wood is a flame of fire': this will effectually prevent me from insisting, that such a God of *India* was *the* JUPITER of *Greece*; such, *the* APOLLO: such, *the* MERCURY: in fact, since all the causes of polytheism contributed largely to the assemblage of *Grecian* divinities (though BACON reduces them all to refined allegories,[5] and NEWTON to a poetical disguise of true history),[6] we find many JOVES, many APOLLOS, many MERCURIES, with distinct attributes and capacities; nor shall I presume to suggest more, than that, in one capacity or another, there exists a striking similitude between the chief objects of worship in ancient *Greece* or *Italy* and in the very interesting country, which we now inhabit.

---

[2] [Zeus, who was believed by the Cretans to have been buried in Crete.]

[3] [Samuel Bochart, *Geographia sacra, seu Phaleg et Chanaan* (Leyden, 1712), 1:447-54.]

[4] [Apuleius, *The Metamorphoses*, IV-VI.]

[5] [As in *De Sapientia Veterum* (London, 1609).]

[6] [In *The Chronology of Ancient Kingdoms Amended* (London, 1728).]

The comparison, which I proceed to lay before you, must needs be very superficial, partly from my short residence in *Hindustan*, partly from my want of complete leisure for literary amusements, but principally because I have no *European* book, to refresh my memory of old fables, except the conceited, though not unlearned, work of POMEY,[7] entitled the *Pantheon*, and that so miserably translated, that it can hardly be read with patience. A thousand more strokes of resemblance might, I am sure, be collected by any, who should with that view peruse HESIOD, HYGINUS,[8] CORNUTUS,[9] and the other mythologists; or, which would be a shorter and a pleasanter way, should be satisfied with the very elegant *Syntagmata* of LILIUS GIRALDUS.[10]

Disquisitions concerning the manners and conduct of our species in early times, or indeed at any time, are always curious at least and amusing; but they are highly interesting to such, as can say of themselves with CHREMES in the play, 'We are men, and take an interest in all that relates to mankind:'[11] They may even be of solid importance in an age, when some intelligent and virtuous persons are inclined to doubt the authenticity of the accounts, delivered by MOSES, concerning the primitive world; since no modes or sources of reasoning can be unimportant, which have a tendency to remove such doubts. Either the first eleven chapters of *Genesis*, all due allowances being made for a figurative Eastern style, are true, or the whole fabrick of our national religion is false; a conclusion, which none of us, I trust, would wish to be drawn. I, who cannot help believing the divinity of the MESSIAH, from the undisputed antiquity and manifest completion of many prophesies, especially those of ISAIAH, in the only person recorded by history, to whom they are applicable, am obliged of course to believe the sanctity of the venerable books, to which that sacred person refers as genuine; but it is not the truth of our national religion, as such, that I have at heart: it is truth itself; and, if any cool unbiassed reasoner will clearly convince me, that MOSES drew his narrative through *Egyptian* conduits from the primeval fountains of *Indian* literature, I shall esteem him as a friend for having weeded my mind from a capital error, and promise to

stand among the foremost in assisting to circulate the truth, which he has ascertained. After such a declaration, I cannot but persuade myself, that no candid man will be displeased, if, in the course of my work, I make as free with any arguments, that he may have advanced, as I should really desire him to do with any of mine, that he may be disposed to controvert. Having no system of my own to maintain, I shall not pursue a very regular method, but shall take all the Gods, of whom I discourse, as they happen to present themselves; beginning, however, like the *Romans* and the *Hindus*, with JANU or GANĒSA.

The titles and attributes of this old *Italian* deity are fully comprized in two choriambick verses of SULPITIUS; and a farther account of him from OVID would here be superfluous:[12]

> *Jane* pater, *Jane* tuens, dive biceps, biformis,
> O cate rerum sator, O principium deorum!

'Father JANUS, all-beholding JANUS, thou divinity with two heads, and with two forms; O sagacious planter of all things, and leader of deities!'[13]

He was the God, we see, of *Wisdom*; whence he is represented on coins with *two*, and, on the *Hetruscan* image found at *Falisci*, with *four*, faces; emblems of prudence and circumspection: thus is GANĒSA, the God of *Wisdom* in *Hindustan*, painted with an *Elephant's* head, the symbol of sagacious discernment, and attended by a favourite *rat*, which the *Indians* consider as a wise and provident animal. His next great character (the plentiful source of many superstitious usages) was that, from which he is emphatically styled *the father*, and which the second verse before-cited more fully expresses, *the origin and founder of all things*: whence this notion arose, unless from a tradition that he first built shrines, raised altars, and instituted sacrifices, it is not easy to conjecture; hence it came however, that his name was invoked before any other God; that, in the old sacred rites, corn and wine, and, in later times, incense also, were first offered to JANUS; that the *doors* or *entrances* to private houses were called *Januæ*, and any pervious passage or thorough-fare, in the plural number, *Jani*, or *with two beginnings*; that he was represented holding a rod as guardian of ways, and a key, as *opening*, not gates only, but *all important works and affairs* of mankind; that he was thought to preside over the morning, or *beginning of*

---

[7] [Frangois-Antoine Pomey's *Pantheon* (trans., London, 1698) discusses various mythological systems.]

[8] [The author of the *Genealogiæ*, or *Fabulæ*.]

[9] [Lucis Annaeus Cornutus, who wrote a summary of Greek mythology.]

[10] [By Lilio Gregario Giraldi in 1515.]

[11] [Terence, *Heauton Timorumenos*, I, i, 25.]

[12] [Janus is discussed in the *Fasti* I.]

[13] [Lines now attributed to Septimius Serenus.]

day; that, although the *Roman* year began regularly with *March*, yet the eleventh month, named *Januarius*, was considered as *first* of the twelve, whence the whole year was supposed to be under his guidance, and opened with great solemnity by the consuls inaugurated in his fane, where his statue was decorated on that occasion with fresh laurel; and, for the same reason, a solemn denunciation of war, than which there can hardly be a more momentous national act, was made by the military consul's opening the gates of his temple with all the pomp of his magistracy. The twelve altars and twelve chapels of JANUS might either denote, according to the general opinion, that he leads and governs twelve months, or that, as he says of himself in OVID, all entrance and access must be made through him to the principal Gods, who were, to a proverb, of the same number. We may add, that JANUS was imagined to preside over infants at their birth, or the *beginning* of life.

The *Indian* divinity has precisely the same character: all sacrifices and religious ceremonies, all addresses even to superiour Gods, all serious compositions in writing, and all worldly affairs of moment, are begun by pious *Hindus* with an invocation of GANĒSA; a word composed of *īsa*, the *governor* or *leader*, and *gana*, or *a company* of deities, *nine* of which companies are enumerated in the *Amarcōsh*.[14] Instances of opening business auspiciously by an ejaculation to the JANUS of *India* (if the lines of resemblance here traced will justify me in so calling him) might be multiplied with ease. Few books are begun without the words *salutation* to GANĒS, and he is first invoked by the *Brāhmans*, who conduct the trial by ordeal, or perform the ceremony of the *hōma*, or sacrifice to fire: M. SONNERAT represents him as highly revered on the Coast of *Coromandel*; 'where the *Indians*', he says, 'would not on any account build a house, without having placed on the ground an image of this deity, which they sprinkle with oil and adorn every day with flowers; they set up his figure in all their temples, in the streets, in the high roads, and in open plains at the foot of some tree; so that persons of all ranks may invoke him, before they undertake any business, and travellers worship him, before they proceed on their journey'.[15] To this I may add, from my own observation, that in the commodious and

useful town, which now rises at *Dharmāranya* or *Gayā*, under the auspices of the active and benevolent THOMAS LAW, Esq. collector of *Rotas*,[16] every new-built house, agreeably to an immemorial usage of the *Hindus*, has the name of GANĒSA superscribed on its door; and, in the old town, his image is placed over the gates of the temples.

- - - - - -   - - - - - -   - - - - - -

This epitome of the first *Indian* History, that is now extant, appears to me very curious and very important; for the story, though whimsically dressed up in the form of an allegory, seems to prove a primeval tradition in this country of the *universal deluge* described by MOSES, and fixes consequently the *time*, when the genuine *Hindu* Chronology actually begins. We find, it is true, in the *Purān*, from which the narrative is extracted, *another deluge* which happened towards the close of the *third* age, when YUDHIST'HIR was labouring under the persecution of his inveterate foe DURYŌDHAN, and when CRISHNA, who had recently become incarnate for the purpose of succouring the pious and of destroying the wicked, was performing wonders in the country of *Mat'hurà*; but the second flood was merely *local* and intended only to affect the people of *Vraja*: they, it seems, had offended INDRA, the God of the firmament, by their enthusiastick adoration of the wonderful child, 'who lifted up the mountain *Gōverdhena*, as if it had been a flower, and, by sheltering all the herdsmen and shepherdesses from the storm, convinced INDRA of his supremacy'. That the *Satya*, or (if we many venture so to call it) the *Saturnian*, age was in truth the age of the *general* flood, will appear from a close examination of the ten *Avatārs*, of *Descents*, of the deity in his capacity of preserver; since of the four, which are declared to have happened in the *Satya yug*, the *three first* apparently relate to some stupendous convulsion of our globe from the fountains of the deep, and the fourth exhibits the miraculous punishment of pride and impiety: first, as we have shown, there was, in the opinion of the *Hindus*, an interposition of Providence to preserve a devout person and his family (for all the *Pandits* agree, that his wife, though not named, must be understood to have been saved with him) from an inundation, by which all the wicked were destroyed; next, the power of the deity descends in the form of a *Boar*, the symbol of strength, to draw up and support on his tusks the whole earth, which had been sunk beneath the ocean; thirdly, the same power is represented as a *tortoise* sustaining the globe, which had been convulsed by the violent assaults

---

[14] [Jones translated his manuscript of the *Amarakosa*, which lists, 12,608 words with etymological comment. See Entry 440 in *Catalogue of the Library of the Late Sir William Jones* (London, 1831).]

[15] [See Pierre Sonnerat's *Voyage aux Indes orientales et à la Chine* (Paris, 1782.]

[16] [Law (1759-1834), with the East India Company since 1773, became a member of the Board of Trade.]

of demons, while the Gods churned the sea with the mountain *Mandar*, and forced it to disgorge the sacred things and animals, together with the water of life, which it had swallowed: these three relate, I think, to the same event, shadowed by a moral, a metaphysical, and an astronomical, allegory; and all three seem connected with the hieroglyphical sculptures of the old *Egyptians*. The fourth *Avatār* was a *lion* issuing from a bursting column of marble to devour a blaspheming monarch, who would otherwise have slain his religious son; and of the remaining six, not one has the least relation to a deluge: the three, which are ascribed to the *Trētāyug*, when tyranny and irreligion are said to have been introduced, were ordained for the overthrow of Tyrants, or, their natural types, Giants with a thousand arms formed for the most extensive oppression; and, in the *Dwāparyug*, the incarnation of CRISHNA was partly for a similar purpose, and partly with a view to thin the world of unjust and impious men, who had multiplied in that age, and began to swarm on the approach of the *Caliyug*, or the age of *contention* and baseness. As to BUDDHA, he seems to have been a reformer of the doctrines contained in the *Vēdas*; and, though his good nature led him to censure those ancient books, because they enjoined sacrifices of cattle, yet he is admitted as the ninth *Avatār* even by the *Brāhmans* of *Cāsì*, and his praises are sung by the poet JAYADĒVA: his character is in many respects very extraordinary; but, as an account of it belongs rather to History than to Mythology, it is reserved for another dissertation. The tenth *Avatār*, we are told, is yet to come, and is expected to appear mounted (like the crowned conqueror in the *Apocalyps*) on a white horse, with a cimeter blazing like a comet to mow down all incorrigible and impenitent offenders, who shall then be on earth.

These four *Yugs* have so apparent an affinity with the *Grecian* and *Roman* ages, that one origin may be naturally assigned to both systems: the first in both is distinguished as abounding in *gold*, though *Satya* mean *truth* and *probity*, which were found, if ever, in the times immediately following so tremendous an exertion of the divine power as the destruction of mankind by a general deluge; the next is characterized by *silver*, and the third, by *copper*; though their usual names allude to proportions imagined in each between vice and virtue: the present, or *earthen*, age seems more properly discriminated than by *iron*, as in ancient *Europe*; since that metal is not baser or less useful, though more common in our times and consequently less precious, than copper; while mere *earth* conveys an idea of the lowest degradation. We may here observe, that the true History of the World seems

obviously divisible into *four* ages or periods; which may be called, first, the *Diluvian*, or purest age; namely, the times preceding the deluge, and those succeeding it till the mad introduction of idolatry at *Babel*; next, the *Patriarchal*, or pure, age; in which, indeed, there were mighty hunters of beasts and of men, from the rise of patriarchs in the family of SEM to the simultaneous establishment of great Empires by the descendants of his brother HÁM: thirdly, the *Mosaick*, or less pure, age; from the legation of MOSES, and during the time, when his ordinances were comparatively well-observed and uncorrupted; lastly, the *Prophetical*, or *impure*, age, beginning with the vehement warnings given by the Prophets to apostate Kings and degenerate nations, but still subsisting and to subsist, until the genuine prophecies shall be fully accomplished. The duration of the Historical ages must needs be very unequal and disproportionate; while that of the *Indian Yugs* is disposed so regularly and artificially, that it cannot be admitted as natural or probable: men do not become reprobate in a geometrical progression or at the termination of regular periods; yet so well-proportioned are the *Yugs*, that even the length of human life is diminished, as they advance, from an hundred thousand years in a subdecuple ration; and, as the number of principal *Avatārs* in each decreases arithmetically from four, so the number of years in each decreases geometrically, and all together constitute the extravagant sum of four million three hundred and twenty thousand years, which aggregate, multiplied by seventy-one, is the period, in which every MENU is believed to preside over the world. Such a period, one might conceive, would have satisfied ARCHYTAS, the *measurer of sea and earth and the numberer of their sands*,[17] or ARCHIMEDES, who invented a notation, that was capable of expressing the number of them;[18] but the comprehensive mind of an *Indian* Chronologist has no limits; and the reigns of fourteen MENUS are only a single day of BRAHMĀ, fifty of which days have elapsed, according to the *Hindus*, from the time of the Creation: that all this puerility, as it seems at first view, may be only an astronomical riddle, and allude to the apparent revolution of the fixed stars, of which the *Brāhmans* made a mystery, I readily admit, and am even inclined to believe; but so technical an arrangement excludes all idea of serious History. I am sensible, how much these remarks will offend the warm advocates for *Indian* antiquity; but we

---

[17] [Quoted from Horace, *Carmina* 1: xxviii: 1-2.]

[18] [The *Sand-Reckoner*, to determine the number of grains of sand that a sphere as large as a 'universe' could contain.]

must not sacrifice truth to a base fear of giving offence: that the *Vēdas* were actually written before the flood, I shall never believe; nor can we infer from the preceding story, that the learned *Hindus* believe it; for the allegorical slumber of BRAHMĀ and the theft of the sacred books mean only, in simpler language, that *the human race was become corrupt*; but that the *Vēdas* are very ancient, and far older than other *Sanscrit* compositions, I will venture to assert from my own examination of them, and a comparison of their style with that of the *Purāns* and the *Dherma Sāstra*. A similar comparison justifies me in pronouncing, that the excellent law-book ascribed to SWĀYAMBHUVA MENU,[19] though not even pretended to have been written by him, is more ancient than the BHĀGAVAT: but that it was composed in the first age of the world, the *Brāhmans* would find it hard to persuade me; and the date, which has been assigned to it, does not appear in either of the two copies, which I possess, or in any other, that has been collated for me: in fact the supposed date is comprized in a verse, which flatly contradicts the work itself; for it was not MENU who composed the system of law, by the command of his father BRAHMĀ, but a holy personage or demigod, named BHRIGU, who revealed to men what MENU had delivered at the request of him and other saints or patriarchs. In the *Mānava Sāstra*, to conclude this digression, the measure is so uniform and melodious, and the style so perfectly *Sanscrit*, or *polished*, that the book must be more modern than the scriptures of MOSES, in which the simplicity, or rather nakedness, of the *Hebrew* dialect, metre, and style, must convince every unbiassed man of their superior antiquity.

I leave etymologists, who decide every thing, to decide whether the word MENU, or, in the nominative case, MENUS, has any connexion with MINOS, the Lawgiver, and supposed son of JOVE: the *Cretans*, according to DIODORUS of Sicily, used to feign, that most of the great men, who had been deified, in return for the benefits which they had conferred on mankind, were born in their island; and hence a doubt may be raised, whether MINOS was really a *Cretan*. The *Indian* legislator was the first, not the seventh, MENU, SATYAVRATA, whom I suppose to be the SATURN of *Italy*: part of SATURN's character, indeed, was that of a great lawgiver,

Qui genus indocile ac dispersum montibus altis
Composuit, *legesque dedit*,[20]

and, we may suspect, that all the fourteen MENUS are reducible to one, who was called NUH by the *Arabs*, and probably by the *Hebrews*, though we have disguised his name by an improper pronunciation of it. Some near relation between the seventh MENU and the Grecian MINOS may be inferred from the singular character of the *Hindu* God, YAMA, who was also a child of the Sun, and thence named VAIVASWATA: he had too the same title with his brother, ŚRĀDDHADĒVA: another of his titles was DHERMARĀJA, or *King of Justice*; and a third, PITRIPETI, or *Lord of the Patriarchs*; but he is chiefly distinguished as *judge of departed souls*; for the *Hindus* believe, that, when a soul leaves its body, it immediately repairs to *Yamapur*, or the city of YAMA, where it receives a just sentence from him, and either ascends to *Swerga*, or the first heaven, or is driven down to *Narac*, the region of serpents, or assumes on earth the form of some animal, unless its offence had been such, that it ought to be condemned to a vegetable, or even to a mineral, prison. Another of his names is very remarkable: I mean that of CĀLA, or *time*, the idea of which is intimately blended with the characters of SATURN and of NOAH; for the name CRONOS has a manifest affinity with the word *chronos*, and a learned follower of ZERĀTUSHT assures me, that, in the books, which the *Behdīns* hold sacred, mention is made of an *universal inundation*, there named the deluge of TIME.[21]

It having been occasionally observed, that CERES was the poetical daughter of SATURN, we cannot close this head without adding, that the *Hindus* also have their *Goddess of Abundance*, whom they usually call LACSHMĪ, and whom they consider as the daughter (not of MENU, but) of BHRIGU, by whom the first Code of sacred ordinances was promulgated: she is also named PEDMĀ and CAMALĀ from the sacred Lotos or *Nymphæa*; but her most remarkable name is SRĪ, or, in the first case, SRĪS, which has a resemblance to the *Latin*, and means *fortune* or *prosperity*. It may be contended, that, although LACSHMĪ may be figuratively called the CERES of *Hindustan*, yet any two or more idolatrous nations, who subsisted by agriculture, might naturally conceive a Deity to preside over their labours, without having the least intercourse with each other; but no reason appears, why two

---

[19] [Two chapters of Jones's translation of Manus's *Institutes of Hindu Law* are included in the Religion Section elsewhere in this book.]

[20] [Virgil, *Aeneid*, 8: 321-22.]
[21] [See the *Bundahis* (*Pahlavi Texts*, trans. E.W. West, *Sacred Books of the East* (London, 1880) 6: 25-28.]

nations should concur in supposing that Deity to be a female: one at least of them would be more likely to imagine, that the *Earth* was a Goddess, and that the God of abundance rendered her fertile. Besides, in very ancient temples near *Gayā*, we see images of LACSHMĪ, with full breasts and a *cord* twisted under her arm like a *horn of plenty*, which look very much like the old *Grecian* and *Roman* figures of CERES . . . .

On the Antiquity of the Indian Zodiack[1]

[Extract]

I engage to support an opinion (which the learned and industrious M. MONTUCLA seems to treat with extreme contempt), that the *Indian* division of the Zodiack was not borrowed from the *Greeks* or *Arabs*, but, having been known in this country from time immemorial, and being the same in part with that used by other nations of the old *Hindu* race, was probably invented by first progenitors of that race before their dispersion. 'The *Indians*', he says, 'have two divisions of the Zodiack; one, like that of the *Arabs*, relating to the moon, and consisting of *twenty-seven* equal parts, by which they can tell very nearly the hour of the night; another relating to the sun, and, like ours, containing twelve signs, to which they have given as many names corresponding with those, which we have borrowed from the *Greeks*'. All that is true; but he adds: 'It is highly probable that they received them at some time or another by the intervention of the *Arabs*; for no man, surely, can persuade himself, that it is the ancient division of the Zodiack formed, according to some authors, by the forefathers of mankind and still preserved among the *Hindus*'. Now I undertake to prove, that the *Indian* Zodiack was not borrowed mediately or directly from the *Arabs* or *Greeks*; and, since the solar division of it in *India* is the same in substance with that used in *Greece*, we may reasonably conclude, that both *Greeks* and *Hindus* re-

[1] [This once-celebrated essay, published in *Asiatick Researches* (2), rejects the usual eighteenth-century view of the origin of the zodiac, with an attack upon the theory of the French mathematician Jean Montucla that the Indian zodiac was borrowed from the Greeks or Arabs. Jones argues that the Indian zodiac derived from an early source, from which both the Greek and Indian zodiac developed. Though he pioneered in defending the originality of Sanskrit astronomy, he was carried too far by the extent of his discoveries to perceive the Greek and Indian resemblances and therefore did not simply amend the standard view. His scholarly reputation and the perceptiveness of his observations stirred a heated controversy that still finds a few modern scholars embracing his view. This essay is yet another striking result of Jones's Sanskrit studies. His two large, complicated astronomical drawings are not included here.]

ceived it from an older nation, who first gave names to the luminaries of heaven, and from whom both *Greeks* and *Hindus*, as their similarity in language and religion fully evinces, had a common descent.[2]

The same writer afterwards intimates, that 'the time, when *Indian* Astronomy received its most considerable improvement, from which it has now, as he imagines, wholly declined, was either the age, when the *Arabs*, who established themselves in *Persia* and *Sogdiana*, had a great intercourse with the *Hindus*, or that, when the successors of CHENGÍZ united both *Arabs* and *Hindus* under one vast dominion'. It is not the object of this essay, to correct the historical errors in the passage last cited, nor to defend the astronomers of *India* from the charge of gross ignorance in regard to the figure of the earth and the distances of the heavenly bodies; a charge, which MONTUCLA very boldly makes on the authority, I believe, of father SOUCIET: I will only remark, that in our conversations with the *Pandits*, we must never confound the system of the *Jyautishicas*, or mathematical astronomers, with that of the *Pauránics*, or poetical fabulists; for to such a confusion alone must we impute the many mistakes of *Europeans* on the subject of *Indian* science. A venerable mathematician of this province, named RĀMA-CHANDRA, now in his eightieth year, visited me lately at *Crishnanagar*, and part of his discourse was so applicable to the inquiries, which I was then making, that, as soon as he left me, I committed it to writing. 'The *Pauránics*', he said, 'will tell you, that our earth is a plane figure studded with eight mountains, and surrounded by seven seas of milk, nectar, and other fluids; that the part, which we inhabit, is one of seven islands, to which eleven smaller isles are subordinate; that a God, riding on a huge *elephant*, guards each of the eight regions; and that a mountain of gold rises and gleams in the centre; but we believe the earth to be shaped like a *Cadamba* fruit, or spheroidal, and admit only four oceans of salt water, all which we name from the four cardinal points, and in which are many great peninsulas with innumerable islands: they will tell you, that a dragon's head swallows the moon, and thus causes an eclipse; but we know, that the supposed head and tail of the dragon mean only the nodes, or points formed by intersections of the ecliptick and the moon's orbit; in short, they have imagined a system, which exists only in their fancy; but we consider

[2] [Here Jones attempts to broaden the linguistic relationship of Greek and Sanskrit so as to link two religions and the 'races' of their speakers as well, in a kind of universalization that he usually aspired toward.]

nothing as true without such evidence as cannot be questioned'. I could not perfectly understand the old Gymnosophist, when he told me, that the *Rāsichacra* or *Circle of Signs* (for so he called the Zodiack) was like a *Dhustūra* flower; meaning the *Datura*, to which the *Sanscrit* name has been softened, and the flower of which is conical or shaped like a funnel: at first I thought, that he alluded to a projection of the hemisphere on the plane of the colour, and to the angle formed by the ecliptick and equator; but a younger astronomer named Vināyaca, who came afterwards to see me, assured me that they meant only the circular mouth of the funnel, or the base of the cone, and that it was usual among their ancient writers, to borrow from fruits and flowers their appellations of several plane and solid figures.

From the two *Brāhmans*, whom I have just named, I learned the following curious particulars; and you may depend on my accuracy in repeating them, since I wrote them in their presence, and corrected what I had written, till they pronounced it perfect. They divide a great circle, as we do, into three hundred and sixty degrees, called by them *ansas* or *portions*; of which they, like us, allot thirty to each of the twelve signs in this order:

Mēsha, the Ram.	*Tulà, the Balance.*
Vrisha, the Bull.	8. *Vrishchica*, the Scorpion.
Mit'huna, the Pair.	Dhanus, the Bow.
4. *Carcat'i*, the Crab.	*Macara*, the Sea-Monster.
Sinha, the Lion.	*Cumbha*, the Ewer.
Canyà, the Virgin.	12. *Mīna*, the Fish.

The figures of the twelve asterisms, thus denominated with respect to the sun, are specified, by SRĪPETI, author of the *Retnamālà*, in *Sanscrit* verses; which I produce, as my vouchers, in the original with a verbal translation:

Mēshādayō nāma samānarūpi
Vināgadād h'yam mit'hunam nriyugmam,
Pradipas'asyē dadhatī carābhyām
Nāvi st'hitā vārin'i canyacaiva.
Tulā tulābhrit pretimānapānir
Dhanur dhanushmān hayawat parāngah,
Mrigānanah syān macarō't'ha cumbhah
Scandhē ñerō rictaghatam dadhānah,
Anyanyapuchch'hābhimuc'hō hi mīnah
Matsyadwayam swast'halachārinōmì.

'The *ram, bull, crab, lion,* and *scorpion,* have the figures of those five animals respectively: the pair are a damsel playing on a *Vīnà* and a youth wielding a mace: the *virgin* stands on a boat in water, holding in one hand a lamp, in the other an ear of ricecorn: the *balance* is held by a weigher with a weight in one hand: the *bow*, by an archer, whose hinder parts are like those of a horse: the *sea-monster* has the face of an antelope: the *ewer* is a waterpot borne on the shoulder of a man, who empties it: the *fish* are two with their heads turned to each others' tails; and all these are supposed to be in such places as suit their several natures'.

To each of the *twenty-seven* lunar stations, which they call *nacshatras*, they allow thirteen *ansas* and one third, or *thirteen degrees twenty minutes*; and their names appear in the order of the signs, but without any regard to the figures of them:

Aświnì	*Maghà*	Mūla
Bharanì.	Pūrva p'halgunì.	Pūrvāshād'ha'.
Criticà	Uttara p'halgunì.	Uttarāshād'hà.
Rōhinī.	Hasta.	*Sravanà.*
Mrigasiras.	*Chitrà.*	Dhanisht'à.
Ārdrà.	Swātì	Satabhishà.
Punarvasu	*Vīsāc'hà.*	Pūrva bhadrapadā.
Pushya.	Anurādhà.	Uttara- bhadrapadā.
9. As'lēshà.	18. *Jyēsht'hà.*	27. Rēvatì.

Between the twenty-first and twenty-second constellations, we find in the plate three called *Abhijit*; but they are the last quarter of the asterism immediately preceding, or the latter *Ashār*, as the word is commonly pronounced. A complete revolution of the moon, with respect to the stars, being made in twenty-seven days, odd hours, minutes, and seconds, and perfect exactness being either not attained by the *Hindus* or not required by them, they fixed on the number twenty-seven, and inserted *Abhijit* for some astrological purpose in their nuptial ceremonies. The drawing, from which the plate was engraved, seems intended to represent the figures of the twenty-seven constellations, together with *Abhijit*, as they are described in three stanzas by the author of the *Retnamālā*:

1. Turagmuc'hasadricsham yōnirūpam cshurābham,
   Sacat'asamam at'haiṇ'asyōttamāngēna tulyam,
   Maṇ'igrihaśara chacrābhāni sālōpaman bham,
   Sayanasadriśamanyachchātra paryancarūpam.

2. Hastācārayutam cha maucticasamam
   chānyat pravālōpamam,

Dhrishyam tōrana sannibham balinibham,
   satcund'alābham param;
   Crudhyatcēsarivicramēna sadriśam,
   śayyāsamānam param,
   Anyad dentivilāsavat st'hitamatah
   śringāt'acavyacti bham.

3. Trivicramābham cha mridangarūpam,
   Vrittam tatōnyadyamalābhwayābham,
   Paryancarūpam murajānucāram,
   Ityēvam aś'wādibhachacrarūpam.

'A horse's head; *yōni* or *bhaga*; a razor; a wheeled carriage; the head of an antelope; a gem; a house; an arrow; a wheel; another house; a bedstead; another bedstead; a hand; a pearl; a piece of coral; a festoon of leaves; an oblation to the Gods; a rich ear-ring; the tail of a fierce lion; a couch; the tooth of a wanton elephant, near which is the kernel of the *śringātaca* nut; the three footsteps of VISHNU; a tabor; a circular jewel; a two-faced image; another couch; and a smaller sort of tabor: such are the figures of *Aświnì* and the rest in the circle of lunar constellations'.

The *Hindu* draughtsman has very ill represented most of the figures; and he has transposed the two *Asharas* as well as the two *Bhadrapads*;[3] but his figure of *Abhijit*, which looks like our ace of hearts, has a resemblance to the kernel of the *trapa*, a curious water-plant described in a separate essay. In another *Sanscrit* book the figures of the same constellations are thus varied:

A horse's head.	A straight tail.	A conch
Yōni or *bhaga*.	Two stars S. to N.	A winnowing fan.
A flame.	Two, N. to S.	Another.
A waggon.	A hand.	An arrow.
A cat's paw.	A pearl.	A tabor.
One bright star.	Red saffron.	A circle of stars.
A bow.	A festoon.	A staff for burdens.
A child's pencil.	A snake.	The beam of a balance.
9. A dog's tail.	18. A boar's head.	27. A fish.

From twelve of the asterisms just enumerated are

---

[3] [Jones's insistence on correct drawings of figures, temples, etc. was constantly frustrated in Calcutta.]

derived the names of the twelve *Indian* months in the usual form of patronymicks; for the *Paurānics*, who reduce all nature to a system of emblematical mythology, suppose a celestial nymph to preside over each of the constellations, and feign that the God SŌMA, or *Lunas*, having wedded twelve of them, became the father of twelve *Genii*, or months, who are named after their several mothers; but the *Jyautishicas* assert, that, when their lunar year was arranged by former astronomers, the moon was at the full in each month on the very day, when it entered the *nacshatra*, from which that month is denominated. The manner, in which the derivatives are formed, will best appear by a comparison of the months with their several constellations:

	Āświna.		Chaitra.
	Cārtica.	8.	Vaisāc'ha.
	Mārgaśīrsha.		Jyaisht'ha.
4.	Pausha.		Āshāra.
	Māgha.		Srāvana.
	P'hālguna.	12.	Bhādra.

The third month is also called *Āgrahāyana* (whence the common word *Agran* is corrupted) from another name of *Mrigaśiras*.

Nothing can be more ingenious than the memorial verses, in which the *Hindus* have a custom of linking together a number of ideas otherwise unconnected, and of chaining, as it were, the memory by a regular measure: thus by putting *teeth* for thirty-two, *Rudra* for eleven, *season* for six, *arrow* or *element* for five, *ocean*, *Vēda*, or *age*, for four. RĀMA, *fire*, or *quality* for three, *eye*, or CUMĀRA for two, and *earth* or *moon* for one, they have composed four lines, which express the number of stars in each of the twenty-seven asterisms.

> Vahni tri ritwishu gunēndu critāgnibhūta,
> Bānāś winētra śara bhūcu yugabdhi rāmāh,
> Rudrābdhirāmagunavēdaśatā dwiyugma,
> Dentā budhairabhihitāh cramaśō bhatārāh.

That is: 'three, three, six; five, three, one; four, three, five; five, two, two; five, one, one; four, four, three; eleven, four and three; three, four, a hundred; two, two, thirty-two: thus have the stars of the lunar constellations, in order as they appear, been numbered by the wise'.

If the stanza was correctly repeated to me, the *two Ashārās* are considered as one asterism, and *Abhijit* as three separate stars' but I suspect an error in the third line, because *dwibāna* or *two and five* would suit the metre as well as *bdhirāma*; and because there were only three

Vēdas in the early age, when, it is probable, the stars were enumerated and the technical verse composed.

Two lunar stations, or *mansions*, and a quarter are co-extensive, we see, with one sign; and nine stations correspond with four signs: by counting, therefore, thirteen degrees and twenty minutes from the first star in the head of the Ram, inclusively, we find the whole extent of *Āświnī*, and shall be able to ascertain the other stars with sufficient accuracy; but first let us exhibit a comparative table of both *Zodiacks*, denoting the mansions, as in the *Vārānes* almanack, by the first letters or syllables of their names:

MONTHS	SOLAR ASTERISMS		MANSIONS		
Āświn	Mēsh		A	+ bh +
Cārtic	Vrish		$\frac{3c}{4}$	+ rò +
Āgrahāyan	Mit'hun		$\frac{M}{2}$	+ ā +
Paush	Carcat'4.		$\frac{P}{4}$	+ p +
Māgh	Sinh		m	+ PU +
P'hālgun	Canyà		$\frac{3U}{4}$	+ h +
Chaitr	Tulà		$\frac{ch}{2}$	+ s +
Vaisāc'h	Vrīschic 8.		$\frac{v}{4}$	+ a +
Jaishth	Dhan		mū	+ pù +
Āshār	Macar		$\frac{3u}{4}$	+ S +
Srāvan	Cumbh		$\frac{dh}{2}$	+ s' +
Bhādr	Mīn 12.		$\frac{pū}{4}$	+ u +

Hence we may readily know the stars in each mansion, as they follow in order:

LUNAR MANSIONS	SOLAR ASTERISM	STARS
Āświnī.	Ram.	*Three*, in and near the head.
Bharanì.	——	*Three*, in the tail.
Criticà.	Bull.	*Six*, of the Pleiads.
Rōhinī.	——	*Five*, in the head and neck.

LUNAR MANSIONS	SOLAR ASTERISM	STARS
Mrigasiras.	Pair.	*Three*, in or near the feet, perhaps in the Galaxy.
Ārdrā.	——	*One*, on the knee.
Punarvasu.	——	*Four*, in the heads, breast and shoulder.
Pushya.	Crab.	*Three*, in the body and claws.
As'lēshà.	Lion.	Five, in the face and mane.
Maghà.	——	*Five*, in the leg and haunch.
Pūrvap'hal-gunì.	——	*Two*, one in the tail.
Uttarap'hal-gunì.	Virgin.	*Two*, on the arm and zone.
Hasta.	——	*Five*, near the hand.
Chitrà.	——	*One*, in the spike.
Swāti.	Balance.	*One*, in the N. Scale.
Visāc'hà.	——	*Four*, beyond it.
Anurādhà.	Scorpion.	*Four*, in the body.
Jyēsht'hà.	——	*Three*, in the tail.
Mūla.	Bow.	*Eleven*, to the point of the arrow.
Pārvāshāra.	——	*Two*, in the leg.
Uttarāshāra.	Sea-Monster.	*Two*, in the horn.
Sravanà.	——	*Three*, in the tail.
Dhanisht'à.	Ewer.	*Four*, in the arm.
Satabhishà.	——	*Many*, in the stream.
Pūrva-bhadrapadà.	Fish.	*Two*, in the first fish.
Uttara-bhadrapadà.		*Two*, in the cord.
Rēvatì.	——	*Thirty-two*, in the second fish and cord.

Wherever the *Indian* drawing differs from the memorial verse in the *Retnamālà*, I have preferred the authority of the writer to that of the painter, who has drawn some terrestrial things with so little similitude, that we must not implicitly rely on his representation of objects merely celestial: he seems particularly to have erred in the stars of *Dhanisht'à*.

For the assistance of those, who may be inclined to re-examine the twenty-seven constellations with a chart before them, I subjoin a table of the degrees, to which the *nacshatras* extend respectively from the first star in the asterism of *Aries*, which we now see near the beginning of the sign *Taurus*, as it was placed in the ancient sphere.

N.	D.	M.	N.	D.	M.	N.	D.	M.
I.	13°.	20'.	X.	133°.	20'.	XIX.	253°.	20'.
II.	26°.	40'.	XI.	146°.	40'.	XX.	266°.	40'.
III.	40°.	0'.	XII.	160°.	0'.	XXI.	280°.	0'.
IV.	53°.	20'.	XIII.	173°.	20'.	XXII.	293°.	20'.
V.	66°.	40'.	XIV.	186°.	40'.	XXIII.	306°.	40'.
VI.	80°.	0'.	XV.	200°.	0'.	XXIV.	320°.	0'.
VII.	93°.	20'.	XVI.	213°.	20'.	XXV.	333°.	20'.
VIII.	106°.	40'.	XVII.	226°.	40'.	XXVI.	346°.	40'.
1X.	120°.	0'.	XVIII.	240°.	0'.	XXVII.	360°.	0'.

The asterisms of the *first* column are in the signs of *Taurus, Gemini, Cancer, Leo*; those of the second, in *Virgo, Libra, Scorpio, Sagittarius*; and those of the *third*, in *Capricornus, Aquarius, Pisces, Aries*: we cannot err much, therefore, in any series of *three* constellations; for, by counting 13° 20' forwards and backwards, we find the spaces occupied by the two extremes, and the intermediate space belongs of course to the middlemost. It is not meaned, that the division of the *Hindu* Zodiack into such spaces is exact to a minute, or that *every* star of each asterism must necessarily be found in the space to which it belongs; but the computation will be accurate enough for our purpose, and no lunar mansion can be very remote from the path of the moon: how Father SOUCIET could dream, that *Visāc'hà* was in the Northern Crown, I can hardly comprehend; but it surpasses all comprehension, that M. BAILLY should copy his dream, and give reasons to support it;[4] especially as four stars, arranged pretty much like those in the *Indian* figure, present themselves obviously near the balance or the scorpion. I have not the boldness to exhibit the individual stars in each mansion, distinguished in BAYER's method by *Greek* letters; because, though I have little doubt, that the five stars of *As'lēshà*, in the form of a wheel, are η, γ, ζ, μ, ε, of the Lion, and those of *Mūla*, γ, ε, δ, ζ, φ, τ, σ, ν, ο, ξ, π, of the *Sagittary*, and though I think many of the others equally clear, yet, where the number of stars in a mansion is less than three, or even than four, it is not easy to fix on them with confidence; and I must wait, until some young *Hindu* astronomer, with a good memory and good

[4] [Jean-Sylvain Bailly (1736-93), an astronomer and Orientalist, had written five books such as *Traite de l'astronomie indienne et orientale* (Paris, 1787).]

eyes, can attend my leisure on serene nights at the proper seasons, to point out in the firmament itself the several stars of all the constellations, for which he can find names in the *Sanscrit* language: the only stars, except those in the *Zodiack*, that have yet been distinctly named to me, are the *Septarshi*, *Dhruva*, *Arundhatì*, *Vishnupad*, *Mātrimandel*, and, in the southern hemisphere, *Agastya*, or *Canopus*. The twenty-seven *Yōga* stars, indeed, have particular names, in the order of the *nacshatras*, to which they belong; and since we learn, that the *Hindus* have determined *the latitude, longitude, and right ascension of each*, it might be useful to exhibit the list of them: but at present I can only subjoin the names of twenty-seven *Yōgas*, or divisions of the Ecliptick.

Vishcambha.	*Ganda.*	*Parigha.*
Prīti.	*Vriddhi.*	*Siva.*
Āyushmat.	*Dhruva.*	*Siddha.*
Saubhāgya.	*Vyāghāta.*	*Sādhya.*
Sōbhana.	*Hershana.*	*Subha.*
Atiganda.	*Vajra.*	*Sucra.*
Sucarman.	*Asrij.*	*Brāhman.*
Dhriti.	*Vyatipāta.*	*Indra.*
Sūla.	*Varīyas.*	*Vaidhriti.*

Having shown in what manner the *Hindus* arrange the *Zodiacal* stars with respect to the sun and moon, let us proceed to our principal subject, *the antiquity of that double arrangement*. In the first place, the *Brāhmans* were always too proud to borrow their science from the *Greeks, Arabs, Moguls*, or any nation of *Mlēchch'has*, as they call those, who are ignorant of the *Vēdas* and have not studied the language of the Gods: they have often repeated to me the fragment of an old verse, which they now use proverbially, *na nichò yavanātparah*, or *no base creature can be lower than a Yavan*: by which name they formerly meant an *Ionian* or *Greek*, and now mean a *Mogul*, or, generally, a *Muselman*. When I mentioned to different *Pandits*, at several times and in several places, the opinion of MONTUCLA, they could not prevail on themselves to oppose it by serious argument; but some laughed heartily; others, with a sarcastick smile, said it was a *pleasant imagination*; and all seemed to think it notion bordering on phrensy. In fact, although the figures of the twelve *Indian* signs bear a wonderful resemblance to those of the *Grecian*, yet they are too much varied for a mere copy, and the nature of the variation proves them to be original; nor is the resemblance more extraordinary than that, which has often been observed, between our

Gothick days of the week and those of the *Hindus*, which are dedicated to the same luminaries, and (what is yet more singular) revolve in the same order: *Ravi*, the Sun; *Sōma*, the Moon; *Mangala*, Tuisco; *Budha*, Woden; *Vrihaspati*, Thor; *Sucra*, Freya; *Sani*, Sater; yet no man ever imagined, that the *Indians* borrowed so remarkable an arrangement from the *Goths* or *Germans*. On the planets I will only observe, that SUCRA, the regent of *Venus*, is, like all the rest, a *male* deity, named also USANAS, and believed to be a sage of infinite learning; but ZOHRAH, the NĀHĪD of the *Persians*, is a goddess like the FREYA of our *Saxon* progenitors: the drawing, therefore, of the planets, which was brought into *Bengal* by Mr. JOHNSON[5] relates to the *Persian* system, and represents the genii supposed to preside over them, exactly as they are described by the poet HÁTIFÍ: 'He bedecked the firmament with stars, and ennobled this earth with the race of men; he gently turned the auspicious new moon of the festival, like a bright jewel, round the ankle of the sky; he placed the *Hindu* Saturn on the seat of that restive elephant, the revolving sphere, and put the rainbow into his hand, as a hook to coerce the intoxicated beast; he made silken strings of sun-beams for the lute of VENUS; and presented JUPITER, who saw the felicity of true religion, with a rosary of clustering Pleiads. The bow of the sky became that of MARS, when he was honoured with the command of the celestial host; for GOD conferred sovereignty on the Sun, and squadrons of stars were his army'.

The names and forms of the lunar constellations, especially of *Bharanì* and *Abhijit*, indicate a simplicity of manners peculiar to an ancient people; and they differ entirely from those of the *Arabian* system, in which the very first asterism appears in the dual number, because it consists only of two stars. *Menzil*, or *the place of alighting*, properly signifies a *station* or *stage*, and thence is used for an ordinary day's *journey*; and that idea seems better applied than *mansion* to so incessant a traveller as the moon: the *menāzilu'l kamar*, or *lunar stages*, or the *Arabs* have *twenty-eight* names in the following order, the particle *al* being understood before every word:

Sharatàn.	Nathrah.	Ghafr.	Dhábih'.
But'ain.	Tarf.	Zubánīyah.	Bulaâ.
Thurayyà.	Jabhah.	Iclìl.	Suûd.
Debaràn.	Zubrah.	Kalb.	Akhbīya.
Hakâah.	Sarfah.	Shaulah.	Mukdim.

[5] [Richard Johnson's knowledge of Persian made him a natural friend and correspondent of Jones.]

Hanâah.	Awwà.	Naâïm.	Múkhir.
7. Dhiráà.	14. Simàc.	21. Beldah.	28. Rishà.

Now, if we can trust the *Arabian* lexicographers, the number of stars in their several *menzils* rarely agrees with those of the *Indians*; and two such nations must naturally have observed, and might naturally have named, the principal stars, near which the moon passes in the course of each day, without any communication on the subject: there is no evidence, indeed, of a communication between the *Hindus* and *Arabs* on any subject of literature or science; for, though we have reason to believe, that a commercial intercourse subsisted in very early times between *Yemen* and the western coast of *India*, yet the *Brāhmans*, who alone are permitted to read the six *Vēdāngas*, one of which is the astronomical *Sāstra*, were not then commercial, and, most probably, neither could nor would have conversed with *Arabian* merchants. The hostile irruption of the Arabs into *Hindustān*, in the eighth century, and that of the *Moguls* under CHENGÍZ, in the thirteenth, were not likely to change the astronomical system of the *Hindus*; but the supposed consequences of *modern* revolutions are out of the question; for, if any historical records be true, we know with as positive certainty, that AMARSINH and CĀLIDĀS composed their works before the birth of CHRIST, as that MENANDER and TERENCE wrote before that important epoch: now the twelve *signs* and twenty-seven *mansions* are mentioned, by the several names before exhibited, in a *Sanscrit* vocabulary by the first of those *Indian* authors, and the second of them frequently alludes to Rōhinī and the rest by name in his *Fatal Ring*, his *Children of the Sun*, and his *Birth* of CUMĀRA; from which poem I produce two lines, that my evidence may not seem to be collected from mere conversation:

Maitrè muhūrtè śaśalānch' hanēna,
Yōgam gatāsūttarap'halganīshu.

'When the stars of *Uttarap'halgun* had joined in a fortunate hour the fawn-spotted moon'.

This testimony being decisive against the conjecture of M. MONTUCLA, I need not urge the great antiquity of MENU's Institutes, in which the twenty-seven asterisms are called the daughters of DACSHA and the consorts of SŌMA, or the Moon, nor rely on the testimony of the *Brāhmans*, who assure me with one voice, that the names of the *Zodiacal* stars occur in the *Vēdas*; three of which I firmly believe, from internal and external evidence, to be more than *three thousand* years old. Having therefore proved what I engaged to prove, I will close my essay

with a general observation. The result of NEWTON's researches into the history of the primitive sphere was, 'that the practice of observing the stars began in *Egypt* in the days of AMMON, and was propagated thence by conquest in the reign of his son SISAC, into *Africk, Europe,* and *Asia*; since which time ATLAS formed the sphere of the *Lybians*; CHIRON, that of the *Greeks*; and the *Chaldeans*, a sphere of their own': now I hope, on some other occasions, to satisfy the publick, as I have perfectly satisfied myself, that 'the practice of observing the stars began, with the rudiments of civil society, in the country of those, whom we call *Chaldeans*; from which it was propagated into *Egypt, India, Greece, Italy,* and *Scandinavia*, before the reign of SISAC or SĀCYA, who by conquest spread a new system of religion and philosophy from the *Nile* to the *Ganges* about a thousand years before CHRIST; but that CHIRON and ATLAS were allegorical or mythological personages, and ought to have no place in the serious history of our species'.[6]

---

[6] [Thus Jones corrected Newton, who had laboriously dated Chiron as having lived about 1170 B.C., in *The Chronology of Ancient Kingdoms Amended* (London, 1728), 89.

Discourse the Eleventh on the Philosophy of the Asiaticks[1]

Delivered 20 February 1794

Had it been of any importance, gentlemen, to arrange these anniversary dissertations according to the ordinary progress of the human mind, in the gradual expansion of its three most considerable powers, *memory, imagination*, and *reason*, I should certainly have presented you with an essay on the *liberal arts* of the five *Asiatick* nations, before, I produced my remarks on their *abstract sciences*; because, from my own observation at least, it seems evident, that *fancy* or the faculty of combining our ideas agreeably by various modes of imitation and substitution, is in general earlier exercised, and sooner attains maturity, than the power of separating and comparing those ideas by the laborious exertions of intellect; and hence, I believe, it has happened, that all nations in the world poets before they had mere philosophers: but, as M. D'ALEMBERT has deliberately placed science before art,[2] as the question of precedence is, on this occasion, of no moment whatever, and as many new facts on the subject of *Asiatick* philosophy are fresh in my remembrance, I propose to address you now on the sciences of *Asia*, reserving for our next annual meeting a disquisition concerning those fine arts, which have immemorially been cultivated, with different success and in

very different modes, within the circle of our common inquiries.

By science I mean an assemblage of transcendental propositions discoverable by human reason, and reducible to first principles, axioms, or maxims, from which they may all be derived in a regular succession; and there are consequently as many sciences as there are general objects of our intellectual powers: when man first exerts those powers, his objects are *himself* and the rest of *nature*; himself he perceives to be composed of *body* and *mind*, and in his *individual* capacity, he reasons on the *uses* of his animal frame and of its parts both exterior and internal, on the *disorders* impeding the regular functions of those parts, and on the most probable methods of preventing those disorders or of removing them; he soon feels the close connexion between his corporeal and mental faculties, and when his *mind* is reflected on itself, he discourses on its *essence* and its *operations*; in his *social* character, he analyzes his various *duties* and *rights* both private and publick; and in the leisure, which the fullest discharge of those duties always admits, his intellect is directed to *nature* at large, to the *substance* of natural bodies, to their several *properties*, and to their quantity both separate and united, finite and infinite; from all which objects he deduces notions, either purely abstract and universal, or mixed with undoubted facts, he argues from phenomena to theorems, from those theorems to other phenomena, from causes to effects, from effects to causes, and thus arrives at the demonstration of a *first intelligent cause*; whence his collected wisdom, being arranged in the form of science, chiefly consists of *physiology* and *medicine*, *metaphysicks* and *logick*, *ethicks* and *jurisprudence*, *natural philosophy* and *mathematicks*; from which the *religion of nature* (since revealed religion must be referred to *history*, as alone affording evidence of it) has in all ages and in all nations been the sublime and consoling result. Without professing to have given a logical definition of science, or to have exhibited a perfect enumeration of its objects, I shall confine myself to those *five* divisions of *Asiatick* philosophy, enlarging for the most part on the progress which the *Hindus* have made in them, and occasionally introducing the sciences of the *Arabs* and *Persians*, the *Tartars*, and the *Chinese*; but, how extensive soever may be the range which I have chosen, I shall beware of exhausting your patience with tedious discussions, and of exceeding those limits, which the occasion of our present meeting has necessarily prescribed.

I. The first article affords little scope; since I have no evidence, that, in any language of *Asia*, there exists one

---

[1] [This was Jones's last anniversary discourse to the Asiatic Society, delivered in pain only two months before his death. It presents an enlightened definition of the then-fuzzy term *science*, seeking to relate the sciences to the arts. With primary attention to classical India, he describes some of the Asian contributions to physiology and medicine, metaphysics and logic, ethics and jurisprudence, astronomy and mathematics, and 'the supremacy of an all-creating and all-preserving spirit'. His praise of the piety and sublimity of Hindu and Persian religious writings and the Koran reveals his wide, comparative sweep. This essay, like most of his essays composed in India, tantalized the West and deepened Europeans' growing interest in East-West intellectual-cultural cooperation. It was published in *Asiatick Researches* 4 (1794).

[2] [Jean Le Rond d'Alembert (1717-1783), French philosopher and mathematician, is remembered for his enunciation of a basic concept in mechanics and for his work on Diderot's *Encyclopédie*.]

original treatise on medicine considered as a *science*: physick, indeed, appears in these regions to have been from time immemorial, as we see it practised at this day by *Hindus* and *Muselmans*, a mere empirical *history* of diseases and remedies; useful, I admit, in a high degree, and worthy of attentive examination, but wholly foreign to the subject before us: though the *Arabs*, however, have chiefly followed the *Greeks* in this branch of knowledge, and have themselves been implicitly followed by other *Mohammedan* writers, yet (not to mention the *Chinese*, of whose medical works I can at present say nothing with confidence) we still have access to a number of *Sanscrit* books on the old *Indian* practice of physick, from which, if the *Hindus* had a theoretical system, we might easily collect it. The *Ayurvēda*, supposed to be the work of a celestial physician, is almost entirely lost, unfortunately perhaps for the curious *European*, but happily for the patient *Hindu*; since a revealed science precludes improvement from experience, to which that of medicine ought, above all others, to be left perpetually open; but I have myself met with curious fragments of that primeval work, and, in the *Vēda* itself, I found with astonishment an entire *Upanishad* on the internal parts of the human body; with an enumeration of nerves, veins, and arteries, a description of the heart, spleen, and liver, and various disquisitions on the formation and growth of the fetus: from the laws, indeed, of MENU, which have lately appeared in our own language, we may perceive, that the ancient *Hindus* were fond of reasoning in their way on the mysteries of animal generation, and on the comparative influence of the sexes in the production of perfect offspring; and we may collect from the authorities adduced in the learned Essay on *Egypt* and the *Nile*, that their physiological disputes led to violent schisms in religion, and even to bloody wars. On the whole, we cannot expect to acquire many valuable truths from an examination of eastern books on the science of medicine; but examine them we must, if we wish to complete the history of universal philosophy, and to supply the scholars of *Europe* with authentick materials for an account of the opinions anciently formed on this head by the philosophers of *Asia*: to know, indeed, with certainty, that so much and no more can be known on any branch of science, would in itself be very important and useful knowledge, if it had not other effect than to check the boundless curiosity of mankind, and to fix them in the straight path of attainable science, especially of such as relates to their duties and may conduce to their happiness. II. We have an ample field in the next division, and a field almost wholly new; since the metaphysicks and

logick of the *Brāhmens*, comprised in their *six* philosophical *Sāstras*, and explained by numerous glosses or comments, have never yet been accessible to *Europeans*; and, by the help of the *Sanscrit* language, we now may read the works of the *Saugatas*, *Bauddhas*, *Ārhatas*, *Jainas*, and other heterodox philosophers, whence we may gather the metaphysical tenets prevalent in *China* and *Japan*, in the eastern peninsula of *India*, and in many considerable nations of *Tartary*: there are also some valuable tracts on these branches of science in *Persian* and *Arabick*, partly copied from the *Greeks*, and partly comprising the doctrines of the *Sùfis* which anciently prevailed, and still prevail in great measure over this oriental world, and which the *Greeks* themselves condescended to borrow from eastern sages.

The little treatise in four chapters, ascribed to *Vyāsa*, is the only philosophical *Sāstra*, the original text of which I have had leisure to peruse with a *Brāhmen* of the *Vēdānti* school: it is extremely obscure, and, though composed in sentences elegantly modulated, had more resemblance to a table of contents, or an accurate summary, than to a regular systematical tract; but all its obscurity has been cleared by the labour of the very judicious and most learned SANCARA, whose commentary on the *Vēdānta*, which I read also with great attention, not only elucidates every word of the text, but exhibits a perspicuous account of all other *Indian* schools, from that of CAPILA to those of the more modern hereticks. It is not possible, indeed, to speak with too much applause of so excellent a work; and I am confident in asserting, that, until an accurate translation of it shall appear in some *European* language, the general history of philosophy must remain incomplete; for I perfectly agree with those, who are of opinion, that one correct version of any celebrated *Hindu* book would be of greater value than all the dissertations or essays, that could be composed on the same subject; you will not, however, expect, that, in such a discourse as I am now delivering, I should expatiate on the diversity of *Indian* philosophical schools, on the several founders of them, on the doctrines, which they respectively taught, or on their many disciples, who dissented from their instructors in some particular points. On the present occasion, it will be sufficient to say, that the oldest head of a sect, whose entire work is preserved, was (according to some authors) CAPILA; not the divine personage, a reputed grandson of BRĀHMA, to whom CRĪSHNA compares himself in the *Gītā*, but a sage of his name, who invented the *Sānc'hya*, or *Numeral*, philosophy, which CRĪSHNA himself appears to impugn in his conversation with ARJUNA, and which, as far as I can recollect it from a few original texts, resembled in part the metaphysicks of PYTHAGORAS, and in part the theology of ZENO: his doctrines were enforced and illustrated, with some additions, by the venerable PATANJALI, who has also left us a fine comment on the grammatical rules of PĀNINI, which are more obscure, without a gloss, than the darkest oracle; and here by the way let me add, that I refer to metaphysicks the curious and important science of *universal grammar*, on which many subtil disquisitions may be found interspersed in the particular grammars of the ancient *Hindus*, and in those of the more modern *Arabs*. The next founder, I believe, of a philosophical school was GŌTAMA, if, indeed, he was not the most ancient of all; for his wife AHALYÁ was, according to *Indian* legends, restored to a human shape by the great RĀMA: and a sage of his name, whom we have no reason to suppose a different personage, is frequently mentioned in the *Vēda* itself; to his rational doctrines those of CANĀDA were in general conformable; and the philosophy of them both is usually called *Nyāya*, or logical a title aptly bestowed; for it seems to be a system of metaphysicks and logick better accommodated than any other anciently known in *India*, to the natural reason and common sense of mankind; admitting the actual existence of *material substance* in the popular acceptation of the word *matter*, and comprising not only a body of sublime dialecticks, but an artificial method of reasoning, with distinct names for the three parts of a proposition, and even for those of a regular syllogism. Here I cannot refrain from introducing a singular tradition, which prevailed, according to the well-informed author of the *Dabistān*,[3] in the *Panjāb* and in several *Persian* provinces, that, 'among other *Indian* curiosities, which CALLISTHENES transmitted to his uncle, was a *technical system of logick*, which the *Brāhmens* had communicated to the inquisitive *Greek*', and which the *Mohammedan* writer supposes to have been the groundwork of the famous *Aristotelean* method: if this be true, it is one of the most interesting facts, that I have met with in *Asia*; and if it be false, it is very extraordinary, that such a story should have been fabricated either by the candid MOHSANI FÁNÌ; or by the simple *Pársìs Pandits*, with whom he had conversed; but, not having had leisure to study the *Nyāya Sāstra*, I can only assure you, that I have frequently seen perfect syllogisms in the philosophical

writings of the *Brāhmens*, and have often heard them used in their verbal controversies.

Whatever might have been the merit or age of GŌTAMA, yet the most celebrated *Indian* school is that, with which I began, founded by VYĀSA, and supported in most respects by his pupil JAIMINI, whose dissent on a few points is mentioned by his master with respectful moderation: their several systems are frequently distinguished by the names of the first and second *Mīmānsā*, a word, which, like *Nyāya*, denotes the operations and conclusions of reason; but the tract of VYĀSA has in general the appellation of *Vēdanta*, or the scope and end of the *Vēda*, on the texts of which, as they were understood by the philosopher, who collected them, his doctrines are principally grounded. The fundamental tenet of the *Vēdantī* school, to which in a more modern age the incomparable SANCARA was a firm and illustrious adherent, consisted, not in denying the existence of matter, that is, of solidity, impenetrability, and extended figure (to deny which would be lunacy), but, in correcting the popular notion of it, and in contending, that it has no essence independent of mental perception, that existence and perceptibility are convertible terms, that external appearances and sensations are illusory, and would vanish into nothing, if the divine energy, which alone sustains them, were suspended but for a moment; an opinion, which EPICHARMUS and PLATO seem to have adopted, and which has been maintained in the present century with great elegance, but with little publick applause; partly because it has been misunderstood, and partly because it has been misapplied by the false reasoning of some unpopular writers, who are said to have disbelieved in the moral attributes of GOD, whose omnipresence, wisdom, and goodness are the basis of the *Indian* philosophy: I have not sufficient evidence on the subject to profess a belief in the doctrine of the *Vēdanta*, which human reason alone could, perhaps, neither fully demonstrate, nor fully disprove; but it is manifest, that nothing can be farther removed from impiety than a system wholly built on the purest devotion; and the inexpressible difficulty, which any man, who shall make the attempt, will assuredly find in giving a satisfactory definition of *material substance*, must induce us to deliberate with coolness, before we censure the learned and pious restorer of the ancient *Vēda*; though we cannot but admit, that, if the common opinions of mankind be the criterion of philosophical truth, we must adhere to the system of GŌTAMA, which the *Brāhmens* of this province almost universally follow.

If the metaphysicks of the *Vēdantīs* be wild and erroneous, the pupils of BUDDHA have run, it is asserted,

---

[3] [Composed in India about the mid-seventeenth century, it is an elaborate treatise on twelve different religions, which Jones thought was 'the most amusing and instructive book I ever read in Persian' (*Letters* 2: 739).]

into an error diametrically opposite; for they are charged with denying the existence of pure spirit, and with believing nothing absolutely and really to exist but *material substance*; a heavy accusation which ought only to have been made on positive and incontestable proof, especially by the orthodox *Brāhmens*, who, as BUDDHA dissented from their ancestors in regard to *bloody sacrifices*, which the *Vēda* certainly prescribes, may not unjustly be suspected of low and interested malignity. Though I cannot credit the charge, yet I am unable to prove it entirely false, having only read a few pages of *Saugata* book, which Captain KIRKPATRICK had lately the kindness to give me;[4] but it begins, like other *Hindu* books, with the word *Ōm*, which we know to be a symbol of the divine attributes: then follows, indeed, a mysterious hymn to the Goddess of Nature, by name of *Āryā*, but with several other titles, which the *Brāhmens* themselves continually bestow on their *Dēvì*; now the *Brāhmens*, who have no idea, that any such personage exists as DĒVÌ, or the *Goddess*, and only mean to express allegorically the *power* of GOD, exerted in creating, preserving and renovating this universe, we cannot with justice infer, that the dissenters admit no deity but *visible nature*: the *Pandit*, who now attends me, and who told Mr. WILKINS, that the *Saugatas* were atheists, would not have attempted to resist the decisive evidence of the contrary, which appears in the very instrument, on which he was consulted, if his understanding had not been blinded by the intolerant zeal of a mercenary priesthood. A literal version of the book just mentioned (if any studious man had learning and industry equal to the talk) would be an inestimable treasure to the compiler of such a history as that of the laborious BRUCKER; but let us proceed to the *morals* and *jurisprudence* of the *Asiaticks*, on which I could expatiate, if the occasion admitted a full discussion of the subject, with correctness and confidence.

III.   That both ethicks and abstract law might be reduced to the *method of science*, cannot surely be doubted; but, although such a method would be of infinite use in a system of universal, or even of national, jurisprudence, yet the *principles* or morality are so few, so luminous, and so ready to present themselves on every occasion, that the practical utility of a scientifical arrangement, in a treatise on ethicks, may very justly be questioned. The moralists of the east have in general chosen to deliver their precepts in short sententious maxims, to illustrate them by sprightly comparisons, or to inculcate them in

the very ancient form of agreeable apologues: there are, indeed, both in *Arabick* and *Persian*, philosophical tracts on ethicks written with sound ratiocination and elegant perspicuity: but in every part of this eastern world, from *Pekin* to *Damascus*, the popular teachers of moral wisdom have immemorially been poets, and there would be no end of enumerating their works, which are still extant in the five principal languages of *Asia*. Our divine religion, the truth of which (if any history be true) is abudantly proved by historical evidence, has no need of such aids, as many are willing to give it, by asserting, that the wisest men of this world were ignorant of the two great maxims, that *we must act in respect of others, as we should wish them to act in respect of ourselves*, and that, *instead of returning evil for evil, we should confer benefits even on those who injure us*; but the first rule is implied in a speech of LYSIAS, and expressed in distinct phrases by THALES and PITTACUS; and I have even seen it word for word in the original of CONFUCIUS, which I carefully compared with the *Latin* translation. It has been usual with zealous men, to ridicule and abuse all those, who dare on this point to quote the *Chinese* philosopher; but, instead of supporting their cause, they would shake it, if it could be shaken, by their uncandid asperity; for they ought to remember, that one great end of revelation, as it is most expressly declared, was not to instruct the wise and few, but the many and unenlightened. If the conversation, therefore, of the *Pandits* and *Maulavis* in this country shall ever be attempted by protestant missionaries, they must beware of asserting, while they teach the gospel of truth, what those *Pandits* and *Maulavis* would know to be false: the former would cite the beautiful *Āryā* couplet, which was written at least three centuries before our era, and which pronounces the duty of a good man, even in the moment of his destruction, to consist *not only in forgiving, but even in a desire of benefiting, his destroyer, as the Sandal-tree, in the instant of its overthrow, sheds perfume on the axe, which fells it*; and the latter would triumph in repeating the verse of SADÌ, who represents *a return of good for good as a slight reciprocity*, but says to the virtuous man, '*Confer benefits on him, who has injured thee*', using an *Arabick* sentence, and a maxim apparently of the ancient *Arabs*. Nor would the *Muselmans* fail to recite four distichs of HÁFIZ, who has illustrated that maxim with fanciful but elegant allusions;

Learn from yon orient shell to love thy foe,
And store with pearls the hand, that brings thee wo:
Free, like yon rock, from base vindictive pride,

---

[4] [William Kirkpatrick (1754-1812), Orientalist, who eventually reached the rank of colonel in the Bengal Army.]

Imblaze with gems the wrist, that rends thy side:
Mark, where yon tree rewards the stony show'r
With fruit nectareous, or the balmy flow'r:
All nature calls aloud: 'Shall man do less
Than heal the smiter, and the railer bless?'

Now there is not a shadow of reason for believing, that the poet of *Shiraz* had borrowed this doctrine from the *Christians*; but, as the cause of *Christianity* could never be promoted by falsehood or errour, so it will never be obstructed by candour and veracity; for the lessons of CONFUCIUS and CHANACYA, of SADÌ and HÁFIZ, are unknown even at this day to millions of *Chinese* and *Hindus*, *Persians* and other *Mahommedans*, who toil for their daily support; nor, were they known ever so perfectly, would they have a divine sanction with the multitude; so that, in order to enlighten the minds of the ignorant, and to enforce the obedience of the perverse, it is evidently *a priori*, that a revealed religion was necessary in the great system of providence: but my principal motive for introducing this topick, was to give you a specimen of that ancient oriental morality, which is comprised in an infinite number of *Persian, Arabick*, and *Sanscrit* compositions.

Nearly one half of *jurisprudence* is closely connected with ethicks; but, since the learned of *Asia* consider most of their laws as positive and divine institutions, and not as the mere conclusions of human reason, and since I have prepared a mass of extremely curious materials, which I reserve for an introduction to the digest of *Indian* laws, I proceed to the fourth division, which consists principally of *science* transcendently so named, or *the knowledge of abstract quantities, of their limits, properties, and relations*, impressed on the understanding with the force of irresistible *demonstration*, which, as all other knowledge depends at best on our fallible senses, and in great measure on still more fallible testimony, can only be found, in pure mental abstractions; though for all the purposes of life, our own senses, and even the credible testimony of others, give us in most cases the highest degree of certainty, physical and moral.

IV.   I have already had occasion to touch on the *Indian* metaphysicks of *natural bodies* according to the most celebrated of the *Asiatick* schools from which the *Phythagoreans* are supposed to have borrowed many of their opinions; and, as we learn from CICERO, that the old sages of *Europe* had an idea of *centripetal force* and a principle of *universal gravitation* (which they never indeed attempted to demonstrate), so I can venture to affirm, without meaning to pluck a leaf from the neverfading laurels of our immortal NEWTON, that the whole of his theology and part of his philosophy may be found in the *Vēdas* and even in the works of the *Sûfis*: that *most subtil spirit*, which he suspected to pervade natural bodies, and, lying concealed in them, to cause attraction and repulsion, the emission, reflection, and refraction of light, electricity, calefaction, sensation, and muscular motion, is described by the *Hindus* as a *fifth element* endued with those very powers; and the *Vēdas* abound with allusions to a force universally attractive, which they chiefly ascribe to the Sun, thence called *Āditya*, or the *Attractor*; a name designed by the mythologists to mean the child of the Goddess ĀDITI; but the most wonderful passage on the theory of attraction occurs in the charming allegorical poem of SHÍRÍN and FERHÁD, or the *Divine Spirit* and *a human Soul disinterestedly pious*; a work which from the first verse to the last, is a blaze of religious and poetical fire.[5] The whole passage appears to me so curious, that I make no apology for giving you a faithful translation of it: 'There is a strong propensity, which dances through every atom, and attracts the minutest particle to some peculiar object; search this universe from its base to its summit, from fire to air, from water to earth, from all below the Moon to all above the celestial spheres, and thou wilt not find a corpuscle destitute of that natural attractibility; the very point of the first thread, in this apparently tangled skein, is no other than such a principle of attraction, and all principles beside are void of a real basis; from such a propensity arises every motion perceived in heavenly or in terrestrial bodies; it is a disposition to be attracted, which taught hard steel to rush from its place and rivet itself on the magnet; it is the same disposition, which impels the light straw to attach itself firmly on amber; it is this quality, which gives every substance in nature a tendency toward another, and an inclination forcibly directed to a determinate point'. These notions are vague, indeed, and unsatisfactory; but permit me to ask, whether the last paragraph of NEWTON's incomparable work goes much farther, and whether any subsequent experiments have thrown light on a subject so abstruse and obscure: that the sublime astronomy and exquisitely beautiful geometry, with which that work is illumined, should in any degree be approached by the Mathematicians of *Asia*, while of all *Europeans*, who ever lived, ARCHIMEDES alone was capable of emulating them, would be a vain expectation; but we must suspend

---

[5] [Nizámi's *Khosrau va Shirin*, a romantic poem in *Khamsa* that treats Khosrau II's love for the Caucasian princess Shírín, with the tragic consequences for Farhád, the kind's rival.]

our opinion of *Indian* astronomical knowledge, till the *Sūrya siddhānta* shall appear in our own language, and even then (to adopt a phrase of CICERO) our *greedy and capacious ears* will by no means be satisfied; for in order to complete an historical account of genuine *Hindu* astronomy, we require verbal translations of at least three other *Sanscrit* books; of the treatise by PARASARA, for the first age of *Indian* science, of that by VARĀHA, with the copious comment of his very learned son, for the middle age, and of those written by BHASCARA, for times comparatively modern. The valuable and now accessible works of the last mentioned philosopher, contain also an *universal*, or *specious*, arithmetick, with one chapter at least on geometry; nor would it, surely, be difficult to procure, through our several residents with the *Pishwā* and with SCINDHYA, the older books on algebra, which BHASCARA mentions, and on which Mr. DAVIS would justly set a very high value;[6] but the *Sanscrit* work, from which we might expect the most ample and important information, is entitled *Cshētrādersa*, or *a View of Geometrical Knowledge*, and was compiled in a very large volume by order of the illustrious JAYASINHA, comprising all that remains on the science in the sacred language of *India*: it was inspected in the west by a *Pandit* now in the service of Lieutenant WILFORD,[7] and might, I am persuaded, be purchased at *Jayanagar*, where Colonel POLIER had permission from the *Rājā* to buy the four *Vēdas* themselves. Thus have I answered, to the best of my power, the three first questions obligingly transmitted to us by professor PLAYFAIR;[8] whether the *Hindus* have books in *Sanscrit* expressly on geometry, whether they have any such on arithmetick, and whether a translation of the *Sūrya siddhānta* be not the great *desideratum* on the subject of *Indian* astronomy: to his three last ques-

tions, whether an accurate summary account of all the *Sanscrit* works on that subject, a delineation of the *Indian* celestial sphere, with correct remarks on it, and a description of the astronomical instruments used by the ancient *Hindus*, would not severally be of great utility, we cannot but answer in the affirmative, provided that the utmost critical sagacity were applied in distinguishing such works, constellations, and instruments, as are clearly of *Indian* origin, from such as were introduced into this country by *Muselman* astronomers from *Tartary* and *Persia*, or in later days by Mathematicians from *Europe*. V.   From all the properties of man and of nature, from all the various branches of science, from all the deductions of human reason, the general corollary, admitted by *Hindus, Arabs*, and *Tartars*, by *Persians*, and by *Chinese*, is the supremacy of an all-creating and all-preserving spirit, infinitely wise, good, and powerful, but infinitely removed from the comprehension of his most exalted creatures; nor are there in any language (the ancient *Hebrew* always excepted) more pious and sublime addresses to the being of beings, more splendid enumerations of his attributes, or more beautiful descriptions of his visible works, than in *Arabick, Persian* and *Sanscrit*, especially in the *Koran*, the introductions to the poems of SADÍ, NIZÁMÍ, and FIRDAUSÌ, the four *Vēdas* and many parts of the numerous *Puranas*: but supplication and praise would not satisfy the boundless imagination of the *Vēdāntī* and *Sùfì* theologists, who blending uncertain metaphysicks with undoubted principles of religion, have presumed to reason confidently on the very nature and essence of the divine spirit, and asserted in a very remote age, what multitudes of *Hindus* and *Muselmans* assert at this hour, that all spirit is homogeneous, that the spirit of GOD is in *kind* the same with that of man, though differing from it infinitely in *degree*, and that, as material substance is mere illusion, there exists in this universe only one generick spiritual substance, the sole primary cause, efficient, substantial and formal of all secondary causes and of all appearances whatever, but endued in its highest degree, with a sublime providential wisdom, and proceeding by ways incomprehensible to the spirits which emane from it; an opinion, which GŌTAMA never taught, and which we have no authority to believe, but which, as it is grounded on the doctrine of an immaterial creator supremely wise, and a constant preserver supremely benevolent, differs as widely from the pantheism of SPINOZA and TOLAND as the affirmation of a proposition differs from the negation of it; though the last named professor of that *insane philosophy* had the baseness to conceal his meaning under the very words of Saint PAUL[9] which are

[6] [Samuel Davis (1760-1819), a district judge in Bengal, whom Jones had strongly interested in Sanskrit astronomy.]

[7] [Francis Wilford (1760/1-1822), a Lieutenant on survey work in India, had done considerable work on Indian geography based on poor translations of the Purāṇas. But Jones required proof before accepting the 'discoveries' of Wilford, whose ignorance of Sanskrit but determination to find Biblical names in the sacred literature led Wilford's pundit to forge a sheet containing such names and thus greatly to mislead Western scholars for years.]

[8] [John Playfair (1748-1819), Edinburgh mathematician and geologist, sent a paper, 'Questions and Remarks on the Astronomy of the Hindus', in response to Jones's Advertisement in *Asiatick Researches* (2) for European scholars to use the Society's research facilities and organization. The paper appeared in *Asiatick Researches* (4).]

cited by NEWTON for a purpose totally different, and has
even used a phrase, which occurs, indeed, in the *Vēda*,
but in a sense diametrically opposite to that, which he
would have given it. The passage, to which I allude, is in
a speech of VARUNA to his son, where he says: 'That
spirit, from which these created beings proceed; through
which having proceeded from it, they live; toward which
they tend and in which they are ultimately absorbed, that
spirit study to know; that spirit is the Great One'.

    The subject of this discourse, gentlemen, is inex-
haustible: it has been my endeavour to say as much on it
as possible in the fewest words; and, at the beginning of
next year, I hope to close these general disquisitions with
topicks measureless in extent, but less abstruse than that,
which has this day been discussed, and better adapted to
the gaiety, which seems to have prevailed in the learned
banquets of the *Greeks*, and which ought, surely, to
prevail in every symposiack assembly.

---

⁹ [See John Toland, *Christianity as Old as the Creation: or,
the Gospel, a Republication of the Religion of Nature* (London,
1730), Chapter 12: 193.]

INSTITUTES
OF
HINDU LAW:
OR,
THE ORDINANCES OF MENU,[1]
ACCORDING TO THE
GLOSS OF CULLŪCA
COMPRISING THE
INDIAN SYSTEM OF DUTIES,
RELIGIOUS AND CIVIL

Verbally Translated From The Original
Sanscrit with A PREFACE
BY SIR WILLIAM JONES.
[Extract]

1 [This celebrated rendering of *Mānava-Dharmaśāstra*, published by the Bengal Government in 1794, has justly received perhaps the largest applause for any of Jones's translations from the Sanskrit, Persian, Arabic, or Greek. Intended as one of several sources for his digest of Indian law, which was designed to permit British officers to govern in accord with Indian laws and religions, this pioneering version may have become the basis for later studies of other *smṛtis* (such as Nārada, Yājñavalkya, Bṛhaspati among many). This monumental first translation of the Brahmanic code opened the stratled eyes of the Christian West to the richness and antiquity of Hinduism, and was reprinted in part as late as 1911 (in Watford). His masterly translation from the Sanskrit is introduced by a Preface distinguished by its polish, humanity, and comparative insights. To prepare his English translation, Jones used the collation of the Bengali commentator Kulluka Bhaṭṭa, checking the version against others. The Preface and the first ('On the Creation') and twelfth ('On Transmigration and Final Beatitude') chapters are included here, to introduce as well as conclude the essential meaning of 'the most sacred book next to the *Vēda*', to which Jones allotted most of his leisure in 1785-94. His version went through numerous editions and reprintings. See Cannon, *Sir William Jones: A Bibliography*, 32-34; and note 3 in *Letters* 2: 927-28.]

The Preface

It is a maxim in the science of legislation and government, that *Laws are of no avail without manners*, or, to explain the sentence more fully, that the best intended legislative provisions would have no beneficial effect even at first, and none at all in a short course of time, unless they were congenial to the disposition and habits, to the religious prejudices, and approved immemorial usages, of the people, for whom they were enacted; especially if that people universally and sincerely believed, that all their ancient usages and established rules of conduct had the sanction of an actual revelation from heaven: the legislature of *Britain* having shown, in compliance with this maxim, an intention to leave the natives of these *Indian* provinces in possession of their own Laws, at least on the titles of *contracts* and *inheritances*, we may humbly presume, that all future provisions for the administration of justice and government in *India*, will be conformable, as far as the natives are affected by them, to the manners and opinions of the natives themselves; an object, which cannot possibly be attained, until those manners and opinions can be fully and accurately known. These considerations, and a few others more immediately within my province, were my principal motives for wishing to know, and have induced me at length to publish, that system of duties, religious and civil, and of law in all its branches, which the *Hindus* firmly believe to have been promulged in the beginning of time by MENU, son or grandson of BRAHMÁ, or, in plain language, the first of created beings, and not the oldest only, but the holiest, of legislators; a system so comprehensive and so minutely exact, that it may be considered as the *Institutes* of *Hindu* Law, preparatory to the copious *Digest*, which has lately been compiled by *Pandits* of eminent learning, and introductory perhaps to a *Code*, which may supply the many natural defects in the old jurisprudence of this country, and, without any deviation from its principles, accomodate it justly to the improvements of a commercial age.

We are lost in an inextricable labyrinth of imaginary astronomical cycles, *Yugas, Mahāyugas, Calpas*, and *Menwantaras*, in attempting to calculate the time, when the first MENU, according to the *Brāhmens*, governed this world, and became the progenitor of mankind, who from him are called *Mānavāh*; nor can we, so clouded are the old history and chronology of *India* with fables and allegories, ascertain the precise age, when the work, now presented to the Publick, was actually composed; but we are in possession of some evidence, partly extrinsick and partly internal, that it is really one of the oldest compositions existing. From a text of PARĀSARA, discovered by MR. DAVIS, it appears, that the vernal equinox had gone back from the *tenth* degree of *Bharanì* to the *first* of *Āswinì*, or *twenty-three degrees and twenty minutes*, between the days of that *Indian* philosopher, and the year of our Lord 499, when it coincided with the origin of the *Hindu* eclipitick; so that PARĀSARA probably flourished near the close of the *twelfth* century before CHRIST: now PARĀSARA was the grandson of another sage, named VASISHT'HA, who is often mentioned in the laws of MENU, and once as contemporary with the divine BHRIGU himself; but the character of BHRIGU, and the whole dramatical arrangement of the book before us, are clearly fictitious and ornamental, with a design, too common among ancient lawgivers, of stamping authority on the work by the introduction of supernatural personages, though VASISHT'HA may have lived many generations before the actual writer of it; who names him, indeed, in one or two places as a philosopher in an earlier period. The style, however, and metre of this work (which there is not the smallest reason to think affectedly obsolete) are widely different from the languages and metrical rules of CĀLIDĀS, who unquestionably wrote before the beginning of our era; and the dialect of MENU is even observed in many passages to resemble that of the *Vēda*, particularly in a departure from the more modern grammatical forms; whence it must at first view seem very probable, that the laws, now brought to light, were considerably older than those of SOLON[2] or even of LYCURGUS,[3] although the promulgation of them, before they were reduced to writing, might have been coeval with the first monarchies established in *Egypt* or *Asia*: but, having had the singular good fortune to procure ancient copies of eleven *Upanishads* with a very perspicuous comment, I am enabled to fix with more exactness the probable age of the work before us, and even to limit its highest possible age, by a mode of reasoning, which may be thought new, but will be found, I persuade myself, satis-

factory; if the Publick shall on this occasion give me credit for a few very curious facts, which, though capable of strict proof, can at present be only asserted. The *Sanscrit* of the three first *Vēdas* (I need not here speak of the fourth), that of the *Mānava Dherma Sāstra*, and that of the *Purānas*, differ from each other in pretty exact proportion to the Latin of NUMA, from whose laws entire sentences are preserved, that of APPIUS, which we see in the fragments of the Twelve Tables, and that of CICERO, or of LUCRETIUS, where he has not affected an obsolete style: if the several changes, therefore, of *Sanscrit* and *Latin* took place, as we may fairly assume, in times very nearly proportional, the *Vēdas* must have been written about 300 years before these Institutes, and about 600 before the *Purānas* and *Itihāsas*, which, I am fully convinced, were not the productions of VYĀSA; so that, if the son of PARĀSARA committed the traditional *Vēdas* to writing in the *Sanscrit* of his father's time, the original of this book must have received its present form about 880 years before CHRIST's birth. If the texts, indeed, which VYĀSA collected, had been actually *written*, in a much older dialect, by the sages preceding him, we must inquire into the greatest possible age of the *Vēdas* themselves: now one of the longest and finest *Upanishads* in the second *Vēda* contains three lists, in a regular series upwards, of at most forty-two pupils and preceptors, who successively received and transmitted (probably by oral tradition) the doctrines contained in that *Upanishad*; and, as the old *Indian* priests were students at *fifteen*, and instructors at *twenty-five*, we cannot allow more than *ten* years on an average for each interval between the respective traditions; whence, as there are *forty* such intervals, in two of the lists, between VYĀSA, who arranged the whole work, and AYĀSYA, who is extolled at the beginning of it, and just as many, in the third list, between the compiler and YĀJNYAWALCYA, who makes the principal figure in it, we find the highest age of the *Yajur-Vēda* to be 1580 years before the birth of our Saviour, (which would make it older than the five books of MOSES) and that of our *Indian* lawtract about 1280 years before the same epoch. The former date, however, seems the more probable of the two, because the *Hindu* sages are said to have delivered their knowledge orally, and the very word *Sruta*, which we often see used for the *Vēda* itself, means *what was heard*; not to insist, that CULLŪCA expressly declares the sense of the *Vēda* to be conveyed in the *language* of VYĀSA. Whether MENU, or MENUS in the nominative and MENŌS in an oblique case, was the same personage with MINOS, let others determine; but he must indubitably have been far older than the work, which

[2] [The celebrated Athenian legislator, born about 638 B.C., whose laws were inscribed on wooden cylinders and triangular tablets and set up in the Acropolis and Prytaueum, had also provided the citizenry with some control over their government, by making officials subject to the courts.]

[3] [A famous Spartan law-giver, said to have visited Crete and there to have studied the laws of Minos.]

contain his laws, and, though perhaps he was never in *Crete*, yet some of his institutions may well have been adopted in that island, whence LYCURGUS a century or two afterwards may have imported them to *Sparta*.

There is certainly a strong resemblance, though obscured and faded by time, between our MENU with his divine Bull, whom he names as DHERMA himself, or the genius of abstract justice, and the MNEUES of Egypt with his companion or symbol, *Apis*; and, though we should be constantly on our guard against the delusion of etymological conjecture, yet we cannot but admit that MINOS and MNEUES, or *Mneuis*, have only Greek terminations, but that the crude noun is composed of the same radical letters both in *Greek* and in *Sanscrit.* 'That APIS and MNEUIS', says Analyst of ancient Mythology, 'were both representations of some personage, appears from the testimony of LYCOPHRON and his scholiast; and that personage was the same, who in *Crete* was styled MINOS, and who was also represented under the emblem of the *Minotaur*: DIODORUS, who confines him to *Egypt*, speaks of him by the title of the bull *Mneuis*, as the first lawgiver, and says, 'That he lived after the age of the gods and heroes, when a change was made in the manner of life among men; that he was a man of a most exalted soul, and a great promoter of civil society, which he benefited by his laws; that those laws were unwritten, and received by him from the chief *Egyptian* deity HERMES, who conferred them on the world as a gift of the highest importance. He was the same', adds my learned friend, 'with MENES, whom the *Egyptians* represented as their first king and principal benefactor, who first sacrificed to the gods, and brought about a great change in diet'. If MINOS, the son of JUPITER, whom the *Cretans*, from national vanity, might have made a native of their own island, was really the same person with MENU, the son of BRAHMĀ, we have the good fortune to restore, by means of *Indian* literature, the most celebrated system of heathen jurisprudence, and this work might have been entitled *The Laws of* MINOS; but the paradox is too singular to be confidently asserted, and the geographical part of the book, with most of the allusions to natural history, must indubitably have been written after the *Hindu* race had settled to the south of *Himālaya*. We cannot but remark that the word MENU has no relation whatever to the *Moon*; and that it was the *seventh*, not the *first*, of that name, whom the *Brāhmens* believe to have been preserved in an ark from the general duluge: him they call the *Child of the Sun*, to distinguish him from our legislator; but they assign to his brother YAMA *the office*

(which the *Greeks* were pleased to confer on MINOS) *of Judge in the shades below.*

The name of MENU is clearly derived (like *menes, mens*, and *mind*) from the root *men* to *understand*; and it signifies, as all the *Pandits* agree, *intelligent*, particularly in the doctrines of the *Vēda*, which the composer of our *Dherma* Sastra must have studied very diligently; since great number of its texts, changed only in a few syllables for the sake of the measure, are interspersed through the work and cited at length in the commentaries: the Publick may, therefore, assure themselves, that they now possess a considerable part of the *Hindu* scripture, without the dullness of its profane ritual or much of its mystical jargon. DĀRA SHUCŪH was persuaded, and not without sound reason, that the first MENU of the *Brāhmens* could be no other person than the progenitor of mankind, to. whom *Jews, Christians*, and *Muselmāns* unite in giving the name of ADAM; but, whoever he might have been, he is highly honoured by name in the *Vēda* itself, where it is declared, that 'whatever MENU pronounced, was a medicine for the soul'; and the sage VRIHASPETI, now supposed to preside over the planet *Jupiter*, says in his own law tract, that 'MENU held the first rank among legislators, because he had expressed in his code the whole sense of the *Vēda*; that no code was approved, which contradicted MENU; that other *Sāstras*, and treatises on grammar or logick, retained splendour so long only, as MENU, who taught the way to just wealth, to virtue, and to final happiness, was not seen in competition with them': VYĀSA too, the son of PARĀSARA before mentioned, has decided, that 'the *Vēda* with its *Angas*, or the six composition deduced from it, the revealed system of medicine, the *Purānas*, or sacred histories, and the code of MENU, were four works of supreme authority, which ought never to be shaken by arguments merely human'.

It is the general opinion of *Pandits*, that BRAHMĀ taught his laws to MENU in a *hundred thousand verses*, which MENU explained to the primitive world in the very words of the book now translated, where he names himself, after the manner of ancient sages, in the third person; but, in a short preface to the lawtract of NĀRED, it is asserted, that 'MENU, having written the laws of BRAHMĀ in a hundred thousand *slōcas* or couplets, arranged under *twenty-four* heads in a *thousand* chapters, delivered the work to NĀRED, the sage among gods, who abridged it, for the use of mankind, in *twelve thousand* verses, and gave them to a son of BHRIGU, named SUMATI, who, for greater ease to the human race, reduced them to *four thousand*; that mortals read only the second abridgement by SUMATI, while the gods of the lower heaven, and the band of celestial musicians, are engaged in studying the primary code, beginning with the fifth verse, a little varied, of the work now extant on earth; but that nothing remains of NĀRED's abridgement, except an elegant epitome of the *ninth* original title *on the administration of justice'*. Now, since these institutes consist only of *two thousand six hundred and eighty-five* verses, they cannot be the whole work ascribed to SUMATI, which is probably distinguished by the name of the *Vriddha*, or ancient, *Mānava*, and cannot be found entire; though several passages from it, which have been preserved by tradition, are occasionally cited in the new digest.

A number of glosses or comments on MENU were composed by the *Munis*, or old philosophers, whose treatises, together with that before us, constitute the *Dhermasāstra*, in a collective sense, or *Body of Law*; among the more modern commentaries, that called *Mēdhātit'hi*, that by GŌVINDARĀJA, and that by DHARANĪ-DHERA, were once in the greatest repute; but the first was reckoned prolix and unequal; the second, concise but obscure; and the third, often erroneous. At length appeared CULLŪCA BHATTA; who, after a painful course of study, and the collation of numerous manuscripts, produced a work, of which it may, perhaps, be said very truly, that it is the shortest, yet the most luminous, the least ostentatious, yet the most learned, the deepest yet the most agreeable, commentary ever composed on any author ancient or modern, *European* or *Asiatick*. The *Pandits* care so little for genuine chronology, that none of them can tell me the age of CULLŪCA, whom they always name with applause; but he informs us himself, that he was a *Brāhmen* of the *Vārēndra* tribe, whose family had been long settled in *Gaur* or *Bengal*, but that he had chosen his residence among the learned on the banks of the holy river at *Cāsi*. His text and interpretation I have almost implicitly followed, though I had myself collated many copies of MENU, and among them a manuscript of a very ancient date: his gloss is here printed in *Italicks*; and any reader who may choose to pass it over as if unprinted, will have in *Roman* letters an exact version of the original, and may form some idea of its character and structure, as well as of the *Sanscrit* idiom, which must necessarily be preserved in a verbal translation; and a translation not scrupulously verbal would have been highly improper in a work on so delicate and momentous a subject as private and criminal jurisprudence.

Should a series of *Brāhmens* omit, for three generations, the reading of MENU, their sacerdotal class, as all the *Pandits* assure me, would in strictness be forfeited; but they must explain it only to their pupils of the three highest classes; and the *Brāhmen*, who read it with me, requested most earnestly, that his name might be concealed; nor would he have read it for any consideration on a forbidden day of the moon, or without the ceremonies prescribed in the second and fourth chapters for a lecture on the *Vēda*: so great, indeed, is the idea of sanctity annexed to this book, that, when the chief native magistrate at *Banares* endeavoured, at my request, to procure a Persian translation of it, before I had a hope of being at any time able to understand the original, the *Pandits* of his court unanimously and positively refused to assist in the work; nor should I have procured it at all, if a wealthy *Hindu* at *Gayà* had not caused the version to be made by some of his dependants, at the desire of my friend Mr. [Thomas] LAW. The *Persian* translation of MENU, like all others from the *Sanscrit* into that language, is a rude intermixture of the text, loosely rendered, with some old or new comment, and often with the crude notions of the translator; and, though it expresses the general sense of the original, yet it swarms with errours, imputable partly to haste, and partly to ignorance: thus where MENU says, *that emissaries are the eyes of a prince*, the *Persian* phrase makes him ascribe *four eyes* to the person of a king; for the word *chār*, which means an *emmissary* in *Sanscrit*, signifies *four* in the popular dialect.

The work, now presented to the *European* world, contains abundance of curious matter extremely interesting both to speculative lawyers and antiquaries, with many beauties, which need not be pointed out, and with many blemishes, which cannot be justified or palliated. It is a system of despotism and priestcraft, both indeed limited by law, but artfully conspiring to give mutual support, though with mutual checks; it is filled with strange conceits in metaphysicks and natural philosophy, with idle superstitions, and with a scheme of theology most obscurely figurative, and consequently liable to dangerous misconception; it abounds with minute and childish formalities, with ceremonies generally absurd and often ridiculous; the punishments are partial and fanciful, for some crimes dreadfully cruel, for others reprehensibly slight; and the very morals, though rigid enough on the whole, are in one or two instances (as in the case of light oaths and of pious perjury) unaccountably relaxed: nevertheless, a spirit of sublime devotion, of benevolence to mankind, and of amiable tenderness to all sentient creatures, pervades the whole work; the style of it has a certain austere majesty, that sounds like the language of legislation and extorts a respectful awe; the sentiments of independence on all beings but GOD, and the harsh admonitions even to kings are truly noble; and the many panegyricks on the *Gāyatrì*, the *Mother*, as it is called, of the *Vēda*, prove the author to have *adored* (not the visible material sun, but) *that divine and incomparably greater light*, to use the words of the most venerable text in the *Indian* scripture, *which illumines all, delights all, from which all proceed to which all must return, and which alone can irradiate* (not our visual organs merely, but our souls and) *our intellects*. Whatever opinion in short may be formed of MENU and his laws, in a country happily enlightened by sound philosophy and the only true revelation, it must be remembered, that those laws are actually revered, as the word of the Most High, by nations of great importance to the political and commercial interests of *Europe*, and particularly by many millions of *Hindu* subjects, whose well directed industry would add largely to the wealth of *Britain*, and who ask no more in return than protection for their persons and places of abode, justice in their temporal concerns, indulgence to the prejudices of their own religion, and the benefit of those laws, which they have been taught to believe sacred, and which alone they can possibly comprehend.

W. JONES.

The Laws of Menu, Son of Brahmā

CHAPTER THE FIRST

On the Creation; with a Summary of the Contents

1. MENU *sat* reclined, with his attention fixed on one object, *the Supreme* GOD; *when* the divine Sages approached *him, and,* after mutual salutations in due form, delivered the following address:

2. 'Deign, sovereign ruler, to apprize us of the sacred laws in their order, as they must be followed by all the *four* classes, and by each of them, in their several degrees, together with the duties of every mixed class;

3. 'For thou, Lord, *and thou* only *among mortals,* knowest the true sense, the first principle, *and* the prescribed ceremonies, of this universal, supernatural *Vēda,* unlimited in extent and unequalled in authority'.

4. HE, whose powers were measureless, being thus requested by the great Sages, whose thoughts were profound, saluted them all with reverence, and gave them a comprehensive answer, *saying*: 'Be it heard!

5. 'This *universe* existed only *in the first divine idea yet unexpanded, as if involved* in darkness, imperceptible, undefinable, undiscoverable *by reason, and* undiscovered *by revelation,* as if it were wholly immersed in sleep:

6. 'Then the *sole* self-existing power, himself undiscerned, but making this world discernible, with five elements and other principles *of nature,* appeared with undiminished glory, *expanding his idea, or* dispelling the gloom.

7. 'HE, whom the mind alone can perceive, whose essence eludes the external organs, who has no visible parts, who exists from eternity, even HE, the soul of all beings, whom no being can comprehend, shone forth in person.

8. 'HE, having willed to produce various beings from his own divine substance, first with a thought created the waters, and placed in them a productive seed:

9. 'That seed became an egg bright as gold, blazing like the luminary with a thousand beams; and in that egg he was born himself, *in the form of* BRAHMĀ, the great forefather of all spirits.

10. 'The waters are called *nārā*, because they were the production of NARA, *or the spirit of* GOD; and, since they were his first *ayana*, or *place of motion*, he thence is named NĀRĀYANA, or *moving on the waters*.

11. 'From THAT WHICH IS, the first cause, not the object of sense, existing *every where in substance*, not existing *to our perception*, without beginning or end, was produced the divine male, famed in all worlds under the appellation of BRAHMĀ.

12. 'In that egg the great power sat inactive a whole year *of the Creator*, at the close of which by his thought alone he caused the egg to divide itself;

13. 'And from its two divisions he framed the heaven *above* and the earth *beneath*: in the midst *he placed* the subtil ether, the eight regions, and the permanent receptacle of waters.

14. 'From the supreme soul he drew forth Mind, existing substantially though unperceived by sense, immaterial; and, *before mind, or the reasoning power, he produced* consciousness, the internal monitor, the ruler;

15. 'And, before them both, he produced the great *principle of the soul, or first expansion of the divine idea*; and all vital forms endued with the three qualities of *goodness, passion,* and *darkness;*[4] and the *five* perceptions of sense, and the five organs of sensation.

16. '*Thus,* having at once pervaded, with emanations from the Supreme Spirit, the minutest portions of six principles immensely operative, *consciousness and the five perceptions,* He framed all creatures;

17. 'And since the minutest particles of visible nature have a dependence on those *six* emanations from GOD, the wise have accordingly given the name of *śarīra,* of *depending on six, that is, the ten organs on consciousness,* and the *five elements on as many perceptions,* to His *image or* appearance in visible nature:

18. 'Thence proceed the great elements, endued with peculiar powers, the Mind with operations infinitely subtil, the unperishable cause of all apparent forms.

19. 'This *universe,* therefore, is compacted from the minute portions of those seven divine and active principles, *the great Soul, or first emanation, consciousness, and five perceptions;* a mutable *universe* from immutable *ideas.*

20. 'Among them each succeeding element acquires the quality of the preceding; and, in as many degrees as each of them is advanced, with so many properties is it said to be endued.

4 [In chapter XII Manu elaborates on these three modes of material nature, which are emphasized by Krishna in the fourth chapter, 'Transcendental Knowledge', of the *Bhagavad-Gita*.]

21. 'HE too first assigned to all creatures distinct names, distinct acts, and distinct occupations; as they had been revealed in the preexisting *Vēda:*

22. 'HE, the supreme Ruler, created an assemblage of inferior Deities, with divine attributes and pure souls; and a number of Genii exquisitely delicate; and he *prescribed* the sacrifice ordained from the beginning.

23. 'From fire, from air, and from the sun he milked out, *as it were,* the three primordial *Vēdas,* named *Rich, Yajush,* and *Sāman,* for the due performance of the sacrifice.

24. 'HE gave being to time and the divisions of time, to the stars also, and to the planets, to rivers, oceans, and mountains, to level plains, and uneven valleys,

25. 'To devotion, speech, complacency, desire, and wrath, and to the creation, which shall presently be mentioned; for He willed the existence of all those created things.

26. 'For the sake of distinguishing actions, He made a total difference between right and wrong, and enured these sentient creatures to pleasure and pain, *cold and heat,* and other opposite pairs.

27. 'With every minute transformable portions, called *mātrās,* of the five elements, all this perceptible world was composed in fit order;

28. 'And in whatever occupation the supreme Lord first employed any vital soul, to that occupation the same soul attaches itself spontaneously, when it receives a new body again and again:

29. 'Whatever quality, noxious or innocent, harsh or mild, unjust or just, false or true, He conferred on any being at its creation, the same quality enters it of course *on its future births;*

30. 'As the *six* seasons of the year attain respectively their peculiar marks in due time and of their own accord, even so the several acts of each embodied spirit *attend it naturally.*

31. 'That the human race might be multiplied, He caused the *Brāhmen,* the *Cshatriya,* the *Vaisya,* and the *Sūdra* (so named from the *scripture, protection, wealth,* and *labour*) to proceed from his mouth, his arm, his thigh, and his foot.

32. 'Having divided his own substance, the mighty power became half male, half female, *or nature active and passive;* and from that female he produced VIRĀJ:

33. 'Know Me, O most excellent of *Brāhmens,* to be that person, whom the male *power* VIRĀJ, having performed austere devotion, produced by himself; Me, the *secondary* framer of all this *visible world.*

34. 'It was I, who, desirous of giving birth to a race of

men, performed very difficult religious duties, and first produced ten Lords of created beings, eminent in holiness,

35. 'MARĪCHI, ATRI, ANGIRAS, PULASTYA, PULAHA, CRATU, PRACHĒTAS, or DACSHA, VASISHT'HA, BHRIGU, and NĀRADA:

36. 'They, abundant in glory, produced seven other *Menus,* together with deities, and the mansions of deities, and *Maharshis,* or great Sages, unlimited in power;

37. 'Benevolent genii, and fierce giants, blood-thirsty savages, heavenly quiristers, nymphs and demons, huge serpents and snakes of smaller size, birds of mighty wing, and separate companies of *Pitirs,* or progenitors of mankind;

38. 'Lightnings and thunder-bolts, clouds and coloured bows of INDRA, falling meteors, earth-rending vapours, comets, and luminaries of various degrees;

39. 'Horse-faced sylvans, apes, fish, and a variety of birds, tame cattle, deer, men, and ravenous beasts with two rows of teeth;

40. 'Small and large reptiles, moths, lice, fleas, and common flies, with every biting gnat, and immovable substances of distinct sorts.

41. 'Thus was this whole assemblage of stationary and moveable bodies framed by those high-minded beings, through the force of their own devotion, and at my command, with separate actions allotted to each.

42. 'Whatever act is ordained for each of those creatures here below, *that* I will now declare to you, together with their order in respect to birth.

43. 'Cattle and deer, and wild beasts with two rows of teeth, giants, and blood-thirsty savages, and the race of men, are born from a secundine:

44. 'Birds are hatched from eggs; *so are* snakes, crocodiles, fish *without shells,* and tortoises, with other animal kinds, terrestrial, *as chameleons,* and aquatick, *as shellfish:*

45. 'From hot moisture are born biting gnats, lice, fleas, and common flies; these, and whatever is of the same class, are produced by heat.

46. 'All vegetables, propagated by seed or by slips, grow from shoots: some herbs, abounding in flowers and fruits, perish when the fruit is mature;

47. 'Other plants, called lords of the forest, have no flowers, but produce fruit; and, whether they have flowers also, or fruit only, *large woody plants* both sorts are named trees.

48. 'There are shrubs with many stalks from the root upwards, and reeds with single roots but united stems, all

of different kinds, and grasses, and *vines or* climbers, and creepers, which spring from a seed or from a slip.

49. 'These *animals and vegetables,* encircled with multiform darkness, by reason of past actions, have internal conscience, and are sensible of pleasure and pain.

50 'All transmigrations, recorded *in sacred books,* from the state of BRAHMĀ, to that of plants, happen continually in this tremendous world of beings; a world *always* tending to decay.

51. 'HE, whose powers are incomprehensible, having thus created both me and this universe, was again absorbed in the supreme Spirit, changing *the* time *of energy* for *the* time *of repose.*

52. 'When that power awakes, (*for, though slumber be not predicable of the sole eternal Mind, infinitely wise and infinitely benevolent, yet it is predicated of* BRAHMĀ, *figuratively, as a general property of life*) then has this world its full expansion; but, when he slumbers with a tranquil spirit, then the whole system fades away;

53. 'For, while he reposes, *as it were,* in calm sleep, embodied spirits, endued with principles of action, depart from their several acts, and the mind itself becomes inert;

54. 'And, when they once are absorbed in that supreme essence, then the divine soul of all beings withdraws his energy, and placidly slumbers;

55. 'Then too this vital soul *of created bodies,* with all the organs of sense and of action, remains long immersed *in the first idea or in* darkness, and performs not its natural functions, but migrates from its corporeal frame:

56. 'When, being *again* composed of minute elementary principles, it enters at once into vegetable of animal seed, it then assumes a *new* form.

57. 'Thus that immutable Power, by waking and reposing alternately, revivifies and destroys in eternal succession this whole assemblage of locomotive and immovable creatures.

58. 'HE, having enacted this code of laws, himself taught it fully to me in the beginning: afterwards I taught it MARĪCHI and the *nine* other holy sages.

59. 'This my *son* BHRIGU will repeat the divine code to you without omission; for that sage learned from me to recite the whole of it'.

60. 'BHRIGU, great and wise, having thus been appointed by MENU to promulge his laws, addressed all the *Rishis* with an affectionate mind, saying: 'Hear!

61. 'FROM this MENU, named SWĀYAMBHUVA, or *Sprung from the self-existing,* came six descendants, other MENUS, *or perfectly understanding the scripture,* each giving birth to a race of his own, all exalted in dignity, eminent in power;

62. 'SWĀRŌCHISHA, AUTTAMI, TĀMASA, RAIVATA likewise and CHĀCSHUSHA, beaming with glory, and VAIVASWATA, child of the sun.

63. 'The seven MENUS, (of *those first created, who are to be followed by seven more*) of whom SWĀYAMBHUVA is the chief, have produced and supported this world of moving and stationary beings, each in his own *Antara*, or *the period of his reign.*

64. 'Eighteen *nimēshas*, or *twinklings of an eye*, are on *cāsht'hā*; thirty *cāsht'hās*, one *calā*; thirty *calās*, one *muhūrta*: and just so many *muhūrtas* let mankind consider as the duration of their day and night.

65. 'The sun causes the distribution of day and night both divine and human; night being *intended* for the repose of *various* beings, and day for their exertion.

66. 'A month *of mortals* is a day and a night of the *Pitris* or *patriarchs inhabiting the moon*; and the division *of a month* being into equal halves, the half beginning from the full moon is their day for actions; and that beginning from the new moon is their night for slumber:

67. 'A year *of mortals* is a day and a night of the Gods, or *regents of the universe seated round the north pole*; and again their division is this: their day is the northern, and their night the southern, course of the sun.

68. 'Learn now the duration of a day and a night of BRAHMĀ, and of the several ages, which shall be mentioned in order succinctly.

69. 'Sages have given the name of *Crita* to an age containing four thousand years of the Gods; the twilight preceding it consists of as many hundreds, and the twilight following it, of the same number:

70. 'In the other three *ages*, with their twilights preceding and following, are thousands and hundreds diminished by one.

71. 'The divine years, in the four *human* ages just enumerated, being added together, their sum, *or* twelve thousand, is called the age of the Gods;

72. 'And, by reckoning a thousand such divine ages, a day of BRAHMĀ may be known: his night has also an equal duration:

73. 'Those persons best know the divisions of days and nights, who understand, that the day of BRAHMĀ, which endures to the end of a thousand such ages, gives rise to virtuous exertions; and that his night endures as long as his day.

74. 'At the close of his night, having long reposed, he awakes, and, awaking, exerts intellect, *or reproduces the great principle of animation*, whose property it is to exist unperceived by sense:

75. 'Intellect, called into action by his will to create

worlds, performs *again* the work of creation; and thence *first* emerges the subtil ether, to which philosophers ascribe the quality of *conveying* sound;

76. 'From ether, effecting a transmutation in form, springs the pure and potent air, a vehicle of all scents; and air is held endued with the quality of touch:

77. 'Then from air, operating a change, rises light *or fire*, making objects visible, dispelling gloom, spreading bright rays; *and* it is declared to have the quality of figure;

78. 'But from light, a change being effected, comes water the quality of taste; and from water is *deposited* earth with the quality of smell: such were they created in the beginning.

79. 'The beforementioned age of the Gods, or twelve thousand *of their* years, being multiplied by seventy-one, *constitutes what* is here named a *Menwantara*, *or the reign of a* MENU.

80. 'There are numberless *Menwantaras*; creations also and destructions of worlds, *innumerable*: the Being supremely exalted performs all this, *with as much ease as* if in sport, again and again *for the sake of conferring happiness.*

81. 'In the *Crita* age *the Genius of* truth and right, *in the form of a Bull*, stands firm on his four feet; nor does any advantage accrue to men from iniquity;

82. 'But in the following ages, by reason of unjust gains, he is deprived successively of one foot; and even just emoluments, through the prevalence of theft, falsehood, and fraud, are *gradually* diminished by a fourth part.

83. 'Men, free from disease, attain all sorts of prosperity and live four hundred years, in the *Crita* age; but, in the *Trētà* and the succeeding ages, their life is lessened gradually by one quarter.

84. 'The life of mortals, which is mentioned in the *Vēda*, the rewards of good works, and the powers of embodied spirits, are fruits proportioned among men to the order of the *four* ages.

85. 'Some duties are performed by *good* men in the *Crita* age; others, in the *Trētà*; some, in the *Dwāpara*; others in the *Cali*; in proportion as those ages decrease in length.

86. 'In the *Crita* the prevailing virtue is declared to be devotion; in the *Trētà*, divine knowledge; in the *Dwāpara*, holy sages call sacrifice the duty chiefly performed; in the *Cali*, liberality alone.

87. 'FOR the sake of preserving this universe, the Being supremely glorious allotted separate duties to those, who sprang respectively from his mouth, his arm, his thigh, and his foot.

88. 'To *Brāhmens* he assigned the duties of reading the *Vēda*, of teaching it, of sacrificing, of assisting others to sacrifice, of giving alms, *if they be rich*, and, *if indigent*, of receiving gifts:

89. 'To defend the people, to give alms, to sacrifice, to read the *Vēda*, to shun the allurements of sensual gratification, are in few words the duties of a *Cshatriya*:

90. 'To keep herds of cattle, to bestow largesses, to sacrifice, to read the scripture, to carry on trade, to lend at interest, and to cultivate land, are prescribed *or permitted* to a *Vaisya*:

91. 'One principal duty the supreme Ruler assigned to a *Sūdra*; namely, to serve the beforementioned classes, without depreciating their worth.

92. 'Man is declared purer above the navel; but the self-existing Power declared the purest part of him to be the mouth:

93. 'Since the *Brāhmen* sprang from the most excellent part, since he was the first born, and since he possesses the *Vēda*, he is by right the chief of this whole creation.

94. 'Him the Being, who exists of himself, produced in the beginning from his own mouth; that, having performed holy rites, he might present clarified butter to the Gods, and cakes of rice to the progenitors of mankind, for the preservation of this world:

95. 'What created being then can surpass Him, with whose mouth the Gods of the firmament continually feast on clarified butter, and the manes of ancestors, on hallowed cakes?

96. 'Of created things the most excellent are those which are animated; of the animated, those which subsist by intelligence; of the intelligent, mankind; and of men, the sacerdotal class.

97. 'Of priests, those eminent in learning; of the learned, those who know their duty; of those who know it, such as perform it virtuously; and of the virtuous, those who seek beatitude from a perfect acquaintance with scriptural doctrine.

98. 'The very birth of *Brāhmens* is a constant incarnation of DHERMA, *God of Justice*; for the *Brāhmen* is born to promote justice, and to procure ultimate happiness.

99. 'When a *Brāhmen* springs to light, he is born above the world, the chief of all creatures, assigned to guard the treasury of duties religious and civil.

100. 'Whatever exists in the universe, is all in effect, *though not in form*, the wealth of the *Brāhmen*; since the *Brāhmen* is entitled to it all by his primogeniture and eminence of birth:

101. 'The *Brāhmen* eats but his own food; wears but his own apparel; and bestows but his own in alms: through

the benevolence of the *Brāhmen*, indeed, other mortals enjoy life.

102. 'To declare the sacerdotal duties, and those of the other classes in due order, the sage MENU, sprung from the self-existing, promulgated this code of laws;

103. 'A code which must be studied with extreme care by every learned *Brāhmen*, and fully explained to his disciples, but *must be taught* by no other man *of an inferior class*.

104. 'The *Brāhmen*, who studies this book, having performed sacred rites, is perpetually free from offence in thought, in word, and in deed;

105. 'He confers purity on his living family, on his ancestors, and on his descendants, as far as the seventh person; and He alone deserves to possess this whole earth.

106. 'This most excellent code produces every thing auspicious; this code increases understanding; this code procures fame and long life; this code leads to supreme bliss.

107. 'In this book appears the system of law in its full extent, with the good and bad properties of human actions, and the immemorial customs of the four classes.

108. 'Immemorial custom is transcendent law, approved in the sacred scripture, and in the codes of divine legislators: let every man, therefore, of the three principal classes, who has a due reverence for the *supreme* spirit *which dwells in him*, diligently and constantly observe immemorial custom:

109. 'A man of the priestly, military, or commercial class, who deviates from immemorial usage, tastes not the fruit of the *Vēda*; but, by an exact observance of it, he gathers that fruit in perfection.

110. 'Thus have holy sages, well knowing that law is grounded on immemorial custom, embraced, as the root of all piety, good usages long established.

111. 'The creation of this universe; the forms of institution and education, with the observances and behaviour of a student in theology; the best rules for the ceremony on his return from the mansion of his preceptor;

112. 'The law of marriage in general, and of nuptials in different forms; the regulations for the great sacraments, and the manner, primevally settled, of performing obsequies;

113. 'The modes of gaining subsistence, and the rules to be observed by the master of a family; the allowance and prohibition of diet, with the purification of men and utensils;

114. 'Laws concerning women; the devotion of hermits, and of anchorets wholly intent on final beatitude,

the whole duty of a king, and the judicial decision of controversies,

115. 'With the law of evidence and examination; laws concerning husband and wife, canons of inheritance; the prohibition of gaming, and the punishments of criminals;

116. 'Rules ordained for the mercantile and servile classes, with the origin of those, that are mixed; the duties and rights of all the classes in time of distress for subsistence; and the penances for expiating sins;

117. 'The several transmigrations in this universe, caused by offences of three kinds, with the ultimate bliss attending good actions, on the full trial of vice and virtue;

118. 'All these titles of law, promulgated by MENU, and *occasionally* the customs of different countries, different tribes, and different families, with rules concerning hereticks and companies of traders, are discussed in this code.

119. 'Even as MENU at my request formerly revealed this divine *Sāstra*, hear it now from me without any diminution or addition.'

CHAPTER THE TWELFTH

On Transmigration and Final Beatitude

1. 'O THOU, who art free from sin', *said the devout sages*, 'thou hast declared the whole system of duties ordained for the four classes of men: explain to us now, from the first principles, the ultimate retribution for their deeds'.

2. BHRIGU, whose heart was the pure essence of virtue, who proceeded from MENU himself, thus addressed the great sages: 'Hear the infallible rules for *the fruit of* deeds in this universe.

3. 'ACTION, either mental, verbal, or corporeal, bears good or evil fruit, *as itself is good or evil*; and from the actions of men proceed their various transmigrations in the highest, the mean, and the lowest degree:

4. 'Of that threefold action, connected with bodily functions, disposed in three classes, and consisting of ten orders, be it known in this world, that the heart is the instigator.

5. 'Devising means to appropriate the wealth of other men, resolving on any forbidden deed, and conceiving notions of atheism or materialism, are the three bad acts of the mind:

6. 'Scurrilous language, falsehood, indiscriminate backbiting, and useless tattle, are the four bad acts of the tongue:

7. 'Taking effects not given, hurting sentient creatures without the sanction of law, and criminal intercourse with the wife of another, are the three bad acts of the body; *and all the ten have their opposites, which are good in an equal degree.*

8. 'A rational creature has a reward or a punishment for mental acts, in his mind; for verbal acts, in his organs of speech; for corporeal acts, in his bodily frame.

9. 'For sinful acts mostly corporeal, a man shall assume *after death* a vegetable or mineral form; for such acts mostly verbal, the form of a bird or a beast; for acts mostly mental, the lowest of human conditions:

10. 'He, whose firm understanding obtains a command over his words, a command over his thoughts, and a command over his whole body, may justly be called a *tridandì*, or *triple commander*; *not a mere anchoret, who bears three visible staves*.

11. 'The man, who exerts this triple self-command with respect to all animated creatures, wholly subduing both lust and wrath, shall by those means attain beatitude.

12. 'THAT substance, which gives a power of motion to the body, the wise call *cshētrajnya*, or *jīvātman*, the vital spirit; and that body, which thence derives active functions, they name *bhūtātman*, or *composed of elements*:

13. 'Another internal spirit, called *mahat*, or *the great soul*, attends the birth of all creatures imbodied, and thence in all mortal forms is conveyed a perception either pleasing or painful.

14. 'Those two, the vital spirit and reasonable soul, are closely united with *five* elements, but connected with the supreme spirit, or divine essence, which pervades all beings high and low:

15. 'From the substance of that *supreme spirit* are diffused, *like sparks from fire*, innumerable vital spirits, which perpetually give motion to creatures exalted and base.

16. 'By the vital souls of those men, who have committed sins in *the body reduced to ashes*, another body, composed of *nerves with* five sensations, in order to be susceptible of torment, shall certainly be assumed after death;

17. 'And, being intimely united with those minute nervous particles, according to their distribution, they shall feel, in that new body, the pangs inflicted in each case by the sentence of YAMA.

18. 'Whence the vital soul has gathered the fruit of sins, which arise from a love of sensual pleasure, but must produce misery, and, when its taint has thus been removed, it approaches again those two most effulgent essences *the intellectual soul and the divine spirit*:

19. 'They two, closely conjoined, examine without re-mission the virtues and vices of that sensitive soul, ac-cording to its union with which it acquires pleasure or pain in the present and future worlds.

20. 'If the vital spirit had practised virtue for the most part and vice in a small degree, it enjoys delight in celestial abodes, clothed with a body formed of pure elementary particles;

21. 'But, if it had generally been addicted to vice, and seldom attended to virtue, then shall it be deserted by those pure elements, and, *having a coarser body of sen-sible nerves*, it feels the pains to which YAMA shall doom it:

22. 'Having endured those torments according to the sentence of YAMA, and its taint being almost removed, it again reaches those five pure elements in the order of their natural distribution.

23. 'Let each man, considering with his intellectual powers these migrations of the soul according to its virtue or vice, *into a region of bliss or pain*, continually fix his heart on virtue.

24. 'BE it known, that the three qualities of the rational soul are a tendency to goodness, to passion, and to dark-ness; and, endued with one or more of them, it remains incessantly attached to all these created substances:

25. 'When any one of the *three* qualities predominates in a mortal frame, it renders the imbodied spirit eminently distinguished for that quality.

26. 'Goodness is declared to be true knowledge; dark-ness, gross ignorance; passion, an emotion of desire or aversion: such is the compendious description of those qualities, which attend all souls.

27. 'When a man perceives in the reasonable soul a disposition tending to virtuous love, unclouded with any malignant passion, clear as the purest light, let him recog-nise it as the quality of goodness:

28. 'A temper of mind, which gives uneasiness and produces disaffection, let him consider as the adverse quality of passion, ever agitating imbodied spirits:

29. 'That indistinct, inconceivable, unaccountable dis-position of a mind naturally sensual, and clouded with infatuation, let him know to be the quality of darkness.

30. 'Now will I declare at large the various acts, in the highest, middle, and lowest degrees, which proceed from those three dispositions of mind.

31. 'Study of scripture, austere devotion, sacred knowl-edge, corporeal purity, command over the organ, perfor-mance of duties, and meditation on the divine spirit, accompany the good quality of the soul:

32. 'Interested motives for acts *of religion or morality*,

perturbation of mind on slight occasions, commission of acts forbidden by law, and habitual indulgence in selfish gratifications, are attendant on the quality of passion:

33. 'Covetousness, indolence, avarice, detraction, athe-ism, omission of prescribed acts, a habit of soliciting favours, and inattention to necessary business, belong to the dark quality.

34. 'Of those three qualities, as they appear in the three times, *past, present* and *future*, the following in order *from the lowest* may be considered as a short *but certain* criterion.

35. 'Let the wise consider, as belonging to the quality of darkness, every act which a man is ashamed of having done, of doing, or of going to do:

36. 'Let them consider, as proceeding from the quality of passion, every act, by which a man seeks exaltation and celebrity in this world, though he may not be much afflicted, if he fail of attaining his object:

37. 'To the quality of goodness belongs every act, by which he hopes to acquire divine knowledge, which he is never ashamed of doing and which brings placid joy to his conscience.

38. 'Of the dark quality, as described, the principal object is pleasure; of the passionate, worldly prosperity; but of the good quality, the chief object is virtue: the last mentioned *objects* are superiour in dignity.

39. 'SUCH transmigrations, as the soul procures in this universe by each of those qualities, I now will declare in order succinctly.

40 'Souls, endued with goodness, attain always the state of deities; those filled with ambitious passions, the condition of men; and those immersed in darkness, the nature of beasts: this is the triple order of transmigration.

41. 'Each of those three transmigrations, caused by the several qualities, must also be considered as threefold, the lowest, the mean, and the highest, according to as many distinctions of acts and of knowledge.

42. 'Vegetable and mineral substances, worms, insects, and reptiles, some very minute, some rather larger, fish, snakes, tortoises, cattle, shakals, are the lowest forms, to which the dark quality leads:

43. 'Elephants, horses, men of the servile class, and contemptible *Mlēch'has*, or *barbarians*, lions, tigers, and boars, are the mean states procured by the quality of darkness:

44. 'Dancers and singers, birds, and deceitful men, giants and bloodthirsty savages, are the highest condi-tions, to which the dark quality can ascend.

45. '*J'hallas*, or cudgelplayers, *Mallas*, or boxers and wrestlers, *Natas*, or actors, those who teach the use of

weapons, and those who are addicted to gaming or drinking, are the lowest forms occasioned by the passionate quality:

46. 'Kings, men of the fighting class, domestick priests of kings, and men skilled in the war of controversy, are the middle states caused by the quality of passion:

47. '*Gandharvas*, or aerial musicians, *Guhyacas* and *Yacshas*, or servants and companions of CUVĒRA, genii attending superiour gods, as the *Vidyāharas* and others, together with various companies of *Apsarases* or nymphs, are the highest of those forms, which the quality of passion attains.

48. 'Hermits, religious mendicants, other *Brāhmens*, such orders of demigods as are wasted in airy cars, genii of the signs and lunar mansions, and *Daityas*, or the offspring of DITI, are the lowest of states procured by the quality of goodness:

49. 'Sacrificers, holy sages, deities of the lower heaven, genii of the *Vēdas*, regents of stars *not in the paths of the sun and moon*, divinities of years, *Pitris* or progenitors of mankind, and the demigods named *Sādhyas*, are the middle forms, to which the good quality conveys *all spirits moderately endued with it*:

50. 'BRAHMĀ with four faces, creators of worlds *under him, as* MARĪCHI *and others*, the genius of virtue, the divinities presiding over (*two principles of nature in the philosophy of* CAPILA) *mahat*, or the *mighty*, and *avyacta*, or *unperceived*, are the highest conditions, to which, by the good quality, souls are exalted.

51. 'This triple system of transmigrations, in which each class has three orders, according to actions of three kinds, and which comprises all animated beings, has been revealed in its full extent:

52. 'Thus, by indulging the sensual appetites, and by neglecting the performance of duties, the basest of men, ignorant of sacred expiations, assume the basest forms.

53. 'WHAT particular bodies the vital spirit enters in this world, and in consequences of what sins here committed, now hear at large and in order.

54. 'Sinners in the first degree, having passed through terrible regions of torture for a great number of years, are condemned to the following births at the close of that period *to efface all remains of their sin.*

55. 'The slayer of a *Brāhmen* must enter *according to the circumstances of his crime* the body of a dog, a boar, an ass, a camel, a bull, a goat, a sheep, a stag, a bird, a *Chandāla*, or a *Puccasa*.

56. 'A priest, who has drunk spirituous liquor, shall migrate into the form of a smaller or larger worm or

insect, of a moth, of a fly feeding on ordure, or of some ravenous animal.

57. 'He, who steals the gold of a priest, shall pass a thousand times into the bodies of spiders, of snakes and cameleons, of *crocodiles and other* aquatick monsters, or of mischievous blood sucking demons.

58. 'He, who violates the bed of his *natural or spiritual* father, migrates a hundred times into the forms of grasses, of shrubs with crowded stems, or of creeping and twining plants, of *vultures and other* carnivorous animals, of *lions and other* beasts with sharp teeth, or of *tigers and other* cruel brutes.

59. 'They, who hurt any sentient beings, are born *cats and other* eaters of raw flesh; they, who taste what ought not to be tasted, maggots or small flies; they, who steal *ordinary things*, devourers of each other: they, who embrace very low women, become restless ghosts.

60. 'He, who has held intercourse with degraded men, or been criminally connected with the wife of another, or stolen *common things* from a priest, shall be changed into a spirit, called *Brahmarācshasa*.

61. 'The wretch, who through covetousness has stolen *rubies of other* gems, pearls, or coral, or precious things of which there are many sorts, shall be born *in the tribe of goldsmiths, or* among *birds* called *hēmacāras, or goldmakers.*

62. 'If a man steal grain in the husk, he shall be born a rat; if a yellow mixed metal, a gander; if water, a *plava*, or diver; if honey, a great stinging gnat; if milk, a crow; if expressed juice, a dog; if clarified butter, an ichneumon weasel;

63. 'If he steal fleshmeat, a vulture; if any sort of fat, the waterbird *madgu*; if oil, a blatta, or oildrinking beetle; if salt, a cicada or cricket; if curds, the bird *valāca*;

64. 'If silken clothes, the bird *tittiri*; if woven flax, a frog; if cotton cloth, the waterbird *crauncha*; if a cow, the lizard *gōdhā*; if molasses, the bird *vāgguda*;

65. 'If exquisite perfumes, a muskrat; if potherbs, a peacock; if dressed grain in any of its various forms, a porcupine; if raw grain, a hedgehog;

66. 'If he steal fire, the bird *vaca*; if a household utensil, an ichneumon-fly; if dyed cloth, the bird *chacōra*;

67. 'If a deer or an elephant, he shall be born a wolf; if a horse, a tiger; if roots or fruit, an ape; if a woman, a bear; if water from a jar, the bird *chātaca*; if carriages, a camel; if small cattle, a goat.

68. 'That man, who designedly takes away the property of another, or eats any holy cakes not first presented *to the deity* at a solemn rite, shall inevitably sink to the condition of a brute.

69. 'Women, who have committed similar thefts, incur a similar taint, and shall be paired with those male beasts in the form of their females.

70. 'IF any of the four classes omit, without urgent necessity, the performance of their several duties, they shall migrate into sinful bodies, and become slaves to their foes.

71. 'Should a *Brāhmen* omit his peculiar duty, he shall be changed into a demon called *Ulcāmuc'ha* or *with a mouth like a firebrand*, who devours what has been vomited; a *Cshatriya*, into a demon called *Cataputana*, who feeds on ordure and carrion;

72. 'A *Vaisya*, into an evil being called *Maitrācshajyōtica*, who eats purulent carcasses; and a *Sūdra*, who neglects his occupations, becomes a foul imbodied spirit called *Chailāsaca*, who feeds on lice.

73. 'As far as vital souls, addicted to sensuality, indulge themselves in forbidden pleasures, even to the same degree shall the acuteness of their senses be raised *in their future bodies, that they may endure analogous pains*;

74. 'And, in consequence of their folly, they shall be doomed as often as they repeat their criminal acts, to pains more and more intense in despicable forms on this earth.

75. 'They shall first have a sensation of agony in *Tāmisra* or *utter darkness*, and in other feats of horrour; in *Asipatravana*, or *the swordleaved forest*, and in different places of binding fast and of rending:

76. 'Multifarious tortures await them: they shall be mangled by ravens and owls, shall swallow cakes boiling hot; shall walk over inflamed sands; and shall feel the pangs of being baked like the vessels of a potter:

77. 'They shall assume the forms of beasts continually miserable, and suffer alternate afflictions from extremities of cold and of heat, surrounded with terrours of various kinds:

78. 'More than once shall they lie in different wombs; and, after agonizing births, be condemned to severe captivity, and to servile attendance on creatures like themselves:

79. 'Then shall follow separations from kindred and friends, forced residence with the wicked, painful gains and ruinous losses of wealth; friendships hardly acquired and at length changed into enmities,

80. 'Old age without resource, diseases attended with anguish, pangs of innumerable sorts, and, lastly, unconquerable death.

81. 'With whatever disposition of mind a man shall perform in this life any act *religious or moral*, in a future body endued with the same quality, shall he receive his retribution.

82. 'THUS has been revealed to you the system of punishments for evil deeds: next learn those acts of a *Brāhmen*, which lead to eternal bliss.

83. 'Studying and comprehending the *Vēda*, practising pious austerities, acquiring divine knowledge *of law and philosophy*, command over the organs of sense and action, avoiding all injury to sentient creatures, and showing reverence to a *natural and spiritual* father, are the chief branches of duty which ensure final happiness'.

84. 'Among all those good acts performed in this world', *said the sages*, 'is no single act held more powerful than the rest in leading men to beatitude'?

85. 'OF all those duties', *answered* BHRIGU, 'the principal is to acquire from the *Upanishads* a true knowledge of one supreme GOD that is the most exalted of all sciences, because it ensures immortality:

86. 'In this life, indeed, as well as the next, the study of the *Vēda*, to acquire a knowledge of GOD, is held the most efficacious of those six duties in procuring felicity to man;

87. 'For in the knowledge and adoration of one GOD, which the *Vēda* teaches, all the rules of good conduct, *beforementioned* in order, are fully comprised.

86. 'THE ceremonial duty, prescribed by the *Vēda*, is of two kinds; *one* connected with this world, and causing prosperity on earth; *the other* abstracted from it, and procuring bliss in heaven.

89. 'A religious act proceeding from selfish views in this world, *as a sacrifice for rain*, or in the next, *as a pious oblation in hope of a future reward*, is declared to be concrete and interested; but an act performed with a knowledge of GOD, and without self love, is called abstract and disinterested.

90. 'He, who frequently performs interested rites, attains an equal station with the regents of the lower heaven; but he, who frequently performs disinterested acts of religion, becomes for ever exempt from *a body composed of* the five elements:

91. 'Equally perceiving the supreme soul in all beings and all beings in the supreme soul, he sacrifices his own spirit by fixing it on the spirit of GOD, and approaches the nature of that sole divinity, who shines by his own effulgence.

92. 'Thus must the chief of the twiceborn, though he neglect the ceremonial rites mentioned in the *Sāstras*, be diligent alike in attaining a knowledge of GOD and in repeating the *Vēda*:

93. 'Such is the advantageous privilege of those, who

have a double birth *from their natural mothers and from the* gāyatrì *their spiritual mother*, especially of a *Brāhmen*; since the twiceborn man by performing this duty but not otherwise, may soon acquire endless felicity.

94. 'To patriarchs, to deities, and to mankind, the scripture is an eye giving constant light; nor could the *Vēda Sāstra* have been made by human faculties; nor can it be measured by human reason *unassisted by revealed glosses and comments*: this is a sure proposition.

95. 'Such codes of law as are not grounded on the *Vēda*, and the various heterodox theories of men, produce no good fruit after death; for they all are declared to have their basis on darkness.

96. 'All systems, which are repugnant to the *Vēda*, must have been composed by mortals, and shall soon perish: their modern date proves them vain and false.

97. 'The three worlds, the four classes of men, and their four distinct orders, with all that has been, all that is, and all that will be, are made known by the *Vēda*:

98. 'The nature of sound, of tangible and visible shape, of taste, and of odour, the fifth object of sense, is clearly explained in the *Vēda* alone, together with the three qualities of mind, the births attended with them, and the acts which they occasion.

99. 'All creatures are sustained by the primeval *Vēda Sāstra*, which the wise therefore hold supreme, because it is the supreme source of prosperity to this creature, man.

100. 'Command of armies, royal authority, power of inflicting punishment, and sovereign dominion over all nations, he only well deserves, who perfectly understands the *Vēda Sāstra*.

101. 'As fire with augmented force burns up even humid trees, thus he, who well knows the *Vēda*, burns out the taint of sin, which has infected his soul.

102. 'He, who completely knows the sense of the *Vēda Sāstra*, while he remains in any one of the four orders, approaches the divine nature, even though he sojourn in this low world.

103. 'They, who have read many books, are more exalted than such, as have seldom studied; they, who retain what they have read, than forgetful readers; they, who fully understand, than such as only remember; and they, who perform their known duty, than such men, as barely know it.

104. 'Devotion and sacred knowledge are the best means by which a *Brāhmen* can arrive at beatitude: by devotion he may destroy guilt; by sacred knowledge he may acquire immortal glory.

105. 'Three modes of proof, ocular demonstration, logi-

cal inference, and the authority of those various books, which are deduced from the *Vēda*, must be well understood by that man, who seeks a distinct knowledge of all his duties:

106. 'He alone comprehends the system of duties religious and civil, who can reason, by rules of logic agreeable to the *Vēda*, on the general heads of that system as revealed by the holy sages.

107. 'These rules of conduct, which lead to supreme bliss, have been exactly and comprehensively declared: the more secret learning of this *Mānava Sāstra* shall now be disclosed.

108. 'IF it be asked, how the law shall be ascertained, when particular cases are not comprised *under any of the general rules, the answer is this*: 'That, which well instructed *Brāhmens* propound, shall be held incontestible law'.

109. 'Well instructed *Brāhmens* are they, who can adduce ocular proof from the scripture itself, having studied, as the law ordains, the *Vēdas* and their extended branches, or *Vēdāngas*, *Mīmānsà*, *Nyāya*, *Dhermasāstra*, *Purānas*:

110. 'A point of law, *before not expressly revealed*, which shall be decided by an assembly of ten such virtuous *Brāhmens* under one chief, or, *if ten be not procurable*, of three such under one president, let no man controvert.

111. 'The assembly of ten under a chief *either the king himself or a judge appointed by him*, must consist of three, each of them peculiarly conversant with one of the three *Vēdas*, of a fourth skilled in the *Nyāya*, and a fifth in the *Mīmānsà* philosophy; of a sixth, who has particularly studied the *Niructa*; a seventh, who has applied himself most assiduously to the *Dhermasāstra*; and of three *universal scholars*, who are in the three first orders.

112. 'One, who has chiefly studied the *Rigvēda*, a second, who principally knows the *Yajush*, and a third best acquainted with the *Sāman*, are the assembly of three under a head, who may remove all doubts both in law and casuistry.

113. 'Even the decision of one priest, *if more cannot be assembled*, who perfectly knows the principles of the *Vēdas*, must be considered as law of the highest authority; not the opinion of myriads, who have no sacred knowledge.

114. 'Many thousands of *Brāhmens* cannot form a legal assembly for the decision of contests, if they have not performed the duties of a regular studentship, are unacquainted with scriptural texts, and subsist only by *the name of* their sacerdotal class.

115. 'The sin of that man, to whom dunces, pervaded by the quality of darkness, propound the law, of which they are themselves ignorant, shall pass, increased a hundredfold, to the wretches who propound it.

116. 'This comprehensive system of duties, the chief cause of ultimate felicity, has been declared to you; and the *Brāhmen*, who never departs from it, shall attain a superiour state above.

117. 'THUS did the allwise MENU, who possesses extensive dominion, and blazes with heavenly splendour, disclose to me, from his benevolence to mankind, this transcendent system of law, which must be kept devoutly concealed *from persons unfit to receive it.*

118. 'LET every *Brāhmen* with fixed attention consider all nature, both visible and invisible, as existing in the divine spirit; for, when he contemplates the boundless universe existing in the divine spirit, he cannot give his heart to iniquity:

119. 'The divine spirit alone is the whole assemblage of gods; all worlds are seated in the divine spirit, and the divine spirit no doubt produces *by a chain of causes and effects consistent with free will*, the connected series of acts performed by imbodied souls.

120. 'He may contemplate the subtil ether in the cavities of his body; the air in his muscular motion and sensitive nerves; the supreme *solar and igneous* light, in his digestive heat and his visual organs; in his corporeal fluids, water; in the terrene parts of his fabrick, earth;

121. 'In his heart, the moon; in his auditory nerves, the guardians, of eight regions; in his progressive motion, VISHNU; in his muscular force, HARA; in his organs of speech, AGNI: in excretion, MITRA; in procreation, BRAHMĀ:

122. 'But he must consider the supreme omnipresent intelligence as the sovereign lord of them all, *by whose energy alone they exist*; a spirit, *by no means the object of any sense*, which can only be conceived by a mind *wholly abstracted from matter, and as it were* slumbering; but which *for the purpose of assisting his meditation*, he may imagine more subtil than the finest conceivable essence, and more bright than the purest gold.

123. 'Him some adore as transcendently present in elementary fire; others, in MENU, lord of creatures, *or an immediate agent in the creation*; some, as more distinctly present in INDRA, *regent of the clouds and the atmosphere*; others, in pure air; others, as the most High Eternal Spirit.

124. 'It is He, who, pervading all beings in five elemental forms, causes them by the gradations of birth, growth,

and dissolution, to revolve in this world, *until they deserve beatitude*, like the wheels of a car.

125. 'Thus the man, who perceives in his own soul the supreme soul present in all creatures, acquires equanimity toward them all, and shall be absorbed at last in the highest essence, even that of the Almighty himself'.

126. HERE ended the sacred instructor; and every twice-born man, who attentively reading this *Mānava Sāstra* promulgated by BHRIGU, shall become habitually virtuous, will attain the beatitude which the seeks.

Extracts from the *Vēdas*[1]

That spirit is free from sin, exempt from age, from death, from sorrow, from thirst, seeking truth only, fancying truth only.

☆   ☆   ☆   ☆

This vital spirit, having risen from this body *and* approaching the Supreme Light, appears brilliant in its proper form.

☆   ☆   ☆   ☆

Without hand or foot he runs rapidly and grasps firmly; without eyes he sees, without ears he hears, *all*; he knows whatever can be known, but there is none, who knows Him: Him the wise call the great, supreme, pervading spirit.

☆   ☆   ☆   ☆

Let us adore the supremacy of *that* divine Sun,[2] the Godhead,[3] who illuminates all, who recreates all, from whom all proceed, to whom all must return; whom we invoke to direct our understandings aright in our progress towards his holy seat.

[Gāyatrī, Mandala III, Sukta 62, Rik 10]

☆   ☆   ☆   ☆

[1] [Jones's vital interest in Indian religion is witnessed by his 'Desiderata (in *Works* 3: xi-xii), a list of vast studies that he hoped to do upon his intended return to England. Among the botanical, linguistic, literary, musical, philosophical, and astronomical tasks enumerated there, he projects translations of the *Mahābhārata*, *Ramayana*, *Purāṇas*, and *Vēdas*. He was collecting materials from the *Yajur-Veda*, the sacred Gāyatrī prayer and other passages from the *Rig-Veda*, and other Vedic passages intended for use in a never-composed essay, 'On the Primitive Religion of the Hindus'. See Cannon, 'Sir William Jones and the Association between East and West', *Proceedings of the American Philosophical Society* 121 (April 1977): 183-87. Teignmouth printed Jones's unpolished translations of some of these texts, with minor omissions and errors, in *Works* (13: 367-81). These are now correctly published from Jones's holographs, which are owned by Cannon. The first two passages are not in *Works*, and only the *Mohamudgara* contains sizable sections that Teignmouth obviously deliberately omitted (without indication of deletion).]

[2] Opposed to the visible luminary.

[3] *Bhargas*, a word consisting of three consonants, derived from *bha*, to shine; *ram*, to delight; *gam*, to move. [It is a synonym of the word Bhaskar for the sun, further attesting to Jones's competence in Sanskrit.]

What the *Sun* and *light* are to this *visible world*, that are the *Supreme Good*, and *truth*, to the intellectual and invisible universe; and, as our corporeal eyes have a distinct perception of objects enlightened by the sun thus our souls acquire certain knowledge by meditating on the light of truth, which emanates from the Being of beings: *that* is the light, by which alone our minds can be directed in the path of beatitude.

☆   ☆   ☆   ☆

Augury by the flight of a bird

1. What relish can there be for enjoyments in this unsound body, filled with bad odours, composed of bones, skin, tendons, membranes, muscles, blood, saliva, tears, ordure and urine, bile and mucus?

2. What relish can there be for enjoyments in this body assailed by desire and wrath, by avarice and illusion, by fear and sorrow, by envy and hate, by absence from those whom we love, and by union with those whom we dislike, by hunger and thirst, by disease and emaciation, by growth and decline, by old age and death?

3. Surely, we see this universe tending to decay, even as these biting gnats, and other insects, even as the grass of the field and the trees of the forest, which spring up and then perish.

4. But what are they? Others, far greater have been archers mighty in battle, and some have been kings of the whole earth.

5. SUDHUMNA, BHURIDHUMNA, INDRADHUMNA, CUVALAYĀSWA, YANVANĀSWA, AVADHYASWA, ASWAPATI, SASABINDU, HAVISEHANDRA, BARISHSHA, SURYATI YAYATI, VICRAVIYA, ACSHAYASENA, PRIYAVRATA, and the rest.

6. MARUTTA likewise, and BHARATA, who enjoyed all corporeal delights, yet left their boundless prosperity, and passed from this world to the next.

7. But what are they? Others yet greater, *Gandharvē Asuras*, *Racshasas*, companies of spirits, *Pisachas*, *Uragas*, and *Grāhas*, have, we see, been destroyed.

8. But what are they? Others, greater still, have been changed; vast rivers, dried; mountains, torn up; the pole itself moved from its place; the cords of the stars rent asunder; the whole earth itself deluged with water; even the *Sufas*, or angels, hurled from their stations.

9. In such a world, then, what relish can there be for enjoyments? Thou alone art able to raise me up; I am in this world like a frog in a dry well; Thou only, O lord, art my refuge; thou only art my refuge.

☆   ☆   ☆   ☆

1. May that soul of mine, which mounts aloft, in my waking hours, as an ethereal spark, and which, even in my slumber, has a like ascent, soaring to a great distance as *an emanation from* the light of lights, be united by devout meditation with the Spirit supremely blest and supremely intelligent!

2. May that soul of mine, by *an agent similar to* which the lowborn perform their menial works, and the wise, deeply versed in sciences, duly solemnize their sacrificial rite, *that soul*, which was itself the primeval oblation *placed* within all creatures, be united by devout meditation with the Spirit supremely blest and supremely intelligent!

3. May that soul of mine, which is *a ray of* perfect wisdom, pure intellect, and permanent existence, which is the unextinguishable light fixed within created bodies, without which no good act is performed, be united by devout meditation with the Spirit supremely blest and supremely intelligent!

4. May that soul of mine, in which, as an immortal essence, may be comprised whatever has past, is present, or will be hereafter; by which the sacrifice, where seven ministers officiate, is properly solemnized, be united by devout meditation with the Spirit supremely blest, and supremely intelligent!

5. May that soul of mine, into which are inserted, like the spokes of a wheel in the axle of a car, the holy texts of the *Rigvēda*, the *Sāman*, and the *Yajush*, into which is inwoven all that belongs to created forms, be united by devout meditation with the Spirit supremely blest, and supremely intelligent!

6. May that soul of mine, which, *distributed in other bodies*, guides mankind, as a skillful charioteer guides his rapid horses with reins, *that soul*, which is fixed in my breast, exempt from old age, and extremely swift in its course, be united by divine meditation with the Spirit supremely blest, and supremely intelligent!

Vēda and 1st *Art.* of our CHURCH

'There is but One living and true GOD, everlasting, without body, parts, or passions, of infinite power, wisdom, and goodness, the maker and preserver of all things both visible and invisible'.

One wd. think this *Article* a translat. of ye *Vēda*.

☆   ☆   ☆   ☆

Īsāvāsyam:
or,
An *Upanishad* from the *Yajurvēda*

1. By one Supreme Ruler is this universe pervaded, even every world in the whole compass of nature: enjoy pure delight, O man, by abandoning *all thoughts of* this perishable world; *and* covet not the wealth of any *creature existing*.

2. He, who in this life continually performs his religious duties, may desire to live a hundred years; but even to the end of that period, O mortal, though shouldst have no other occupation here below; since a deviation from duty is imputed to a man like thee.

3. To those regions, where evil spirits dwell and which utter darkness involves, will such men surely go after death, as destroy *the purity of* their own souls.

4. There is one supreme spirit, which nothing can shake, more swift than the thought of man: that primeval mover even divine intelligences cannot reach: that spirit, though unmoved, infinitely transcends others, how rapid soever their course.

5. That *supreme spirit* moves at pleasure, but in itself is immovable; it is distant from us, yet very near us; it pervades this whole system of worlds, yet is infinitely beyond it.

6. The man, who considers all beings as existing even in the Supreme Spirit, and the Supreme Spirit as pervading all beings, thenceforth views no creature with contempt.

7. In him, who knows that all spiritual beings are the same *in kind* with the supreme spirit, what *room can there be for* delusion of mind, when he reflects on the identity of spirit?

8. The pure enlightened soul assumes a luminous form, with no gross body, with no perforation, with no veins or membranes, tendons, unblemished, untainted by sin; *itself being a ray from the infinite Spirit*, which knows the past and the future, which pervades all, which existed with no cause but itself, which created all things as they are, in ages very remote.

9. They, who are ignorantly devoted to the mere ceremonies of religion, are fallen into thick darkness, but

they surely have a thicker gloom around them, who are solely attached to speculative science.

10.   A distinct reward, they say, is reserved for ceremonies, and a distinct reward, they say, for divine knowledge; *adding*: 'This we have heard from sages, who declared it to us'.

11.   He alone is acquainted with the nature of ceremonies and with that of speculative science, who is acquainted with both at once: by religious ceremonies he passes the gulph of death, and by divine knowledge he attains immortality.

12.   They, who adore only the appearances and forms of the deity are fallen into thick darkness; but they surely have a thicker gloom around them, who are solely devoted to the abstract nature of the divine essence.

13.   A distinct reward, they say, is obtained by adoring the forms and attributes, and a distinct reward, they say, by adoring the abstract essence; *adding*: 'This we have heard from sages, who declared it to us'.

14.   He only knows the forms and the essence of the deity, who adores both at once; by adoring the appearances of the deity, he passes the gulph of death, and by adoring his abstract essence he attains immortality.

15.   Unveil, O thou, who givest sustenance to the world, that face of the true Sun, which is now hidden by a vase of golden light; so that we may see the truth and know our whole duty!

16.   O thou, who givest sustenance to the world, thou sole mover of all, thou who restrainest sinners, who pervadest yon great luminary, who appearest as the son of the creator; hide thy dazzling beams and expand thy spiritual brightness, that I may view thy most auspicious, most glorious, real form.

'OM. Remember me, divine spirit;
'OM. Remember my deeds'.

17.   With that all-pervading spirit, with that spirit which gives light to the visible Sun, even the same *in kind* am I, *though how infinitely distant in degree*! Let my soul return to the immortal spirit of GOD, and then let my body, which ends in ashes, return to dust!

18.   O spirit, who pervadest fire, lead us in a straight path to the riches of beatitude! Thou, O GOD , possessest all the treasures of knowledge; remove each foul taint from our souls: we continually approach thee with the highest praise and the most fervid adoration.

☆   ☆   ☆   ☆

From the *Yajurvēda*

1.   As a tree, the lord of the forest, even so, without fiction, is man: his hairs *are as* leaves; his skin, as exterior bark.

2.   Through the skin flows blood; through the rind, sap: from a wounded man, therefore, blood gushes, as the vegetable fluid from a tree, *that is* cut.

3.   His muscles *are as* interwoven fibres; the membrane round his bones, as interior bark, *which is* closely fixed: his bones *are as* the hard pieces of wood within; their narrow is compared to the pith.

4.   Since a tree, *when* felled, springs again, still fresher, from the root, from what root springs mortal man, when felled by the hand of death?

5.   Say not, 'He springs from seed': seed, surely, comes from the living. A tree, no doubt, rises from seed, and, after death, *has* a visible renewal.

6.   *But* a tree, which they have plucked up by the root, flourishes *individually* no more. From what root then springs mortal man, when felled by the hand of death?

7.   Say not: *He was* born before; *he is* now born: who can make him spring again to birth?

8.   GOD, *who is* perfect wisdom, perfect happiness. *He is* the final refuge of the man, who has liberally bestowed his wealth, who has been firm in virtue, who knows *and adores* that Great One.

☆   ☆   ☆   ☆

A Hymn to the Night from the *Vēda*

Night approaches illumined *with stars and planets*, and, looking on all sides with numberless eyes, overpowers all meaner lights:

The immortal goddess pervades the firmament, covering the low *valleys and shrubs*, and the lofty *mountains and trees*, but *soon* she disturbs the gloom with celestial effulgence. Advancing with brightness at length she recalls her sister the Morning; and the nightly shade gradually melts away.

May she at this time be propitious! She, in whose early watch we may calmly recline in our mansion, as birds repose on the tree!

Mankind now sleep in their towns; now herds and flocks peacefully slumber, and winged creatures, even swift falcons and vultures.

O Night, avert from us the she-wolf and the wolf; and oh! suffer us to pass thee in soothing rest!

O Morn, remove *in due time* this black, yet visible,

overwhelming darkness, which at present enfolds me, as thou enablest me to remove the cloud of their debts.

Daughter of heaven, I approach thee with praise, as the cow approaches her milker: accept, O Night, not the hymn only, but the oblation of thy suppliant, who prays, that his foes may be subdued.

☆   ☆   ☆   ☆

Translation of *The Mōhamudgara*, or The Ignorant Instructed[1]
[by Śankara Āchārya]

Stanzas rhymed

1. Restrain, O ignorant man, thy desire of wealth, and become a hater of it in body, understanding, and mind: let the riches, which thou possessest, acquired by thy own good actions, with those gratify thy soul.

2. Day after day, reflect on the vanity of wealth; not the smallest pleasure arises from them: this is certain; but even from the son of the wealthy fear proceeds: this moral lesson is inculcated for every man.

3. The boy so long delights in his play, the youth so long pursues his beloved, the old so long brood over melancholy thoughts, that no man meditates on the Supreme Being.

4. As long as man is able to acquire wealth, so long his own family will be courteous and attentive to him; afterwards, when his body is weakened by age, no creature in his house will convene with him.

5. Who is thy wife; and who, thy son; how great and wonderful is this world; whose thou art, and whence thou comest—meditate on this, my brother, and again on this.

6. Be not proud of wealth, attendants, and youth; since Time destroys all of them in the twinkling of an eye: check thy attachment to all these illusions, like *Māyà*, fix thy heart on the foot of *Brehmē*, and thou wilt soon know him.

---

[1] [This Sanskrit poem attributed to the great philosopher Śankara Āchārya was edited in Darjeeling in 1888 and then published in an English version (Calcutta, 1896). Jones first translated the poem into interlinear Latin, which he corrected, before rendering his Latin into English prose that he extensively revised for intended publication in *Asiatick Researches*. He intended it to be introduced by this note: 'The following version of the poem is disengaged from the stiffness of a verbal translation; but all the images in the original, which would bear an *English* dress without deformity, are carefully preserved; and, though much is omitted, nothing is added'. Omitting whole sections and making numerous errors, Teignmouth published Jones's prose version in *Works* 13: 382-84. Cannon owns the holograph, the source of this first correct printing.]

7.   As a drop of water moves on the leaf of the lotus, thus or more slippery is human life: the company of the virtuous endures here but for a moment; that is the vehicle to bear thee over land and ocean.

8.   Restrain thy lust, wrath, avarice, folly, guile; and conserve thyself, *saying* 'Who am I?' 'They, who know not GOD, are ignorant indeed; and, when their bones are rotten, they are bound fast in a place of punishment.

9.   What mean these numerous meditations? Thou mad man, why hast thou not a vigilant chastiser, who might bind thy hands close, and give thee a lesson in confinement for a whole watch?

10.   To dwell under the vaulted roof in the mansion of Gods at the foot of a tree, to have the ground for a bed, and a hide for vesture, to renounce all ties of family or connexion—who would not receive delight from this devout abhorrence of the world?

11.   Set not thy affections of foe, or friend, on a son, or a relation, in war or in peace! Bear an equal mind towards all; if thou desirest it, thou wilt soon be like Vishnu.

12.   Day and night, evening and morn, winter and spring depart and return! Time sports, age passes on, desire and the wind continue unrestrained!

13.   When the body is tottering, the head grey, and the mouth toothless, when the smooth stick trembles in the hand, which it supports, yet the vessel of covetousness remains unemptied.

14. So soon born! so soon dead! so long lying in thy mother's womb! so great crimes are committed in the world! How then, O man, canst thou live here below with complacency?

15.   There are eight original mountains and seven seas— *Brehmē*, *Indra*, the *Sun*, and *Rudra*—these are permanent, not Thou, not I, not this or that people: what therefore should occasion our sorrow?

16.   In thee, in me, in every other, *Vishnu* resides: in vain art thou angry with me, not bearing my approach: this is perfectly true; all must be esteemed equal: be not, therefore, proud of a magnificent palace.

17.   This is the instruction of learners, delivered in twelve various measures: what more can be done with those, whom this work doth not fill with devotion?

Thus ends the book named *Mōhamudgara*, or, the *Ignorant Instructed* (properly, the *Mallet of the Ignorant*) composed by the Holy, Devout, and Prosperous, *Sancar Āchārya*.

SELECTED BIBLIOGRAPHY

Aarsleff, Hans. 'Sir William Jones and the New Philology'. In *The Study of Language in England, 1780-1860*. Princeton: Princeton University Press, 1967.

Arberry, Arthur J. *Asiatic Jones: The Life and Influence of Sir William Jones*. London: Longmans, 1946.

Bearce, George D. *British Attitudes towards India, 1784-1858*. London: Greenwood Press, 1961.

Brown, Peter. 'Sir William Jones'. In *The Chathamites*. London: Macmillan, 1967.

Cannon, Garland. *Oriental Jones: A Biography of Sir William Jones*. Bombay and London: Asia Publishing House, 1964.

———. *Sir William Jones: A Bibliography of Primary and Secondary Sources*. Amsterdam: John Benjamins B.V., 1979.

Drew, John. 'Sir William Jones: *Asiatic Researches* and the Platonizing of India'. In *India and the Romantic Imagination*. Delhi and Oxford: Oxford University Press, 1987.

Figueria, Dorothy, '*Śakuntalā*'s Reception in Nineteenth Century Europe'. *South Asian Review* 8, no. 5 (1984): 30-37.

Jones, Sir William. *Catalogue of the Library of the Late Sir William Jones*. Ed. R.H. Evans. London: W. Nicol, 1831.

———. *The Letters of Sir William Jones*. 2 vols. Ed. Garland Cannon. Oxford: Clarendon, 1970.

———. *The Works of Sir William Jones*. 6 vols. Ed. Anna Maria Jones. London: G.G. & J. Robinson, 1799.

———. *The Works of Sir William Jones*. 13 vols. London: J. Stockdale, 1807.

Karimi-Hakkak, Ahmad. 'Sir William Jones's Attitude toward the Persian Epic'. *South Asian Review* 8, no. 5 (1984): 71-80.

Kejariwal, Om Prakash. *The Asiatic Society of Bengal and the Discovery of India's Past, 1784-1838*. New York: Oxford University Press, 1988.

Kopf, David. *British Orientalism and the Bengal Renaissance*. Berkeley: University of California Press, 1969.

Lehmann, Winfred P. 'The Impact of India on Jones'. *South Asian Review* 8, no. 5 (1984): 18-21.

Marshall, Peter J. *The British Discovery of Hinduism in the Eighteenth Century*. Cambridge: Cambridge University Press, 1970.

Meisami, Julie Scott. 'Sir William Jones and the Reception of Persian Literature'. *South Asian Review* 8, no. 5 (1984): 61-70.

Mojumder, Abu Taher. *Sir William Jones and the East*. Dacca, 1978.

———. 'Three New Letters by Sir William Jones'. *India Office Library and Records Report 1981* (1982): 24-35.

Morris, Henry. *Sir William Jones, the Learned Oriental Scholar*. London: Christian Literature Society for India, 1901.

Moussa-Mahmoud, Fatma, *Sir William Jones and the Romantics*. Cairo: Anglo-Egyptian Bookshop, 1962.

Mukherjee, S.N. *Sir William Jones: A Study in Eighteenth-Century British Attitudes to India*. Cambridge: Cambridge University Press, 1968.

Pachori, Satya S. 'Sir William Jones's Perspectives on Sanskrit Literature'. *Michigan Academician* 16, no. 3 (spring 1984): 401-10.

———. 'Shelley's 'Hymn to Intellectual Beauty' and Sir William Jones'. *The Comparatist* 9 (May 1987): 54-63.

———. 'Shelley's 'Indian Serenade': Hafiz and Sir William Jones'. *Osmania Journal of English Studies* 11, no. 1 (1974-75): 11-26.

———. 'Tennyson's Early Poems and Their Hindu Imagery'. *Literature East & West* 9, nos. 1-4 (1978-79): 132-38.

Priolkar, Anant K. *The Printing Press in India : Its Beginnings and Early Development*. Bombay: Marathi Samshodhana Mandala, 1958.

Proceedings of the Sir William Jones Bicentenary Conference. London: Royal India Society, 1946.

Rankin, Sir George Claus. *Background to Indian Law*. Cambridge: Cambridge University Press, 1946.

Raychaudhuri, Durgaprasanna. *Sir William Jones and His Translation of Kālidāsa's Śakuntalā*. Calcutta: Oxford Mission, 1928.

Rocher, Ludo, trans. and ed. *Philipp Wesdin Dissertation on the Sanskrit Language*. [1790]. Amsterdam: Benjamins, 1977.

———. 'Vans Kennedy (1784-1846): A Preliminary Bio-Bibliography'. *Journal of the American Oriental Society* 109 (1989): 621-625.

Rocher, Rosane, 'Nathaniel Brassey Halhed, Sir William Jones, and Comparative Indo-European Linguistics'. *Recherches de Linguistique* (1980): 173-80.

———. *Orientalism, Poetry, and the Millennium: The Checkered Life of Nathaniel Brassey Halhed, 1751-1830*. Delhi: Motilal Banarsidass, 1983.

Salus, Peter H. 'Bringing India to Europe: The Asiatic Society's Bicentenary'. *South Asian Review* 8, no. 5 (1984): 6-17.

———. '*Śakuntalā* in Europe: The First Thirty Years'.

Journal of the American Oriental Society 84 (1964): 417.

Sankhdher, Brijendra M. *Press, Politics and Public Opinion in India*. New Delhi: Deep & Deep Publications, 1984.

Sir William Jones: Bicentenary of His Birth Commemoration Volume, 1746-1946. Calcutta: Asiatic Society of Bengal, 1948.

South Asian Review. Ed. Satya S. Pachori. Jacksonville,

Florida: South Asian Literary Association of Modern Language Association of America, 1984.

Teignmouth, Baron, John Shore. *Memoirs of the Life, Writings and Correspondence of Sir William Jones*. London: John Hatchard, 1806. [Reprinted as vols. 1-2 of the 1807 edition of Jones's *Works*.]

The Works of the English Poets. Ed. Alexander Chalmers. London: J. Johnson et al., 1810. 18:425-511.

INDEX

Men and women are listed under their family name (women, if married, under their husbands') with a cross-reference under their title. References to Jones are omitted—except for his writings. Certain entries are subdivided only when there are numerous references such as *Hinduism, The Ramayana*, and *The Mahabharata*. A single reference may well suggest more than one occurrence on a particular page. Entries also consist of prominent titles, places, events, subjects, and themes.

Flavours of the
Middle Kingdom
The origins of traditional Chinese dishes

TIMES EDITIONS

The publisher wishes to thank **Lim's Arts and Living**, **Robinson & Co. (S) Pte Ltd** and **The Life Shop** for the loan and use of their tableware.

Managing Editor : Jamilah Mohd Hassan
Editor : Lydia Leong
Art Direction/Designer : Geoslyn Lim
Photographer : Cedric Lim
Food Preparation & Prop Styling : Amy Van and Sharon Soh
Production Co-ordinator : Nor Sidah Haron

Published by Times Editions – Marshall Cavendish
An imprint of Marshall Cavendish International (Asia) Private Limited
A member of Times Publishing Limited
Times Centre, 1 New Industrial Road, Singapore 536196
Tel: (65) 6213 9288   Fax: (65) 6285 4871
E-mail: te@sg.marshallcavendish.com
Online Bookstore: http://www.timesone.com.sg/te

Malaysian Office:
Federal Publications Sdn Berhad
(General & Reference Publishing) (3024-D)
Times Subang, Lot 46, Persiaran Teknologi Subang
Subang Hi-Tech Industrial Park
Batu Tiga, 40000 Shah Alam
Selangor Darul Ehsan, Malaysia
Tel: (603) 5635 2191   Fax: (603) 5635 2706
E-mail: cchong@tpg.com.my

National Library Board (Singapore) Cataloguing in Publication Data

Flavours of the Middle Kingdom :- the origins of traditional Chinese dishes. – Singapore :- Times Editions,- c2004.
p. cm.
ISBN : 981-232-710-X

1. Cookery, Chinese. 2. Food habits – China. 3. China – Social life and customs.

TX724.5.C5
641.5951 — dc21   SLS2004009369

Printed in Singapore by Times Printers Pte Ltd

CONTENTS

POULTRY

SEAFOOD

China is the most populous country in the world, and it is also the one with the longest history. Archaeological findings testify that the Chinese had already established patriarchal societies some 5,000 years ago, and that villages existed and agriculture was practiced. This, perhaps, explains in part why China boasts such a rich and varied culinary heritage today.

To the Chinese, food is representative of life, health and prosperity. It is one's good fortune and privilege to be able to savour good food, and thus, food is always treated with respect. Every part of the animal or vegetable is utilised for food as far as possible, and wastage is minimal. This mindset also led cooks to experiment creatively with new flavours and ingredients to produce dishes that brought out the best in terms of taste, texture and colour, even when resources were limited.

Given the vastness of China's geographical expanse, the cooking styles and the ingredients used are inevitably varied. Common cooking techniques include deep-frying, stir-frying, braising, stewing, steaming and smoking, among a host of other variations and methods. In the north of China, wheat products such as steamed pancakes, bread and noodles are eaten in place of rice; while in the south, rice is the staple. Despite the differences, a well-prepared Chinese meal is fragrant, and the ingredients would have been cut to a uniform size to be visually pleasing, while fulfilling the practical reason of an even cooking time.

In *Flavours of the Middle Kingdom*, recipes from throughout China have been included, so readers will have the opportunity to try out the various cooking methods and savour the unique flavours of each region. Each recipe has also been carefully and clearly written in step-by-step format, so the dishes can be easily reproduced in the home kitchen. Additional useful tips will also enable readers to be better acquainted with the intricacies of Chinese cooking techniques. With practise, preparing and serving a Chinese meal will be an effortless process.

Some of these recipes date back to antiquity, while others were created more recently. Many of these dishes were also not created by professional or imperial chefs, but by ordinary home cooks and unlikely poets and scholars. A handful even came about by accident. Both the beautiful tales and the more mundane stories behind each of these recipes are recorded in *Flavours of the Middle Kingdom* in the hope that readers will understand the rationale behind the dishes and from there, learn to appreciate the flavours and techniques of Chinese cooking.

SPECIAL DELICACIES

Prawns on Rice Crusts
(The Best Dish in the World)

The 10 Emperors of the Qing Dynasty (1644–1911) were the last to occupy the Forbidden City as China's rulers. Emperor Qianlong (1711–1799) was the sixth emperor. In his reign (1736–1796), he engaged in several wars and expanded China's territorial limits extensively. The feudal ruler tried to sustain the power and influence of the dynasty established by his ancestors. He visited southern China six times on tours of inspection.

In 1762, during his third trip to southern China, Emperor Qianlong visited a small restaurant in Wuxi. He was served a dish of deep-fried rice crusts drizzled with a thick, glistening prawn and shredded chicken sauce. The chef brought out the rice crusts and sauce in two separate containers and poured the sauce over the rice crusts at the table. A sizzling sound was created and a pleasant aroma wafted through the air. The emperor enjoyed the dish so much that he commended it as being the best dish in the world. This is how the dish got its name.

Ingredients

Ingredient	Amount
Cooked rice	450 g
Prawns (shrimps)	160 g, shelled and deveined
Egg white	1
Cornflour (cornstarch)	1 Tbsp
Cooking oil	
Chicken	120 g, boiled and shredded
Water	250 ml
Salt	1 tsp
Sugar	3 tsp
Tomato sauce	2 Tbsp
Chinese cooking wine (hua tiao)	2 Tbsp
White vinegar	2 Tbsp
Coriander (cilantro) leaves	

Tip: Pour the prawn and chicken mixture over the rice crusts just before serving so the rice crust remains crisp.

Method

* Press cooked rice into a pre-oiled baking tray to form a 1-cm thick layer. Bake in a preheated oven at 120°C for 30–40 minutes until rice is dry and slightly crisp.

* Remove rice crust from tray and break into pieces. Set aside.

* Coat prawns with egg white and cornflour.

* Heat 2 Tbsp oil in a wok and stir-fry prawns for 3 minutes. Add shredded chicken, water, salt, sugar, tomato sauce, cooking wine and vinegar. Bring to the boil then thicken with 1 Tbsp cornflour mixed with 2 Tbsp water. Remove from heat.

* In the same wok, heat more oil for deep-frying. Lower in rice crusts and fry until crisp. Remove and drain.

* Place rice crusts on a serving plate and pour prawns and chicken mixture over. Garnish with coriander and serve immediately.

Prawns with Egg Rolls
(Prawns and Gold Coins)

The first emperor of the Song Dynasty (960–1279), Zhao Kuangyin was also known as Emperor Taizu of the Song (927–976). During his reign, he successfully restored the country's flagging economy and stabilised the political unrest. His government became one of the most accomplished in China's history.

Once, during a game of chess, the emperor lost a lot of money, including Mount Huashan, which was used as a stake in the game. As a result, he was sad and lost his appetite. To help stir up the emperor's palate, the imperial chef used chicken, eggs and black hair moss to create a dish to resemble gold coins and served it with prawns. (Black hair moss is called *fa cai* in Mandarin, which sounds like the Mandarin term for "acquiring great fortune". Thus, it is always welcomed as a symbol of good fortune). The chef named the dish "Prawns and Gold Coins". The dish has been popular in Chinese cuisine ever since and is a specialty in Henan, Song Dynasty's former capital.

Ingredients

Minced chicken	120 g
Salt	1 tsp
Chinese cooking wine (hua tiao)	1 Tbsp
Ginger	1-cm knob, peeled and chopped
Black hair moss (fa cai)	50 g, soaked and finely chopped
Egg whites	2
Eggs	4, beaten
Cornflour (cornstarch)	2 Tbsp, mixed with 3 Tbsp water
Dried prawns (shrimps)	160 g, soaked
Cooking oil	2 Tbsp
Water chestnuts	50 g, peeled and diced
Green peas	50 g
Water	2 Tbsp

Method

❀ Combine minced chicken with 1/2 tsp salt, 2 tsp cooking wine, half the ginger, black hair moss and 1 egg white. Set aside.

❀ Steam half the beaten eggs for 5 minutes. Set aside to cool before cutting into long, thick strips.

❀ Add half the cornflour mixture to the remaining beaten eggs and make two thin omelettes.

❀ Spread half of minced chicken mixture over an omelette. Place a strip of steamed egg in the middle and roll omelette up. Repeat for other omelette. Steam for 8 minutes.

❀ Remove from heat and cut into 1-cm thick slices. Set aside.

❀ Mix prawns with remaining egg white and cornflour mixture.

❀ Heat oil and stir-fry prawns with remaining salt, cooking wine and ginger, water chestnut and peas. Add water and bring to the boil. Arrange on a serving plate with omelette rounds and serve.

Tip: When steaming the egg rolls, sit them on the open flap so they do not open up. Alternatively, wrap them with muslin and remove before serving.

Hot and Sour Squid
(Chao Heng's Squid)

In 717, a Japanese student, Abei, arrived in Chang'an, capital of the Tang Dynasty, to further his studies. There, he adopted the Chinese name, Chao Heng. Chao Heng was an excellent scholar and he was made governor of Annan after he graduated.

While serving as governor, Chao Heng made friends with several renowned poets such as Wang Wei and Li Bai, and he often treated them to his favourite dish of squid. This dish was later improved upon by Tang cooks who added chilli and vinegar—two favoured ingredients of the people of Chang'an.

Another legacy that Chao Heng left behind is a poem entitled "Looking West at Chang'an", which he wrote in 755 before returning home to Japan.

Ingredients

Squid tubes	420 g
Vinegar	2 Tbsp
Chinese cooking wine (hua tiao)	2 Tbsp
Salt	1 tsp
Ground white pepper	1/2 tsp
Water	3 Tbsp
Cornflour (cornstarch)	1 Tbsp, mixed with 2 Tbsp water
Cooking oil	500 ml
Dried red chilli	1, soaked to soften and finely cut
Spring onion (scallion)	1, finely cut
Ginger	1.5-cm knob, peeled and shredded

Method

❊ Cut open squid tubes and pull away purplish skin. Cut into 5 x 4-cm pieces and score with criss-cross cuts. Scald briskly in boiling water then drain and pat dry.

❊ Combine vinegar, cooking wine, salt, pepper, water and cornflour mixture for the sauce.

❊ Heat oil in a wok until it smokes and quickly fry squid. Remove and drain.

❊ Leave 2 Tbsp oil in the wok and stir-fry chilli, spring onion and ginger. Add squid and sauce. Stir quickly and remove from heat. Serve.

Tip: Cook the squid quickly over high heat to ensure that it remains tender. Overcooking will cause the squid to become tough and chewy.

Squid with Chicken Strips
(Playful Dragon and Phoenix)

Emperor Wuzong who reigned from 1506–1522 during the Ming Dynasty (1491–1522) was an irresponsible ruler. He constantly ignored state affairs, preferring to hunt, sightsee and feast. This caused great unhappiness among the peasants and resulted in many peasant uprisings in Sichuan and Jiangxi.

Once, Emperor Wuzong toured Meilong town, southern China in disguise. He had lunch at a restaurant run by a brother and sister. Intrigued by the sister, Fengjuan's beauty, the emperor requested her to prepare dishes for him. She cooked a dish of chicken and squid which the emperor aptly named "Playful Dragon and Phoenix"—the dragon referred to the squid and the phoenix to the chicken, as well as to Fengjuan. (In Mandarin, the word for "phoenix" is *feng*, the same character in Fengjuan's name.) When the Emperor returned to the palace, he had this dish included in the imperial kitchen's menu.

Ingredients

Squid tubes	250 g
Chicken breast	160 g, cut into 3-cm pieces
Salt	1 tsp
Chinese cooking wine (*hua tiao*)	3 tsp
Egg white	1
Cornflour (cornstarch)	4 Tbsp + 1 Tbsp, mixed with 2 Tbsp water
Spring onions (scallions)	2, shredded
Ginger	1-cm knob, peeled and shredded
Garlic	1 clove, peeled and sliced
Vinegar	1 tsp
Water	4 Tbsp
Dried wood ear fungus (*hei mu er*)	20 g, soaked to soften and shredded
Bamboo shoot (*zhu sun*)	20 g, shredded
Spinach	3–4 stalks, cut into 4-cm lengths
Cooking oil	

Method

❁ Cut open squid tubes and pull away purplish skin. Cut into 4 x 2-cm pieces and score with criss-cross cuts. Scald briskly in boiling water then drain.

❁ Marinate chicken with $1/2$ tsp salt and 1 tsp cooking wine. Leave for 5 minutes, then mix in egg white and cornflour.

❁ Heat oil in wok and deep-fry chicken breast pieces until golden brown and crisp. Remove and drain. Set aside. Leave 2 Tbsp oil in wok and keep hot.

❁ Combine remaining salt and cooking wine, cornflour mixture, spring onions, ginger, garlic, vinegar, water, wood ear fungus, bamboo shoot and spinach. Pour into wok and stir. Add squid and mix well.

❁ Dish out and arrange on a serving plate with fried chicken pieces. Serve.

Stir-fried Clam Meat
(The Most Delicious Dish on Earth)

Nantong is a coastal city located on the northern bank of the Yangtze River estuary in Jiangsu Province. It has been one of the major clam producing areas since the Sui Dynasty (581–618). Clams from Nantong were highly favoured by the emperors and were often used as imperial tributes. Today, these clams account for two-thirds of China's total clam exports and are still known to the Chinese as "The Most Delicious Food on Earth".

In one of his poems, Wang Anshi (1021–1086), a politician, poet and prose writer of the Northern Song described how these clams were favoured by diners because of their irresistible taste. Ouyang Xiu (1007–1072), another well-known writer and leader of the Northern Song School of Classic Literature, also commented that no other meat or seafood could match up to the taste of these clams.

Ingredients

Clams	1.5 kg, shelled and washed
Salt	1 tsp
Chinese cooking wine (hua tiao)	1 Tbsp
Ginger	1-cm knob, peeled and chopped
Cooking oil	5 Tbsp
Spring onions (scallions)	2, sliced
Water chestnuts	50 g, peeled and sliced
Dried Chinese mushrooms	30 g, soaked to soften and sliced
Ground white pepper	1/2 tsp
Water	3 Tbsp
Cornflour (cornstarch)	1 Tbsp, mixed with 2 Tbsp water

Method

* Marinate clams with 1/2 tsp salt, cooking wine and ginger.

* Heat 2 Tbsp oil in a wok and stir-fry clams briskly for 2–3 minutes. Remove from heat.

* Add remaining oil to wok and heat. Stir-fry spring onions, water chestnuts and mushrooms. Sprinkle in pepper and remaining salt. Add water and stir. Bring to the boil.

* Thicken sauce with cornflour mixture. Stir in clams briskly, mixing well. Garnish with spring onions and serve.

Tip: *When shelling the clams, carefully pry them open with a small knife. Wear thick gloves to prevent hurting yourself. Cook clams briskly or they will turn tough and chewy.*

SPECIAL DELICACIES

Duck Webs with Quail Eggs
(Pearl in the Palm)

During Emperor Qianlong's 60-year reign (1736–1796), duck dishes were served as delicacies in the imperial palace. Duck webs were especially popular because of their unique crunchy texture. About a century later, some chefs in Jiangsu created a special duck web dish which they named "Pearl in the Palm". To prepare this dish, the duck webs (palms) were boiled and deboned. They were then topped with mashed garlic and quail eggs (pearls).

Tip: *Do not overcook the bean sprouts. They should still retain their crunch and juiciness after cooking.*

Ingredients

Duck webs	12, outer skin discarded, cleaned
Chicken stock	625 ml
Prawns (shrimps)	150 g, shelled and deveined
Chinese cooking wine (hua tiao)	2 Tbsp
Salt	1 tsp
Cornflour (cornstarch)	2 Tbsp, mixed with 4 Tbsp water
Quail eggs	12, hardboiled and shelled
Cooking oil	3 Tbsp
Bean sprouts	120 g

Method

* Boil duck webs in 500ml chicken stock and 125 ml water for about 1 hour. Remove duck webs and plunge into cold water to cool.

* Mince prawns and mix well with 1 Tbsp cooking wine, $1/2$ tsp salt and two-thirds of the cornflour mixture. Stir until mixture is sticky.

* Arrange cooled duck webs on a heatproof (flameproof) plate. Spread a spoonful of prawn mixture on each web and top with a quail egg. Steam for 5 minutes.

* Heat oil in a wok and lightly stir-fry bean sprouts. Season with $1/4$ tsp salt.

* Dish bean sprouts onto a serving plate and arrange with duck webs.

* Bring remaining chicken stock, cooking wine and salt to the boil. Thicken with remaining cornflour mixture and pour over webs and beans sprouts before serving.

MEAT

Braised Streaky Pork
(First Official Pork)

Kublai Khan (1215–1294), the leader of the Mongol tribes, began his drive against the Southern Song (1127–1279) in the mid-13th century. During this time, Jia Sidao (1213–1275), a prime minister of Southern Song was to lead the army against the Mongols. Instead, he betrayed the kingdom by secretly signing a humiliating treaty and agreed to pay indemnities in return for peace. Jia then concealed his betrayal by proclaiming that he had won the war.

Several years later, the Mongols launched an attack again and seized major towns in Song territory. The people fought valiantly against the invaders for eight years. During this period, they sent urgent requests to Jia, asking for reinforcements which he conveniently ignored. The people were so infuriated with Jia's behavior that they renamed the popular braised streaky pork dish "First Official Pork", with reference to Jia's title to humiliate him. In 1275, the Mongols attacked again and Jia was forced to fight. He was defeated and sent into exile but was killed on the way.

Ingredients

Belly pork (with skin)	480 g, thickly sliced
Spring onions (scallions)	5, cut into 2.5-cm lengths
Ginger	1.5-cm knob, peeled and sliced
Chinese cooking wine (hua tiao)	2 Tbsp
Light soy sauce	2 tsp
Sugar	30 g
Preserved red bean curd	20 g, minced
Salt	1/2 tsp
Cooking oil	2 Tbsp
Green vegetables	150 g
Sesame oil	1 tsp
Cornflour (cornstarch)	1 tsp, mixed with 2 Tbsp water (optional)

Method

❀ Submerge pork in water and bring to the boil. Remove pork and rinse.

❀ Put pork, spring onions, ginger, cooking wine, light soy sauce, sugar, preserved red bean curd and 1/2 tsp salt in a pot. Add 500 ml water and bring to the boil. Lower heat and simmer for 1 hour.

❀ Carefully remove pork and arrange in a heatproof (flameproof) bowl, skin side down. Drain sauce from pot and pour into bowl. Steam over high heat for 30 minutes.

❀ Meanwhile, heat oil and stir-fry green vegetables. Add remaining salt and sesame oil. Dish out onto a serving plate.

❀ Remove pork from bowl and arrange with cooked vegetables. Reserve sauce.

❀ Heat sauce over high heat until it thickens or stir in cornflour mixture to thicken. Pour sauce over and serve.

Tip: Cut the meat into even-sized pieces so it cooks evenly and will be tender.

MEAT

Soy Sauce Pork
(Home-style Salted Pork)

To preempt an invasion by the Kin army, General Zong Ze (1060–1128) of the Song Dynasty (960–1279) left his hometown of Yiwu for the capital. The people of Yiwu were greatly concerned for the general and decided to support his mission by sending the troops some pork and mutton for food. As the distance between the town and the capital was great, the meat would spoil along the way. Thus, the people found a solution by marinating the meat with large amounts of salt to preserve it. The meat was delivered successfully and the troops enjoyed eating it. In gratitude, General Zong named the meat "Home-style Salted Pork".

Tip: Simmering the meat in the water, wine and soy sauce mixture helps to remove the excess salt. The slow cooking process also ensures that the meat remains tender.

Ingredients

Belly pork (with skin)	1 kg, marinated overnight with 3–4 tsp salt
Chinese cooking wine (*hua tiao*)	200 ml
Dark soy sauce	1 Tbsp
Spring onion (scallion)	1, chopped
Ginger	3.5-cm knob, peeled and roughly chopped

Method

❖ Rinse marinated pork with hot water. Slice in half and place in wok. Fill wok with water until meat is fully submerged. Pour in cooking wine and dark soy sauce and simmer over low heat for 2 hours.

❖ Transfer pork and remaining sauce to a large heatproof (flameproof) bowl. Top with spring onion and ginger and steam for 30 minutes.

❖ Carefully remove meat and slice thickly before serving.

Shredded Pork with Bamboo Shoots

Emperor Daoguang (1782–1850) came to power at a time when Qing rule had begun to deteriorate and corruption was rampant. The imperial treasury was greatly depleted and Western powers had begun to invade China. Opium had also made its inroads into the country.

This dish was first prepared by Emperor Daoguang's chefs when the emperor lost his appetite. The imperial chefs tried their best to rouse his appetite, but he still did not want to eat. One day, fresh bamboo shoots were delivered as a tribute to the emperor and the chefs decided to shred and stir-fry them with shredded pork. When the dish was presented to the emperor, he tasted it and regained his appetite.

Tip: *When a recipe calls for shredding the ingredients, shred them to the same size. This will ensure that the ingredients cook evenly and the final dish is aesthetically pleasing.*

Ingredients

Bamboo shoot (zhu sun)	150 g
Chicken stock or water	450 ml
Cooking oil	3 Tbsp
Pork loin	200 g, cut into long thin strips
Spring onion (scallion)	1, chopped
Ginger	1-cm knob, peeled and shredded
Garlic	1 clove, peeled and sliced
Light soy sauce	1 Tbsp
Chinese cooking wine (hua tiao)	3 tsp
Salt	1/2 tsp
Vinegar	2 tsp
Cornflour (cornstarch)	1 tsp, mixed with 2 Tbsp water

Method

❀ Place bamboo shoots and half the chicken stock or water in a pot and simmer for 20 minutes. Remove bamboo shoots and leave to cool slightly before shredding.

❀ Heat 1 1/2 Tbsp cooking oil in a wok until just smoking. Stir-fry pork strips until dry, then add remaining oil while still stir-frying.

❀ Add spring onion, ginger, garlic and shredded bamboo shoots and mix well. Stir in light soy sauce, cooking wine, salt and vinegar and bring to the boil.

❀ Lower heat and stir in cornflour mixture to thicken sauce. Transfer to a serving plate and serve hot.

Stir-fried Pork
(Golden Meat)

The Later Kin, subsequently called the Qing Dynasty (1644–1911), was founded by the Nuzhen people led by Nuerhachi (1559–1626). As ruler, Nuerhachi created an administrative system of eight banners and also invented the script for the Manchu language.

Before his rise to power, Nuerhachi was an errand boy in the chief commander's palace. One day, the cook fell ill and several palace maids had to help with the cooking. Based on the palace rule, meals were always served with eight different dishes. Although they tried their best, the maids could only produce seven dishes. Fortunately, Nuerhachi came up with a dish which he named "Golden Meat". In memory of their founding emperor, future Qing emperors often ate this dish. Today, it is a national dish in northeastern China.

Ingredients

Pork loin	300 g, sliced
Salt	1/2 tsp
Egg	1, beaten
Cornflour (cornstarch)	2 Tbsp, mixed with 8 Tbsp water
Chinese cooking wine (hua tiao)	1 3/4 Tbsp
Vinegar	3 tsp
Light soy sauce	1 1/2 Tbsp
Sugar	2 tsp
Chicken stock	3 Tbsp
Cooking oil	250 ml
Spring onion (scallion)	1, chopped
Ginger	1-cm knob, peeled and shredded
Garlic	1 clove, peeled and sliced
Coriander (cilantro)	1 sprig
Sesame oil	1 tsp
Ground white pepper	1/2 tsp

Method

❁ Place pork slices, salt, egg and two-thirds of the cornstarch mixture in a bowl. Mix well and leave to marinate.

❁ In another bowl, prepare the sauce by combining cooking wine, vinegar, light soy sauce, sugar, chicken stock and remaining cornstarch mixture. Set aside.

❁ Heat oil in a wok and slowly lower pork slices in one at a time. Deep-fry for 2 minutes or until golden brown. Remove pork slices and drain oil from wok.

❁ Put the pork back into the wok and add spring onion, ginger, garlic and sauce. Stir-fry for a few minutes to heat and cook sauce. Garnish with coriander and serve.

> **Tip:** Placing the pork slices into the hot oil one at a time ensures that they will not stick together when cooking.

MEAT

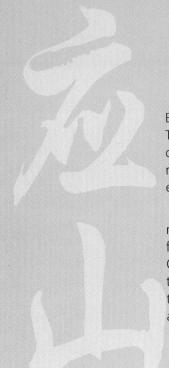

Fried and Steamed Pork
(Yingxian-style Smooth Pork)

Born Li Shimin, Emperor Taizong (599–649) was the second emperor of the Tang Dynasty (618–907). He was an outstanding military strategist and is, today, considered to be one of the greatest Chinese emperors who ever lived. His 23-year rule was characterised by rapid social development, political progress and economic growth.

Once, Emperor Taizong fell seriously ill and lost his appetite. He offered a reward to anybody who could make a dish to help him regain his appetite. A chef from Yingxian County, Hubei, learnt of this and made his way to the capital of Chang'an. There, he prepared a dish of pork for the emperor. When the emperor tasted it, he took an immediate liking to the dish and praised it. The chef was hired to work in the imperial kitchen, and the dish he prepared became widely acclaimed as "Yingxian-style Smooth Pork".

Ingredients

Streaky pork	500 g, cut into 2.5-cm cubes
Ginger	1.5-cm knob, peeled and finely chopped
Salt	1 tsp
Egg	1, beaten
Cornflour (cornstarch)	100 g
Cooking oil for deep-frying	
Chicken stock	150 ml
Light soy sauce	1¹/₂ Tbsp
Ground white pepper	1 tsp
Spring onions (scallions)	2, finely sliced

Method

* Leave pork cubes to soak in water for 10 minutes. Drain well and set aside.

* Combine ginger, salt, egg and cornflour and coat pork cubes.

* Heat oil and deep-fry pork cubes until golden brown. Remove and drain well.

* In a heatproof (flameproof) bowl, combine fried pork with 50 ml chicken stock and 1 Tbsp light soy sauce. Leave to steam for 1 hour.

* Carefully remove bowl from steamer and drain juices into a pan. Set aside.

* Turn bowl onto a serving plate to empty pork onto it. Set aside.

* Heat pan with juice and stir in remaining chicken stock, light soy sauce and pepper. Bring to the boil then lower heat to simmer.

* Mix 1 tsp cornflour with 1 Tbsp water and stir into sauce to thicken it slightly. To serve, pour sauce over pork cubes and garnish with spring onions.

Tip: In order that the resulting dish is smooth and tender, deep-fry and steam the pork well.

MEAT

Pork Tenderloin with Rice Cake (Life Saver)

The last emperor of the Ming Dynasty (1368–1644), Emperor Yongli (1623–1662), fought a losing war against the Qing army throughout his reign.

After the Ming Dynasty fell, the victorious Qing forces forced the emperor to flee southwards to Kunming. In hiding from the Manchus, the emperor and his entourage put up in a small village. Seeing how hungry and worn-out they were, a local offered them some stir-fried pork slices with rice cake. The emperor was so touched by the gesture he exclaimed: "What a great life saver!" This was how a humble, peasant dish became famous throughout Chinese food history.

Tip: *Ensure that the oil is smoking hot before adding the pork. The heat will cook the pork quickly while sealing in the juices. Cooking the pork for too long will make it tough.*

Ingredients

Pork loin	120 g, thinly sliced
Egg white	1
Salt	1 tsp
Cornflour (cornstarch)	3 Tbsp
Cooking oil	2 Tbsp
Foochow rice cake* (*nian gao*)	200 g, soaked overnight
Spring onion (scallion)	1, cut into 2.5-cm lengths
Chinese ham	50 g, thinly sliced
Tomato	50 g, thinly sliced
Spinach	50 g, cut into 3-cm lengths
Light soy sauce	2 Tbsp
Pork or chicken stock	120 ml
Hardboiled egg	1, shelled and sliced

* Foochow rice cakes are available from Chinese supermarkets or grocery stores from the dried goods section. They are translucent white in colour and have a flat longish shape.

Method

❀ Coat pork slices with egg white and sprinkle with salt and cornflour.

❀ Heat oil in a wok and stir-fry pork for 2 minutes. Add rice cake and stir-fry for another 3 minutes. Remove from heat.

❀ Using the same wok, stir-fry spring onion until fragrant. Add remaining salt, ham, tomato, spinach, light soy sauce and stock. Return stir-fried pork and rice cake to the wok and mix well. Add the egg slices in for a final toss. Serve.

MEAT

Minced Pork and Fish Rolls
(The Dragon Dish)

This dish is a specialty of Hubei Province. Before embarking on his trip to the capital to ascend the throne, Zhu Houcong (1507–1566), the future Emperor Jiajing of the Ming Dynasty (1368–1644), visited his teacher in Zhongxiang County, Hubei Province. As a tribute to the future emperor, the teacher had a sumptuous meal prepared. One of these dishes was created in the shape of a dragon to symbolise that the emperor, the "son of heaven" or the "true dragon" would soon ascend the throne. Zhu enjoyed the dish immensely and named it "The Dragon Dish". He also introduced the recipe into the imperial menu. When he made inspection tours of Hubei in later years, local officials always welcomed the emperor with this dish.

Tip: Knead the meat mixture thoroughly so it acquires a light, springy texture. This will ensure that the resulting dish is soft and succulent.

Ingredients

Minced lean pork	360 g
Salt	1 1/2 tsp
Egg whites	4
Cornflour (cornstarch)	150 g
Spring onion (scallion)	1, finely cut
Ginger	1.5-cm knob, peeled and chopped
White fish meat	160 g, minced
Eggs	3, beaten
Chicken stock	60 ml

Method

* Soak minced pork in 250 ml water for 30 minutes. Drain pork and knead in 1 tsp salt, egg whites, cornflour, spring onion and ginger. Add minced fish and continue kneading into a paste. Set aside.

* Heat a large pan and make a thin omelette with beaten eggs.

* Spread meat paste all over omelette and roll. Steam meat roll for 20 minutes, then leave to cool. Cut into 0.25-cm thick slices. Set aside.

* Heat stock and add remaining salt. Bring to the boil and thicken with 1 tsp cornflour mixed with 2 tsp water. Pour sauce over egg roll slices and serve.

Stuffed Eggs
(The Immortals' Eggs)

A long time ago, a gentleman in Liyang County, Jiangsu Province invited a group of his friends for a party. He requested his cook to prepare an egg for each of the guests. Amid the hustle and bustle in the kitchen, the cook overcooked the eggs, causing the shells to crack. Afraid of enraging his employer, the cook peeled the shells away, created an opening in each egg and removed the yolk. He filled the void with minced meat and sealed the opening with cornflour. He then steamed the eggs and served them. The guests were delighted and impressed with the innovative dish. When asked what the name of the dish was, the cook told them that it was something that the Eight Immortals used to eat. This dish was subsequently named "The Immortals' Eggs".

Ingredients

Eggs	8, medium
Minced pork	160 g
Chinese cooking wine (hua tiao)	2 Tbsp
Light soy sauce	2 Tbsp
Salt	1 tsp
Ginger	1-cm knob, peeled and chopped
Cornflour (cornstarch)	2 Tbsp, mixed with 3 Tbsp water
Cooking oil	250 ml
Chicken stock	60 ml
Sugar	1 tsp

Tip: When boiling eggs in their shells, use room temperature eggs to prevent the shells from cracking. Put the eggs into the water from the start, before turning on the heat. The eggs will be heated gradually and the shells will not crack as easily.

Method

❀ Place eggs in cold water and slowly bring to a simmer. Cover and remove from heat. Allow eggs to rest for about 10 minutes. Soak in cold water to cool eggs then carefully make a small hole in each eggshell to remove the yolk.

❀ Mix pork with 2 tsp cooking wine, 1 Tbsp light soy sauce, $1/2$ tsp salt and ginger. Stuff pork into hollowed out eggs.

❀ Smoothen the opening and brush with cornstarch mixture. Steam stuffed eggs for another 5 minutes before shelling them.

❀ Heat oil and deep-fry eggs until golden brown in colour. Remove and drain.

❀ Combine remaining cooking wine, soy sauce, salt, chicken stock, sugar and egg in a pot and bring to the boil. Lower heat and thicken sauce with remaining cornflour mixture. Pour over eggs and serve.

Steamed Meat Balls
(Filial Piety Meat Balls)

It used to be customary to serve steamed meat balls at weddings, birthday parties and funeral dinners in China. This tradition started in Yanyang, Hubei Province during the Five Dynasties (907–960). An official named Zhou Gao, who was known for his filial piety, would always bring home for his parents some of the choice items from the banquets he attended. Others soon followed his example and from there, bringing food home for one's parents became a norm. As such, banquet hosts began to prepare steamed meat balls wrapped up in lotus leaves for guests to take home. Steamed meat balls soon came to symbolise filial piety.

Tip: For a successful dish, knead the meat mixture well until it feels light and springy. As a variation to this dish, you can reduce the amount of pork and replace it with a similar amount of minced prawns.

Ingredients

Minced pork	220 g
White fish meat	120 g, minced
Egg	1, beaten
Salt	1 tsp
Ground white pepper	a pinch
Sesame oil	1 tsp
Spring onion (scallion)	1, finely cut
Ginger	1-cm knob, peeled and finely cut
Cornflour (cornstarch)	3 Tbsp

Method

* Knead pork and fish with the rest of the ingredients until it become a paste. Divide mixture into four portions. Shape each portion into large balls.

* Place meat balls on a heatproof (flameproof) dish and steam for 20 minutes. Remove from heat and serve with vegetables of your choice.

MEAT

Braised Pig's Trotters
(Fengjing Pig's Trotters)

This recipe originated from the Dingwenxing restaurant in Fengjing, a small town in Jinshan, Shanghai. In 1850, the owner of this small eatery planned to improve his business by introducing a new dish of pig's trotters cooked with ingredients that he specially ordered in from other towns. The ingredients were slowly stewed over a medium fire which resulted in very innovative flavours. All his guests grew very fond of the dish and people subsequently named it "Fengjing Pig's Trotters".

Tip: Simmering the pig's trotters on low heat will ensure that the meat absorbs the flavours and becomes tender. Top up with more water as necessary when simmering.

Ingredients

Pig's trotters	1.8 kg
Light soy sauce	120 ml
Chinese cooking wine (hua tiao)	120 ml
Sugar	50 g
Cloves	5 g
Cinnamon sticks	3
Ginger	2-cm knob, peeled and sliced
Coriander (cilantro)	1 sprig

Method

* Clean trotters and submerge in boiling water for 5 minutes. Remove and drain.

* Place trotters in a pot and cover with water. Add light soy sauce, cooking wine, sugar, cloves, cinnamon sticks and ginger. Bring to the boil then lower heat to simmer for 2 hours until trotters are well done.

* Carefully lift trotters out of pot and remove cloves, cinnamon sticks and ginger. Replace trotters and leave to boil until sauce is reduced. Remove from heat and leave to cool.

* When cool, debone trotters and slice. Pour sauce over and garnish with coriander before serving.

Deep-fried Pouches

In China during the 1800s, many noblemen carried tobacco in gold silk pouches embroidered with images of birds, animal and flowers. Drawing his inspiration from this trend, an imperial chef created a dish of minced meat, mushrooms, ham and bamboo shoot encased in an egg wrapping, like a pouch. When deep-fried, the pouches looked very attractive and were delicious. This dish was hence included in the imperial menu.

Tip: Seal the pouches well so they do not split open when frying. Lowering the pouches into the hot oil one by one will prevent them from sticking together when cooking.

Ingredients

Minced pork loin	150 g
Dried Chinese mushrooms	50 g, soaked to soften and diced
Bamboo shoot (*zhu sun*)	25 g, diced
Chinese cooking wine (*hua tiao*)	1¹/₂ Tbsp
Salt	1 tsp
Sesame oil	1 tsp
Eggs	3, beaten
Cornflour (cornstarch)	3 Tbsp, mixed with 2¹/₂ Tbsp water
Plain (all-purpose) flour	3 Tbsp, mixed with 3 Tbsp water
Chinese ham	1 slice, cut into small dices
Spring onion (scallion)	1, finely cut
Cooking oil for deep-frying	

Method

- Combine minced pork, mushrooms and bamboo shoots in a bowl. Add cooking wine, ³/₄ tsp salt and sesame oil. Mix well and shape into 2-cm balls. Set aside.

- Combine eggs, remaining salt and cornflour mixture to make a batter. Heat a large non-stick pan until hot. Pour 1 Tbsp of batter into the pan to make a pancake about 8-cm in diameter. Remove and place on a plate to cool. Repeat with the rest of the batter.

- Place a ball of filling on each pancake and brush the edges with some flour mixture. Fold pancake to enclose filling. Brush pouches with flour mixture and sprinkle with minced ham and spring onion.

- Heat oil for deep-frying and slowly lower pouches in one by one until crisp and golden brown. Drain and serve.

MEAT

Stir-fried Pork and Chicken
(Gold and Silver Strips)

In spite of its grand-sounding name, this dish is simple fare, comprising finely shredded meat, leek, spinach and bean sprouts. The word "gold" in the Mandarin name of the dish refers to the shredded omelette sprinkled on top.

This dish was traditionally served on the first day of spring and was eaten wrapped in thin pancakes. To the people in northern China, the dish was auspicious and symbolic of a good harvest.

Tip: Ensure that the wok is well-heated through before stir-frying. Stir the ingredients around to cook them quickly and evenly. Overcooking will cause the meat to toughen and the vegetables to lose their flavour.

Ingredients

Cooking oil	5 Tbsp
Spring onions (scallions)	2, cut into 3-cm lengths
Ginger	2-cm knob, peeled and shredded
Pork loin	80 g, cut into thin strips
Chicken breast	80 g, cut into thin strips
Chinese cooking wine (hua tiao)	2 Tbsp
Salt	1 tsp
Spinach	50 g, cut into 3-cm lengths, blanched
Glass noodles (fen si)	50 g, soaked in warm water to soften
Bean sprouts	120 g
Vinegar	1 tsp
Chinese chives (jui cai)	50 g, cut into 3-cm lengths
Eggs	2, lightly beaten

Method

* Heat 2 Tbsp oil in a wok and stir-fry spring onions and ginger. Add pork and chicken and cook for 3 minutes. Add cooking wine, 1/2 tsp salt, spinach and glass noodles. Mix well and set aside.

* In the same wok, heat another 2 Tbsp oil and stir-fry bean sprouts. Stir in remaining salt, vinegar and chives. Return stir-fried meat to wok and mix well. Remove from heat and transfer to a serving dish.

* In a pan, heat some of the remaining oil and make thin omelettes with the beaten egg. Roll omelette up and cut into thin shreds. Sprinkle over meat and vegetables.

Braised Ox Tail

The first emperor of the Qin Dynasty (221–206 B.C.), Qin Shi Huangdi (259–210 B.C.) was considered to be the first ruler of a unified China. He expanded the military's strength, building an army of one million soldiers. He also standardised weights and measures, the currency and the written Chinese language during his autocratic reign.

On one of his inspection tours of the country, Emperor Qin ordered a beef dish at a restaurant. Unfortunately, all the beef had been sold and all that was left was an ox tail which the owner had already cooked for his own meal. Nevertheless, the owner brought it out to serve the emperor and the emperor enjoyed it immensely. He then rewarded the restaurant's owner handsomely.

Tip: You may also add other vegetables such as baby corn and potatoes to this dish. Cut them to the same size as the carrots and bamboo shoots. Pre-cook them before adding to the stew if desired.

Ingredients

Ox tail	1 kg, cut into 3-cm lengths
Light soy sauce	3 Tbsp
Chinese cooking wine (hua tiao)	3 Tbsp
Salt	1 tsp
Cooking oil	500 ml
Spring onions (scallions)	2, chopped
Ginger	2-cm knob, peeled and sliced
Sichuan peppercorns	1 tsp
Dried red chillies	3–4, soaked to soften and sliced
Sugar	1 Tbsp
Ground white pepper	3/4 tsp
Licorice	10 g
Star anise	2
Chicken stock	400 ml
Carrots	80 g, peeled and cut into chunks
Bamboo shoot (zhu sun)	80 g, cut into chunks

Method

* Marinate ox tail with 1 1/2 Tbsp light soy sauce, 1 Tbsp cooking wine and 1/2 tsp salt. Leave for 10 minutes.

* Heat oil in a wok and deep-fry ox tail until light red in colour. Remove and drain.

* Leave 2 Tbsp of oil in the wok and stir-fry spring onions and ginger. Add remaining soy sauce, cooking wine, salt, peppercorns, dried red chillies, sugar, pepper, licorice, star anise, chicken stock and ox tail. Simmer on low heat for 2 hours.

* Remove ox tail and strain stewing liquid. Discard spices and herbs. Place ox tail back into wok with strained stewing liquid. Add carrots and bamboo shoots and cook for another 8–10 minutes. Remove and serve.

Hu-style Mutton

In 139 B.C., Chinese explorer Zhang Qian travelled to the western region as an envoy of the Han Dynasty to gather information relating to lands in Central Asia. One of his roles was to create relations with the Yue-chi people against the Xiongnu people. There, he met some wealthy locals who treated him to a dish they called Hu-style mutton. Zhang enjoyed the dish so much that he brought the recipe back to the Han Dynasty emperor. Soon, the recipe spread to eastern China. Those who travelled along the Silk Road regarded Hu-style mutton as one of the most delightful dishes. Centuries later, Hu-style mutton became even more popular when Emperor Renzong (1010–1063) of the Song Dynasty tasted the dish and highly commended it.

Tip: *Use a tender cut of mutton or lamb to get the best results from this dish. Slice the meat evenly so it cooks evenly and will be uniformly tender.*

Ingredients

Stewing mutton	600 g
White peppercorns	1 tsp
Fennel seeds	1 tsp
Ginger	2-cm knob, peeled and sliced
Spring onion (scallion)	1, cut into 2.5-cm lengths
Salt	1 tsp
Ground white pepper	1 tsp
Dried wood ear fungus (hei mu er)	15 g, soaked to soften, shredded
Dried lily buds (golden needles)	10 g, soaked to soften
Light soy sauce	1 Tbsp
Cornflour (cornstarch)	1½ Tbsp, mixed with 2 Tbsp water
Coriander (cilantro)	1 sprig

Method

❀ Cook mutton in boiling water for 30 minutes.

❀ Meanwhile, place peppercorns, fennel seeds and half the sliced ginger in a small muslin bag and secure. Add to boiling mutton stock. Cook for another 30 minutes until the meat is almost done. Remove and leave to cool. Reserve mutton stock.

❀ Cut cooled mutton into strips about 6-cm long. Combine mutton with remaining ginger, spring onion, salt, pepper and 60 ml mutton stock in a heatproof (flameproof) bowl. Steam for 40 minutes. Drain stock from steamed mutton then turn mutton onto a plate. Set aside.

❀ Bring remaining mutton stock to the boil and cook wood ear fungus and lily buds. Stir in light soy sauce. Thicken with cornstarch mixture.

❀ To serve, pour sauce over mutton and garnish with coriander.

Spicy Braised Mutton

This dish has been a popular delicacy in Longxi, Gansu Province since the 700s. One of the stories behind this dish concerns Li He (790–816), a talented poet of the Tang Dynasty (618–907). A native of Longxi, Li had a particular fondness for this specialty of his hometown. Although he was related to the imperial family, Li lived a life of poverty and suffered from poor health. He died of illness when he was just 27, leaving his young wife devastated. One night, Li appeared to his wife in a dream to comfort her and told her that he missed his favourite braised spicy mutton dish.

Tip: *Cutting the ribs into smaller pieces only after it is cooked ensures that the meat adheres to the bone while cooking.*

Ingredients

Red yeast rice (hong qu)	10 g
Mutton rib with bones	2 kg, cut into 4 pieces
Cinnamon sticks	3
Sichuan peppercorns	2 tsp
Cloves	1 tsp
Salt	1 Tbsp
Chinese cooking wine (hua tiao)	4 Tbsp
Ginger	4-cm knob, peeled and cut into chunks

Method

- Put red yeast rice in a pot with 250 ml water and boil for 5 minutes. When the grains break, transfer to a bowl to cool. When cool, spread on top of mutton. Set aside.

- Place cinnamon sticks, peppercorns and cloves in a small muslin bag and secure.

- In a large pot, heat enough water to submerge mutton. Place mutton, salt, cooking wine, ginger and bag of spices in. Bring to the boil, then lower heat and simmer for 2 hours until meat is tender.

- Remove mutton from pot and place on a serving plate. Carefully cut into small pieces before serving.

MEAT

Spicy Mutton

In the late 1800s, a small restaurant selling mutton dishes stood on a quiet lane in Linxia, Gansu Province. Although the dishes were delicious and reasonably priced, business was slow due to the restaurant's location. The owner tried to relocate to a busier area but could not afford the higher rent. He then decided to carry the cooked mutton in wooden buckets to the city and sell it by the roadside. There, his business picked up tremendously. Today, this mutton dish remains a popular snack in Gansu.

Tip: *The garlic can be added whole and unpeeled into the stock. The skin will help to hold the garlic together even as it softens after boiling. Strain off any impurities before serving the stock.*

Ingredients

Mutton leg	2 kg, cut in half
Salt	100 g
Ginger	4-cm knob, peeled and cut into chunks
Garlic	30 g, unpeeled and leave whole
Ground white pepper	40 g
Sesame oil	100 ml

Method

✤ Boil mutton for 5 minutes then drain.

✤ In a large pot, heat enough water to submerge mutton. Put in mutton, salt, ginger, garlic and pepper. Bring to the boil and leave on high heat for 1 hour.

✤ Remove mutton from stock, drain and drizzle with sesame oil. Set aside.

✤ Strain scum off stock and put mutton back in. Stew over low heat for 1 hour.

✤ Remove mutton from stock and cut into slices. Serve with stock and a garnish of your choice.

MEAT

Braised Mutton in Carp

An ancient Chinese legend tells of a man named Peng Zu. He was an imperial official and also an expert in health. During the time of the great floods, Emperor Tang Yao, who is believed to have reigned during the Xia Dynasty, worked hard to protect the people against the floods. As a result, his health suffered. When Peng Zu learnt of this, he prepared some soup for the emperor and the emperor's health improved. In gratitude, the emperor appointed Peng Zu the Lord of Peng city, which is today Xuzhou, Jiangsu Province.

The story behind this dish involves Peng Zu's youngest son. The boy loved to go fishing and he returned home one day with some fish he caught. His mother was preparing some mutton at the time and she put the mutton inside the fish to cook them together. When the fish was done, she took the mutton out and gave the fish to the boy. When Peng Zu ate the mutton, he realised that it was tastier than the usual. The recipe then began to spread.

Ingredients

Stewing mutton	1 kg
Chinese cooking wine (hua tiao)	5 Tbsp
Salt	2 tsp
Spring onions (scallions)	3, cut into 2.5-cm lengths
Ginger	3-cm knob, peeled and finely sliced
Sichuan peppercorns	1 tsp
Carp	1, about 300g
Coriander (cilantro)	1 sprig

Tip: *Slice the mutton accordingly so it fits snugly into the belly of the carp. Do not use too thick a slice of mutton as it will take longer to cook and the fish might break up if overcooked.*

Method

* Season mutton with 2 Tbsp cooking wine, 1 tsp salt, a quarter of the spring onions, one-third of the ginger and one-third of the peppercorns. Set aside to marinate for 6 hours.

* Boil enough water to submerge mutton and cook for 5 minutes. Drain well and set aside.

* Make a slit in the belly of the carp then score carp on both sides. Season with 2 tsp cooking wine and a pinch of remaining salt.

* Stuff mutton into carp's belly and place carp into a wok with remaining cooking wine, salt, spring onions, ginger, peppercorns and 250 ml water. Bring to the boil. When water boils, lower heat and skim off any scum that surfaces. Simmer until mutton and carp are well done.

* Carefully transfer stuffed carp to a serving dish and garnish with coriander to serve.

POULTRY

Fragrant Chicken
(Mr Zhuang's Hot Oil Sprinkled Chicken)

This is one of the most famous dishes of Hunan Province, created in the early 1900s. The story behind this dish involves an official by the name of Zhuang Gengliang.

Zhuang enjoyed eating at the many classy restaurants in Hunan's capital, Changsha. At one of these restaurants, he requested for a chicken dish. The chef decided to prepare a special dish by combining two cooking methods—stewing and the sprinkling of hot oil. Zhuang was very impressed with the dish after taking just one bite. The dish later became known as "Mr Zhuang's Hot Oil Sprinkled Chicken".

Ingredients

Chicken	1, about 750 g
Spring onions (scallions)	2, chopped
Ginger	5-cm knob, peeled and cut into chunks
Salt	1 tsp
Ground white pepper	1 tsp
Light soy sauce	3 Tbsp
Chinese cooking wine (hua tiao)	3 Tbsp
Sugar	1 Tbsp
Cooking oil for deep-frying	
Coriander (cilantro)	1 sprig

Method

* In a pot, marinate chicken with spring onions, ginger, salt, $3/4$ tsp pepper, $1^1/_2$ Tbsp light soy sauce and $1^1/_2$ Tbsp cooking wine. Leave for 10 minutes before adding remaining pepper, light soy sauce, cooking wine and 1 Tbsp water. Simmer for 30 minutes.

* Remove chicken and drain. Dry chicken well.

* Heat oil in a wok and slowly lower in chicken. Continuously scoop oil over chicken to coat completely and allow chicken skin to turn crisp and golden brown. Remove and drain.

* When chicken has cooled, cut meat into bite-sized pieces. Garnish with coriander and serve.

Chicken Parcel (Imperial Tablet Chicken)

This dish originated from the time of the Ming Dynasty (1368–1644). Sun Chengzong (1563–1638) was a righteous official in the Ming imperial court. He devoted much of his time and energy in building the Great Wall and opening up wasteland. There was however, a treacherous official, eunuch Wei Zhongxian (1568–1627), who was unfortunately trusted by the emperor. Wei abused his authority to advance his own interests, killing and imprisoning anybody who opposed him. Sun kept a record of Wei's misdeeds on his imperial tablet or *huban* (a wooden tablet which court officials held in both hands when attending court), awaiting the opportunity to bring Wei to justice.

There was a time when Sun was so preoccupied with his plans to impeach Wei that he put the imperial tablet into his mouth, mistaking it for the fried fish laid out for his meal by his family. Deeply moved by Sun's commitment to justice, his family prepared a dish of chicken moulded into the shape of an official tablet. Since then, this dish has been passed down from generation to generation to become a traditional local delicacy of Hebei, Sun's native province.

The traditional recipe uses pig's caul but as that is difficult to obtain today, it has been substituted with dried bean curd skin in this recipe.

Ingredients

Minced chicken breast	300 g
Spring onion (scallion)	1, finely shredded
Ginger	1.5-cm knob, peeled and finely chopped
Egg	1, lightly beaten
Salt	3/4 tsp
Chinese cooking wine (hua tiao)	3 tsp
Dried bean curd skin (fu pi)	1 sheet
Egg whites	4
Cornflour (cornstarch)	50 g
Cooking oil for deep-frying	
Coriander (cilantro)	1 sprig

Method

* Combine minced chicken with spring onion, ginger, half the egg, salt and cooking wine.

* Brush remaining egg over dried bean curd skin and spoon minced chicken onto it. Bring edges of bean curd skin up to enclose filling and fold into a neat rectangular parcel. Cut away any excess skin. Even out and flatten parcel by pressing down with a chopping board.

* Steam parcel for 15 minutes over high heat. Remove and leave to cool.

* Beat egg whites until light and fluffy. Stir in cornflour to make a thick paste. Brush over steamed parcel.

* Heat oil and deep-fry parcel until golden. Drain well. Slice into 2-cm strips and transfer to a serving plate. Garnish with coriander and serve.

POULTRY

Tea-smoked Chicken
(The Magistrate's Chicken)

This dish is said to be the creation of Zhou Guisheng, a magistrate who ruled several counties in the last years of the Qing Dynasty (1644–1911). When the Qing Dynasty toppled, Zhou's career as an official also ended. He and his family then moved to Guangzhou in southern China where he made a living operating a street-side food stand serving special meat dishes. Drawing on his familiarity with the cooking styles of the different counties which he oversaw as an official, Zhou combined the best of each county's cooking techniques to create the aromatic tea-smoked chicken of Guangdong. The locals in Guangdong also fondly refer to the dish as "The Magistrate's Chicken" in honour of Zhou.

Tip: *Ensure that the chicken is well cooked and drained before steaming. Use quality Chinese tea leaves to get a very fragrant final dish.*

Ingredients

Chicken stock	1 litre
Chicken	1, about 1 kg
Cooking oil	4 Tbsp
Chinese tea leaves	80 g
Brown sugar	150 g
Sesame oil	1 tsp
Coriander (cilantro)	1 sprig

Method

❀ Heat stock in a large pot and boil chicken for 20 minutes. Drain and reserve stock.

❀ Heat oil and stir-fry tea leaves until fragrant. Add brown sugar and stir-fry until it smokes. The smoke will be yellowish in colour.

❀ Pour tea leaves mixture onto a sheet of aluminium foil. Place this in a wok.

❀ Place a steamer rack over the tea leaves and sit the chicken on it. Cover wok tightly and heat for a few minutes.

❀ Remove wok from heat and leave chicken in wok for 5 minutes. The chicken is ready when it turns a light brown colour. Cut chicken into chunks and arrange on a plate.

❀ Combine 5 Tbsp stock, 3 Tbsp water and sesame oil to make a sauce. Drizzle over chicken, garnish with coriander and serve.

Deep-fried Claypot Chicken
(Imperial Concubine's Chicken)

Yang Yuhuan (719–756) was made an imperial concubine in 745 and conferred the title Yang Gui Fei. She was beautiful and also talented in music, dancing and singing. In the palace, she gained the special favour of Emperor Xuanzong (685–763) of the Tang Dynasty (618–907).

Once, the emperor requested that Yang Gui Fei serve him dinner. When the time came, however, he had dinner with another imperial concubine instead. Disappointed, Yang Gui Fei became drunk. After hearing of this, a cook prepared a dish of chicken in wine and named it "Imperial Concubine's Chicken". The dish spread and is today a traditional dish served in Beijing.

Tip: Simmering the chicken over low heat ensures that the flesh remains succulent and tender.

Ingredients

Light soy sauce	2 Tbsp
Chinese cooking wine (hua tiao)	2 Tbsp
Chicken	1, about 750 g
Cooking oil for deep-frying	
Chicken stock	1 litre
Salt	1 tsp
Red wine	90 ml
Spring onions (scallions)	3, chopped

Method

* Combine 1 Tbsp soy sauce and cooking wine and brush inside and outside of chicken.

* Heat oil in a wok and deep-fry chicken until it turns golden brown. Remove from heat and plunge into cold water to remove excess oil. Drain.

* Put chicken in a claypot and add chicken stock, remaining soy sauce, salt and half the red wine. Simmer over low heat for 2 hours. Remove from heat and add remaining red wine. Garnish with spring onions and serve.

Dong'an Tender Chicken

The story behind this dish took place during the reign of Emperor Xuanzong (685–763) of the Tang Dynasty (618–907). It is said that several businessmen went to a restaurant in Dong'an County, Hunan Province one evening for dinner. At the time, the restaurant had run out of other ingredients, except for two chickens. Thus, the cook cut the chickens into small pieces, marinated them and quickly stir-fried them in very hot oil. He then lowered the heat and left the dish to simmer. The result was chicken that was both tender and aromatic. The businessmen were very satisfied with the dish and they spoke about the restaurant wherever they went, prompting many people to visit it. When the county magistrate heard of it, he also went to the restaurant to try the dish for himself. He was so impressed he named the dish "Dong'an Tender Chicken".

Tip: Boiling the chicken briskly in water before stir-frying helps make it tender. Do not leave it to boil too long or the flavour will be lost and the chicken will also be tough.

Ingredients

Chicken	1, about 750 g
Cooking oil	50 ml
Spring onions (scallions)	2, cut into 3-cm lengths
Dried red chillies	5, soaked to soften and thinly sliced
Chinese cooking wine (hua tiao)	2 Tbsp
Salt	1 tsp
Ground white pepper	1 tsp
Vinegar	2 Tbsp
Water	125 ml
Ginger	2.5-cm knob, peeled and shredded
Cornflour (cornstarch)	1 Tbsp, mixed with 2 Tbsp water

Method

* Bring a large pot of water to the boil and cook chicken for about 8 minutes. Carefully remove chicken and leave to cool slightly. Debone chicken and cut meat into 3 x 1-cm pieces.

* Heat oil in a wok and stir-fry chicken strips, spring onions and dried red chillies for about 3 minutes.

* Add cooking wine, salt, pepper, vinegar and water, stir and bring to the boil. Then lower heat and simmer for 3 minutes. Add ginger.

* Slowly stir in cornflour mixture to thicken sauce. Dish out and serve.

Three Cup Chicken

This dish, called *San Pei Ji* (Three Cup Chicken), is a well-known dish in Jiangxi. It was first prepared for Wen Tianxiang (1236–1283) of the Southern Song Dynasty (1127–1279), a prominent writer and military general famous for his loyalty and heroism.

In 1275, General Wen received news that enemy Yuan troops were heading southward into Southern Song territory. He immediately organised a volunteer army in Gangzhou, Jiangxi Province and led them to Lin'an, capital of Southern Song, to protect the city. For his patriotism, General Wen was made prime minister and sent to negotiate with the Yuan troops. There, he was detained but managed to escape. In the following year, General Wen's resistance army battled with the Yuan troops again but this time, General Wen was captured. He refused to yield to the enemy and was eventually put to death.

Prior to his death, an old lady visited him in prison with the assistance of a prison guard. She brought along a dish of chicken that had been slowly stewed with three cups of ingredients—pork fat, cooking oil and light soy sauce. After the general was executed, the prison guard went home to Jiangxi where he prepared the dish in memory of General Wen every year on the date of his execution.

Ingredients

Spring chicken	1, about 500 g, cut into 3-cm pieces
Cooking oil	90 ml
Chinese cooking wine (hua tiao)	90 ml
Light soy sauce	90 ml
Spring onions (scallions)	2, cut into 3-cm lengths
Ginger	3-cm knob, peeled and sliced
Water	500 ml
Coriander (cilantro)	1 sprig

Method

* Bring a pot of water to the boil and lower chicken pieces in briefly. Remove and rinse chicken with cold water.

* Put chicken in a claypot. Add oil, cooking wine, light soy sauce, spring onions, ginger and water. Cover and leave to simmer over low heat for 45 minutes. Remove spring onions and ginger before serving, if preferred. Garnish with coriander.

Tip: Cut the chicken into similarly sized pieces so that the cooking time required will be more even.

POULTRY

Chicken Cubes with Pears

This traditional dish from Yunnan Province was first prepared in 1666. It is said that Li Dingguo (1621–1662), a military general who helped the Ming emperor escape from the invading Manchu army in the late 1650s to 1660s, was being attacked by the Qing troops when he fled to Kunming where a kind lady took him into her home. She prepared a dish of chicken with freshly-picked pears for his meal. The general enjoyed the dish very much and from then on, this dish became a specialty of Kunming.

Tip: Cut the chicken, pear and ham into cubes or squares of similar size. Choose pears that are not too ripe as they will become mushy when cooked, spoiling the resulting dish.

Ingredients

Chicken breast	280 g, cut into 1.5-cm cubes
Egg	1, small, beaten
Cornflour (cornstarch)	2^1/$_2$ Tbsp
Cooking oil	250 ml
Chinese pears	150 g, peeled, cored and cut into cubes
Spring onions (scallions)	2, sliced
Ginger	1.5-cm knob, peeled and sliced
Chinese ham	3 slices, each cut into small squares
Salt	^{1}/$_2$ tsp

Method

* Mix chicken cubes with egg and cornflour.

* Heat 2 Tbsp oil in a wok and stir-fry chicken breast. Add pear cubes, stir quickly and remove from heat.

* In the same wok, stir-fry half the spring onions and ginger. Add ham, chicken and pear and stir-fry to mix. Sprinkle in salt.

* Remove from heat and garnish with remaining spring onions to serve.

Chicken with Yam
(Sima Yi's Braised Chicken)

The Three Kingdoms (Wei, Shu and Wu, which had overlapping reigns from 220–280) was romanticised and touted as the golden age of chivalry in China's history. Although wars continued to be fought during this period, each of the kingdoms focused mainly on the reorganisation of their respective governments and rebuilding their economies.

Sima Yi (179–251) was a general of Wei. He was a cunning strategist who led the resistance against the Shu army headed by another great strategist Zhuge Liang (181–234). In 204, Sima Yi took control of the court affairs in Wei.

Sima Yi enjoyed eating chicken and yam as he believed that these two ingredients were nutritious and nourishing. He had his cook prepare a dish using both chicken and yam which he enjoyed very much. This dish soon became a favourite of the people of Hennan.

Ingredients

Chicken breast	280 g, cut into 2.5-cm cubes
Salt	1 tsp
Egg	1, beaten
Cornflour (cornstarch)	4 Tbsp
Cooking oil for deep-frying	
Yam	180 g, peeled and cut into 2.5-cm cubes
Light soy sauce	1 Tbsp
Spring onions (scallions)	2, cut into 2.5-cm lengths
Ginger	2.5-cm knob, peeled and sliced
Sugar	1 tsp
Star anise	2
Chinese cooking wine (hua tiao)	2 Tbsp
Water	150 ml

Method

* Mix chicken with ¹/₂ tsp salt, egg and 3 Tbsp cornflour.

* In a wok, heat oil and deep-fry chicken cubes until golden brown. Drain and place in a heatproof (flameproof) bowl.

* In the same oil, deep-fry yam cubes until golden brown. Drain and place on top of chicken in bowl.

* Pour remaining salt, light soy sauce, spring onions, ginger, sugar, star anise, cooking wine and water over chicken and yam. Steam for 15–20 minutes until meat is soft.

* Pour sauce from the bowl into a clean wok and bring to the boil. Reduce slightly and thicken with remaining cornflour mixed first with 1 Tbsp water.

* Turn chicken and yam out onto a plate. Pour sauce over and serve.

Tip: *Deep-fry the chicken lightly to seal in the juices and make it fragrant, but do not deep-fry too long or the meat will be dry and tough. The steaming time for the chicken and yam depends on your preference. Steam for a longer time period if you prefer them softer.*

Chicken with Prawns (Li Bai's Chicken)

Li Bai (701–762) was a famous poet of the Tang Dynasty. He is remembered as one of the greatest romantic poets in Chinese history. In 742, Li Bai was invited to serve in the imperial academy. He was delighted as he believed he could contribute much as a politician. To his great dismay, however, Emperor Xuanzong (685–763) only gave him the role of a palace poet. Unwilling to stay on with such a trivial role, Li Bai left the post soon after. He then wandered around the country aimlessly and was looked upon as a political outcast. During his last years, Li Bai lived a life of poverty and died of illness at Dangtu (today's Anhui).

A story is told about a time in Li Bai's life, when he stayed in Anlu County, Hubei Province. He drowned his sorrows by eating, drinking and composing poems. Despite the sorry end to his life, this dish of chicken with prawns was named after Li Bai.

Ingredients

Chicken	1, about 1 kg
Salt	1 1/2 tsp
Chinese cooking wine (hua tiao)	3 Tbsp
Spring onions (scallions)	2, chopped
Ginger	1.5-cm knob, peeled and sliced
Prawns (shrimps)	100 g, shelled and minced
Cornflour (cornflour)	2 1/2 Tbsp
Egg whites	2
Dried Chinese mushrooms	30 g
Cooking oil	3 Tbsp
Chinese ham	3 slices, shredded
Bamboo shoot (zhu sun)	20 g, shredded
Ground white pepper	1 tsp
Light soy sauce	2 tsp
Rice wine	3 Tbsp
Chicken stock	3 Tbsp
Coriander (cilantro)	1 sprig

Method

* Marinate chicken with 3/4 tsp salt, 2 Tbsp cooking wine, spring onions and ginger for 2 hours.

* Steam for 1 hour then and remove from heat and cool slightly. Cut chicken into bite-sized pieces and arrange on a serving plate.

* Combine minced prawns, 1/2 tsp salt, cornflour and egg whites. Mix well and shape into small balls. Steam together with mushrooms for 5 minutes. Arrange with chicken.

* Heat oil and cook ham and bamboo shoots. Add pepper, soy sauce, rice wine, remaining salt and chicken stock.

* Bring to the boil and thicken sauce with 1 tsp cornflour mixed 1/2 Tbsp water.

* Pour sauce over chicken, prawn balls and mushrooms. Garnish with coriander and serve.

Sichuan Diced Chicken with Peppercorns

The story behind this dish involves a rich, affable man who lived in Sichuan Province during the Qing Dynasty (1644–1911). He was so well-liked he always had friends visiting him from near and far. As is in the Chinese custom, the man showed his hospitality by preparing meals for all his friends each time they visited. This became such a regular affair that the man thought of a way to facilitate making these meals. He created a cold dish of diced chicken and peppercorn which could be prepared beforehand in huge quantities and kept over several days. The guests enjoyed the dish so much it eventually became a hallmark dish of Sichuan.

Ingredients

Chicken breast	450 g, cut into 2-cm cubes
Salt	1 tsp
Chinese cooking wine (hua tiao)	2 Tbsp
Spring onions (scallions)	2, finely sliced
Ginger	1-cm knob, peeled and finely sliced
Cooking oil	for deep-frying
Sichuan peppercorns	4 tsp
Dried red chillies	4, soaked to soften and cut into 2-cm lengths
Water	160 ml
Sugar	1 Tbsp

Method

❀ Combine chicken, 1/2 tsp salt, 1 Tbsp cooking wine, spring onions and ginger. Leave to marinate for 10 minutes.

❀ Heat oil in a wok and deep-fry diced chicken until golden brown. Remove and drain. Set aside.

❀ Leave 2 Tbsp oil in the wok and fry peppercorns until aromatic. Add dried chillies, chicken, water, remaining salt and sugar. Simmer until sauce is reduced to about one-third.

Diced Chicken with Chilli and Peanuts
(Imperial Guard Diced Chicken)

This is one of the most popular dishes in Chinese cuisine. Its name, *Gong Bao Ji Ding* roughly translates as "Imperial Guard Diced Chicken" and there are many stories about the origin of the name. The most well-known one is that the dish was named after Ding Baozhen (1820–1886) who had a predilection for this dish. Ding, a native of Guizhou, served as governor of Yuezhou in Hunnan Province and was promoted to the post of touring governor of Shandong Province. In 1876, he was made governor of Sichuan. Under him, the administrative system was reformed and machinery works of the province established. The imperial court gave him the title, *Tai Zi Shao Bao*—Junior Guardian of the Heir Apparent—in recognition of his merits and services. Thereafter, people often referred to him as *Ding Gong Bao* or Ding, The Imperial Guardian.

When Ding first arrived in Sichuan, his subordinates invited him for dinner. Aware of his fondness for diced chicken and chillies, they requested for this dish to be specially prepared and they dubbed the dish *Gong Bao Ji* after Jing's title.

Ingredients

Chicken breast	300 g, cut into 1-cm cubes
Egg	1, beaten
Light soy sauce	2 tsp
Salt	$\frac{1}{2}$ tsp
Chinese cooking wine (*hua tiao*)	3 tsp
Cornflour (cornstarch)	3 Tbsp, mixed with 5 Tbsp water
Vinegar	1 tsp
Sugar	1 tsp
Cooking oil	2 Tbsp
Sichuan peppercorns	1 tsp
Dried red chillies	5, soaked to soften, cut into 2-cm lengths
Spring onion (scallion)	1, finely cut
Ginger	1-cm knob, peeled and sliced
Garlic	1, peeled and sliced
Peanuts (groundnuts)	80 g, deep-fried until crisp

Method

✿ Combine chicken, egg, $\frac{1}{2}$ tsp light soy sauce, $\frac{1}{4}$ tsp salt, 1 tsp cooking wine, and 2 Tbsp cornflour mixture. Set aside.

✿ Make a sauce with the remaining light soy sauce, salt, cooking wine, cornflour mixture, vinegar, sugar and additional 1 Tbsp water. Set aside.

✿ Heat oil in a wok and deep-fry peppercorns until aromatic. Add dried red chillies, spring onion, ginger, garlic and diced chicken. Stir-fry until chicken is cooked.

✿ Add in sauce and stir-fry until sauce is reduced. Quickly stir in peanuts and serve.

Chicken with Fish Cakes
(Lady Li's Chicken)

In the final years of the Qing Dynasty (1644–1911), Empress Dowager Cixi (1835–1908) was the supreme leader during the reign of two emperors: her son, Tongzhi (1856–1875), and nephew, Guangxu (1871–1908). When Emperor Guangxu took over the throne, he was just four years old. The empress dowager, dubbed "The Dragon Lady", ensured that she was still the power behind the scene.

In 1900, when the empress dowager and Emperor Guangxu travelled westward to Xi'an, they felt hungry and decided to stop for a snack in the village of Shicun in Shanxi. The chief eunuch ordered the imperial cooks travelling with them to prepare the food. As the cooks were not familiar with the cuisine of the place, a local woman named Li gave them one of her recipes for a dish of tender chicken casserole with fish meat balls. After finishing the meal, the empress was very satisfied and named the dish "Lady Li's Chicken" in honour of the lady.

Ingredients

Chicken	1, about 750 g
Spring onions (scallions)	2, cut into 2.5-cm lengths + 1, finely cut
Ginger	2-cm knob, peeled and sliced + 1-cm knob, peeled and finely cut
Star anise	2–3
Chinese cooking wine (hua tiao)	2 Tbsp
Salt	1 1/2 tsp
Water	500 ml
White fish fillet	180 g, minced
Egg whites	2
Cooking oil	2 Tbsp
Green peas	20 g
Chinese ham	20 g, thinly sliced
Coriander (cilantro)	1 sprig

Method

❀ Scald chicken in boiling water then place in a claypot.

❀ Add sectioned spring onions, sliced ginger, star anise, 1 Tbsp cooking wine, 1 tsp salt and water and steam for 40 minutes. Remove spring onion, ginger and star anise.

❀ Combine minced fish meat with finely cut spring onions, finely cut ginger, egg whites, remaining cooking wine, remaining salt and mix well until mixture is slightly sticky to the touch.

❀ Take six small containers, oil them and fill with fish paste. Press green peas lightly into paste and steam gently for 5 minutes. Remove from heat and pop fish cakes out.

❀ Place fish cakes into claypot with chicken. Garnish with ham and coriander to serve.

Tip: When preparing the fish paste, use a spoon to mix it in one direction to achieve a well-mixed paste with a sticky consistency.

Braised Goose
(The Duke's Goose)

Duke Huan ruled the State of Qi from 685–643 B.C. His main goal was to form an alliance with the states of Chen, Cai, Zhu, Song, Lu, Wei, Zheng and Cao, and have Qi be the dominant state. There was resistance from some of the states initially but through the good advice of his chief minister, Guan Zhong, Huan made strategic military moves and managed to form the alliance without much warfare.

Once, Duke Huan was tired and famished while out in the battlefield when a soldier brought him a goose. His cook braised the goose and Duke Huan enjoyed it very much. Thereafter, people named the dish "Duke Huan's Goose". Today, this dish remains a famous specialty in central Shanxi.

Tip: For the goose to be fragrant, it has to be well-marinated. Do not cut back on the marinating time. Substitute the goose with duck as desired.

Ingredients

Goose	1, about 1.2 kg
Light soy sauce	4 Tbsp
Chinese cooking wine (*hua tiao*)	3 Tbsp
Water	1 litre
Salt	1 tsp
Spring onions (scallions)	3, chopped
Ginger	1.5-cm knob, peeled and sliced
Star anise	3
Fennel seeds	2 tsp
Cinnamon sticks	2
Chicken stock	500 ml
Chinese cabbage	400 g, thinly cut
Sugar	2 Tbsp
Cornflour (cornstarch)	2 Tbsp, mixed with 3 Tbsp water
Dried Chinese mushrooms	8, soaked and sliced
Bamboo shoot (*zhu sun*)	20 g, sliced

Method

* Cut goose in half and marinate with 3 Tbsp light soy sauce and 1^1/$_2$ Tbsp cooking wine. Leave for about 20 minutes.

* Heat enough oil for deep-frying in a wok and deep-fry goose half at a time until crispy and golden brown. Drain well.

* Put deep-fried goose in a large pot, add water, ^{1}/$_2$ tsp salt, spring onions, ginger, star anise, fennel seeds and cinnamon sticks. Cover and cook for 30 minutes.

* Remove goose, cut into bite-sized pieces and arrange neatly on a heatproof (flameproof) plate. Pour some sauce from pot over goose and steam for 10 minutes. Reserve remaining sauce.

* Heat 400 ml chicken stock and cook cabbage. Set aside.

* Heat remaining chicken stock, soy sauce, cooking wine, salt and sugar in a wok and bring to the boil.

* Add cooked cabbage and simmer for 2 minutes. Thicken sauce by adding a third of the cornflour mixture. Dish out onto a serving plate. Arrange steamed goose on top, reserving juice.

* Using the same wok, boil juice and cook mushrooms and bamboo shoots. Thicken with remaining cornflour mixture. Pour over goose and cabbage and serve.

SEAFOOD

Deep-fried Carp (Squirrel-shaped Fried Fish)

While travelling through Suzhou, Emperor Qianlong (1736–1795) of the Qing Dynasty (1644–1911) visited a restaurant known as the Pine and Crane Tower Restaurant. Much to the cook's alarm, he asked for the live carp offering on the altar table to be cooked for his meal. This was taboo at that time as the people believed that offerings made to gods were not to be consumed by mortals. The emperor was adamant, however, on having the carp.

The cook had no choice but to relent. He then thought of preparing the carp such that it did not look like a carp, hoping that this would exempt him from spiritual wrong-doing. The emperor enjoyed the dish no less and it soon became one of the most renowned dishes in Chinese cuisine.

Tip: Drain fish well to ensure that it remains crisp when serving.

Ingredients

Carp	1, about 700 g
Salt	1¹/₂ tsp
Chinese cooking wine (hua tiao)	3 Tbsp
Egg	1, beaten
Cornflour (cornstarch)	6 Tbsp
Tomato sauce	4 Tbsp
Sugar	150 g
White vinegar	6 Tbsp
Stock or water	250 ml
Spring onions (scallions)	2, finely cut
Ginger	1-cm knob, peeled and finely cut
Prawns (shrimps)	50 g, boiled, shelled and diced
Bamboo shoot (zhu sun)	20 g, diced
Dried Chinese mushrooms	20 g, soaked to soften and diced
Green peas	20 g

Method

- Gut fish and wash thoroughly. Cut off head and set head aside.

- Cut fish along the spine until about 4 cm from tail. Remove spine bone until the uncut tail portion. Spread open fish, flesh side up and carefully remove other remaining bones.

- Score flesh with criss-cross cuts. Marinate with 1 tsp salt, 1¹/₂ Tbsp cooking wine and egg. Leave for 20 minutes, then dust with cornflour.

- Make sauce. Combine tomato sauce, sugar, white vinegar, stock or water, remaining salt and cooking wine. Set aside.

- Heat enough oil for deep-frying in a wok. Spread fish out flat, flesh side up, on a large percolated ladle. Slowly lower fish into oil. Deep-fry until crisp and golden brown. Remove and drain.

- In the same oil, deep-fry fish head until golden brown. Remove and drain. Arrange fish and fish head on a serving plate.

- Leaving 3 Tbsp oil in the wok, stir-fry spring onions and ginger briefly until aromatic. Add prawns, bamboo shoots, mushrooms and green peas. Stir-fry briefly and pour in sauce. Bring to the boil and thicken with 1 Tbsp cornflour mixed with 2 Tbsp water. Pour over fish and serve.

SEAFOOD

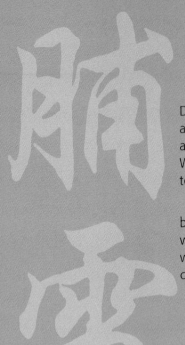

Deep-fried Yellow Croaker

During his inspection tour of the south, Emperor Qianlong (1736–1796) discovered a huge rock that produced melodious sounds when struck. Convinced that it was a rare treasure, he ordered that the rock be transported back to the imperial palace. With tremendous difficulty, the workers eventually succeeded in getting the rock to the capital.

The emperor wanted the rock to be placed in a particular garden in the palace but as the gate to that garden was too narrow for the rock to go through, the gate was dismantled. When the emperor's mother learned of this, she chided him for wasting time and money. To appease her anger, the emperor personally created this deep-fried yellow croaker dish for her.

Tip: *Dip fish slices into hot oil one at a time to prevent them from sticking together when frying.*

Ingredients

Yellow croaker	1, about 750 g
Chinese cooking wine (hua tiao)	2 Tbsp
Salt	1 1/2 tsp
Ground white pepper	1 tsp
Egg whites	4
Cornflour (cornstarch)	4 Tbsp
Plain (all-purpose) flour	6 Tbsp
Egg yolks	2
Tomato sauce	2 Tbsp
Spring onion (scallion)	1, shredded
Ginger	1-cm knob, peeled and shredded
Garlic	1 clove, peeled and sliced
Green capsicum (bell pepper)	1, thinly sliced
Bamboo shoot (zhu sun)	20 g, thinly sliced
Chinese ham	1 slice, thinly sliced
Stock or water	125 ml
Vinegar	1 Tbsp

Method

❀ Gut fish and cut off head and tail. Wash and set aside. Debone and skin body, then cut into 1-cm thick slices.

❀ Marinate fish head and tail with 1 1/2 Tbsp cooking wine, and 1/2 tsp each of salt and pepper. Leave for 10 minutes.

❀ Whip egg whites and add 2 1/2 Tbsp cornflour and 1 1/2 Tbsp flour. Mix into a paste. Set aside.

❀ Combine egg yolks, 1/2 tsp salt and remaining cornflour. Mix into a paste. Set aside.

❀ Heat enough oil for deep-frying in a wok. Dust fish slices evenly with remaining plain flour, dip into egg white paste and deep-fry until golden. Drain and arrange neatly on a serving plate.

❀ Dip fish head in egg yolk paste and deep-fry. Remove and arrange on the same serving plate.

❀ Add tomato sauce to remaining egg white paste and mix well. Dip fish tail in and deep-fry. Drain and arrange on the plate.

❀ Leave 2 Tbsp oil in the wok and stir-fry spring onion, ginger and garlic until fragrant. Add green capsicum, bamboo shoot, ham, remaining cooking wine and salt, stock or water and vinegar. Bring to the boil then thicken sauce with 1 Tbsp cornflour mixed with 2 Tbsp water. Pour sauce over fish to serve.

SEAFOOD

Sweet and Sour Fish
(Fish from West Lake)

This renowned dish from Hangzhou has a bitter-sweet story behind it. Once upon a time, there were two fishermen brothers who lived near West Lake. A despotic official, named Zhao, coveted the older brother's wife and plotted the death of the brother. He then framed the younger brother for the death. Knowing the truth, the latter's sister-in-law urged him to flee. Before he left home, she cooked him a fish caught from the lake, seasoning it with sugar and vinegar to symbolise the sweetness and bitterness (sourness) of life. She then told him not to forget the oppression of the common people when life became good for him.

Years passed and the young man became an official. He had the opportunity to punish Zhao for his evil deed but he could not locate his sister-in-law. Then one day, he was served fish at a dinner that tasted exactly like the one his sister-in-law prepared for him. He asked to meet the cook and it was indeed his sister-in-law. She had disguised herself to escape from Zhao. They were finally reunited.

Ingredients

Carp	1, about 750 g
Light soy sauce	2 Tbsp
Chinese cooking wine (hua tiao)	1 Tbsp
Ginger	2-cm knob, peeled and finely chopped
Sugar	3 Tbsp
Red vinegar	4 Tbsp
Cornflour (cornstarch)	1 Tbsp, mixed with 2 Tbsp water

Method

* Gut fish, cut into two along the spine and wash.

* Boil water in a wok and carefully place fish in, skin side up. Boil for 8 minutes until fish is just cooked.

* Discard half of the water, add light soy sauce, cooking wine, ginger and sugar. Simmer for 10 minutes.

* Remove fish and place on a serving plate, skin side up.

* Stir vinegar into the water in the wok. Thicken with cornflour mixture. Bring to the boil and pour over fish. Serve.

Palace Gate Fish

Emperor Kangxi (1654–1722) of the Qing Dynasty (1644–1911) is often lauded as one of the wisest rulers in Chinese history. He ascended the throne when he was just eight years old and his six-decade reign was a peaceful one. He greatly improved the people's standard of living and encouraged the development of arts and cultural activities.

A caring emperor, Kangxi often made arduous inspection tours to the south incognito. On one of these tours, he visited a village by the name of Palace Gate. Here, he chanced upon a small restaurant by a pond. He ordered a fish dish and enjoyed it with wine. He asked for the name of the dish, but the unsuspecting restaurant owner said it had no name. Thus, the emperor gamely suggested that the dish be named "Palace Gate Fish". He wrote the name down and signed his name next to it. Not long after, the governor of Zhejiang passed by the restaurant and recognised the emperor's signature. Word spread and business at the restaurant flourished.

Ingredients

Mandarin fish	1, about 700 g
Light soy sauce	2 Tbsp
Spring onions (scallions)	2, roughly broken
Ginger	1-cm knob, peeled and sliced
Egg whites	2, mixed with 4 Tbsp cornflour
Chinese ham	1 slice, cut into diamond shapes
Green peas	30 g
Beef	50 g, diced
Dried prawns (shrimps)	20 g
Pickled mustard stems (*mei cai*)	20 g, diced
Bamboo shoot (*zhu sun*)	50 g, diced
Salt	1 tsp
Sugar	1 Tbsp
Vinegar	1 Tbsp
Stock or water	250 ml

Method

* Gut, wash and dry fish. Cut it into three parts—the head, body and tail. Marinate head and tail with 1 tsp light soy sauce, spring onions and ginger. Leave for 30 minutes.

* Debone and skin body. Cut into large slices. Dip slices into egg white mixture and arrange a ham slice on top. Set aside.

* Heat enough oil for deep-frying in a wok and lower in fish head and tail. Cook until golden brown. Drain and set aside.

* In the same oil, deep-fry fish slices until crispy. Drain and arrange fish slices in the centre of a serving plate. Top each slice with a green pea.

* Leaving 2 Tbsp in the wok, stir-fry beef, dried prawns, pickled mustard stems and bamboo shoot. Add remaining light soy sauce, salt, sugar, vinegar and 200 ml stock or water. Add fish head and tail, and bring to the boil. Lower heat and simmer for another 20 minutes.

* Bring remaining stock to the boil and stir in 1 Tbsp cornflour mixed with 2 Tbsp water to thicken slightly. Pour over fish slices.

* Arrange fish head and tail on serving plate with fish slices. Pour stir-fried beef mixture over. Serve.

Braised Fish Tails (Fans on Water)

While serving as an official in Huangzhou, Hubei Province, famous poet Su Dongpo (1037–1101) befriended many local fishermen. He often went fishing with them and shared meals with them. Sometimes he even cooked for them.

Once, while he was cooking a fish dish for his fishermen friends, he thought of how graceful fishes were as they swam, gliding their tails in the water. Immediately, he named the dish he was preparing "Fans on Water".

Tip: *When selecting the fish tails, choose those with more flesh and trim them to a similar size.*

Ingredients

Cooking oil	3 Tbsp
Spring onions (scallions)	2, cut into 3-cm lengths
Ginger	2-cm knob, peeled and sliced
Fish tails	2–4, about 1.2 kg in total
Light soy sauce	2 Tbsp
Salt	1 tsp
Chinese cooking wine (hua tiao)	2 Tbsp
Sugar	3 Tbsp
Chicken stock or water	250 ml
Vinegar	2 tsp
Ground white pepper	1 tsp
Cornflour (cornstarch)	1 Tbsp, mixed with 2 Tbsp water
Egg	1, fried into a thin omelette and shredded
Oranges	2, peeled and segmented

Method

* Heat oil and stir-fry spring onions and ginger for a few minutes. Discard spring onion and ginger.

* Carefully lower fish tails into wok and fry for 3 minutes. Add light soy sauce, salt, cooking wine, sugar and stock or water. Simmer for 5 minutes.

* Add vinegar and pepper, and continue to cook for another 3 minutes. Thicken sauce with cornflour mixture.

* Remove fish tails and arrange on a plate in the shape of a fan.

* Use orange segments to create the bottom edge of the fan and arrange shredded omelette to serve as tassels of the fan.

SEAFOOD

Fish Slices with Pickled Mustard Leaves

This classic dish is attributed to a servant girl named Peihong who lived in Shaoxing, Zhejiang Province in the 1700s. Seeing that her fellow servants had to endure eating old and yellowed vegetables daily, Peihong devised a more palatable solution by soaking the vegetables in salt water for a few days to pickle them. The pickled vegetables were delicious and Peihong was praised by the other servants.

In memory of Peihong's ingenious idea, the people began to refer to pickled mustard leaves as "Peihong vegetables". Cooks also began to use these pickled vegetables with mandarin fish slices, creating this popular and classic dish.

Ingredients

Mandarin fish fillet	250 g, sliced
Egg white	1, lightly beaten
Salt	1/2 tsp
Cornflour (cornstarch)	1 1/2 Tbsp, mixed with 3 Tbsp water
Pickled mustard leaves (*mei cai*)	70 g, soaked in cold water
Cooking oil	
Bamboo shoot (*zhu sun*)	50 g, sliced thinly
Chinese cooking wine (*hua tiao*)	1 Tbsp
Stock or water	125 ml

Method

* Marinate fish slices with egg white, salt and half the cornflour mixture.

* Drain the mustard leaves and chop finely.

* Heat 3 Tbsp oil in a wok and lightly scald fish slices until cooked and white. Remove and set aside.

* In the same wok, leave about 1 1/2 Tbsp oil and stir-fry pickled mustard leaves and bamboo shoots for 1 minute.

* Add cooking wine, stock or water and fish slices. Bring to the boil then thicken sauce with remaining cornstarch mixture. Dish out and serve.

Stir-fried Fish Slices

The story behind this dish is a simple one. It is said that Empress Dowager Cixi (1835–1908) often had a banquet table laden with countless delicacies for her dinning pleasure. One day, she singled out a fish dish and tasted it. She was immediately impressed by its exquisite flavour and found out that the chef simply called it "Stir-fried Fish Slices". Thereafter, this dish was always included in the imperial menu.

Tip: When stir-frying the fish, cook lightly and with quick, brisk strokes. This will ensure that the fish doesn't break up but will be well-coated with the sauce.

Ingredients

White fish fillet	250 g, cut into 5 x 1-cm slices
Egg	1, small, beaten
Cornflour (cornstarch)	4 Tbsp, mixed with 6 Tbsp water
Light soy sauce	1 tsp
Vinegar	2 Tbsp
Sugar	3 Tbsp
Chinese cooking wine (*hua tiao*)	1 Tbsp
Cooking oil for deep-frying	
Spring onion (scallion)	1, shredded
Ginger	1-cm knob, peeled and shredded

Method

❀ Marinate fish with egg and two-thirds of cornflour mixture. Set aside.

❀ Make sauce. Combine light soy sauce, vinegar, sugar, cooking wine and remaining cornflour mixture. Set aside.

❀ Heat oil and deep-fry fish slices until crisp. Remove and drain.

❀ Leaving 2 Tbsp oil in the wok, stir-fry spring onion and ginger until fragrant. Add sauce and bring to the boil. Add fish slices and stir-fry briskly while being careful not to break fish. Dish out and serve.

Dongpo's Fish Chips

Su Dongpo (1037–1101) was a prolific poet and statesman of the Northern Song Dynasty. He held a number of government positions throughout China, with the most notable one being in Hangzhou where he was responsible for constructing a pedestrian causeway across the West Lake.

While Su was in Hangzhou, a friend came to visit. Knowing that his friend enjoyed eating fish, Su prepared some deep-fried fish chips specially for him. The dish was well-received and Su would always prepare the same dish whenever this friend came to visit. Soon, many people came to know of the dish and called on Su hoping to sample it.

The generous Su then wrote the recipe down for those who wanted to try cooking it for themselves. "Dongpo's Fish Chips" is said to be the first dish to be named after the famous scholar-official.

Ingredients

White fish fillet	250 g, cut into 6 x 2-cm strips
Spring onion (scallion)	1, finely cut
Ginger	1-cm knob, peeled and chopped
Salt	1 tsp
Chinese cooking wine (hua tiao)	3 tsp
Ground white pepper	1/2 tsp
Sesame oil	1 tsp
Cornflour (cornstarch)	200 g
Cooking oil for deep-frying	
Sesame seeds	2 Tbsp

Method

❀ Marinate fish with spring onion, ginger, salt, cooking wine, pepper and sesame oil. Leave for 30 minutes.

❀ Roll marinated fish strips in cornflour.

❀ Heat oil and deep-fry fish slices until golden brown. Drain and sprinkle with sesame seeds before serving.

Tip: *Coat the fish strips evenly with cornflour then dust off any excess before deep-frying. Lower the coated fish strips into the oil one by one to prevent them from sticking together.*

SEAFOOD

Jade Belt Fish Rolls

Lu Duo was a conscientious young man who lived in China during the 1600s. His ambition was to be a court official and as he drew nearer the imperial exams, he often studied all day and night. His mother was very worried for his health and she nourished him with the highest quality fish meat. She would mould the fish meat into jade belt-like rolls and steam them.

After Lu passed the exams and became an official, he continued to eat fish prepared the same way his mother did to remind himself of his mother's goodness. Today, the dish is a popular delicacy in Hubei Province.

Tip: Slice the ham, mushroom, bamboo shoot, ginger and spring onion into strips of about the same length, just slightly longer or shorter than the width of the fish slice. When the fish is cooked, it shrinks a little and the filling will be exposed. Thus, having the filling about the same length will make it visually pleasing.

Ingredients

White fish fillet	320 g, sliced
Salt	1 tsp
Chinese cooking wine (hua tiao)	1 Tbsp
Cornflour (cornstarch)	5 Tbsp
Chinese ham	50 g, thinly sliced
Dried Chinese mushrooms	50 g, soaked to soften and thinly sliced
Bamboo shoot (zhu sun)	50 g, thinly sliced
Ginger	3.5-cm knob, peeled and thinly sliced
Spring onions (scallions)	3, cut into 4-cm lengths + 2, finely cut
Cooking oil	2 Tbsp
Chicken stock	120 ml
Ground white pepper	1/2 tsp
Cornflour (cornstarch)	1 Tbsp, mixed with 2 Tbsp water

Method

❀ Marinate fish slices with 1/2 tsp salt and 3 tsp cooking wine. Leave for about 5 minutes.

❀ Dust fish slices with cornflour and top each one with a slice of ham, mushroom, bamboo shoot, ginger and length of spring onion. Roll fish slices up to enclose filling and steam for 5 minutes.

❀ Heat oil in a wok. Stir-fry finely cut spring onions until fragrant. Add chicken stock, pepper and remaining salt and wine. Thicken sauce with cornflour mixture. Pour sauce over fish rolls to serve.

Fried Dough Strips with Prawns
(Fallen Autumn Leaves)

In the last years of the Qing Dynasty (1644–1911), the troubled imperial family led such a secluded and confused existence in the palace that they did not even notice the changes in the season. Hence, to remind the imperial family of the arrival of autumn, palace chefs lovingly prepared a dish with fried dough shaped into fallen autumn leaves.

Restaurants in Beijing have since adapted this dish, serving it during the months of late autumn and early winter.

Tip: You may choose to pour the fried prawns, mushrooms and bamboo shoot mixture over the fried dough pieces at the table so the dough pieces will remain crisp longer.

Ingredients

Plain (all-purpose) flour	150 g
Water	
Prawns (shrimps)	150 g
Egg white	1
Cornflour (cornstarch)	1 Tbsp, mixed with 2 Tbsp water
Spring onion (scallion)	1, finely cut
Ginger	1-cm knob, peeled and chopped
Garlic	1 clove, peeled and sliced
Dried Chinese mushrooms	20 g, soaked to soften and sliced
Bamboo shoot (*zhu sun*)	20 g, sliced
Salt	1 tsp
Chinese cooking wine (*hua tiao*)	2 tsp
Chicken stock	2 Tbsp

Method

- Combine flour and a little water to make a soft, pliable dough. Add water a little at a time until the right consistency is achieved. Add more water if necessary. Roll dough into a thin sheet and cut into leaf shapes.
- Heat enough oil for deep-frying in a wok and deep-fry dough shapes until lightly browned. Drain and set aside on a serving plate.
- Coat prawns with egg white and cornflour mixture. Deep-fry briefly in the same oil and drain.
- Leaving 2 Tbsp oil in the wok, stir-fry spring onion, ginger and garlic until fragrant.
- Add fried prawns, mushrooms, bamboo shoot, salt, cooking wine and stock. Mix well and pour on top of fried dough pieces. Serve immediately.

VEGETABLES & BEAN CURD

Stir-fried Spicy Bean Curd
(Pocked-faced Lady's Tofu)

A lady who ran a restaurant at Wanfu Bridge in Chengdu during the 1800s created this famous Sichuan dish. Her customers were so fond of the spicy bean curd dish that they affectionately called it "Pock-faced Lady's Tofu", with reference to her complexion as it was less than perfect.

The reputation of the dish spread far and wide and it remains as one of the gastronomic icons of Sichuan today.

> **Tip:** For this dish to be full of flavour, stir-fry the beef until it is fragrant before adding other ingredients. You can also vary the recipe by using minced chicken or pork.

Ingredients

Cooking oil	75 ml
Minced beef	120 g
Spicy bean paste	1¹/₂ Tbsp
Black bean paste	2 tsp
Chilli powder	1 tsp
Salt	1 tsp
Light soy sauce	2 tsp
Ginger	2-cm knob, peeled and chopped
Garlic	2 cloves, peeled and finely cut
Soft bean curd	400 g, cut into 2-cm squares
Cornflour (cornstarch)	2 Tbsp, mixed with 2 Tbsp water
Spring onion (scallion)	1, finely cut
Sichuan peppercorn oil	1 tsp

Method

❈ Heat oil in a wok and stir-fry minced beef.

❈ Add spicy bean paste, black bean paste, chilli powder, salt, light soy sauce, ginger, garlic and bean curd. Sprinkle in some water and bring to the boil.

❈ Lower flame and simmer for another 5 minutes. Add cornflour mixture and spring onion. Sprinkle peppercorn oil over and serve.

VEGETABLES & BEAN CURD

Bean Curd Boxes with Minced Meat

In Boshan of Shandong Province, bean curd is made using a special technique that gives it a delicate fragrance. This variety of bean curd is so highly regarded that it is always served during festive, auspicious and other special occasions.

When Emperor Qianlong (1736–1796) of the Qing Dynasty (1644–1911) visited Boshan, he was served a special bean curd dish that was shaped like a treasure chest. He was very pleased with it, and since then, this dish has been replicated by restaurants even beyond Boshan.

Tip: The bean curd pieces must be well-fried in order for them to keep their shape and hold the filling. Deep-fry them until they take on a rich golden brown colour.

Ingredients

Firm bean curd	500 g, cut into 4 x 2-cm pieces
Spring onion (scallion)	1, finely cut
Ginger	1-cm knob, peeled and chopped
Minced pork	250 g
Dried prawns (shrimps)	50 g, finely chopped
Bamboo shoot (*zhu sun*)	20 g, sliced
Dried wood ear fungus (*hei mu er*)	20 g, soaked and finely chopped
Light soy sauce	2 Tbsp
Salt	1 tsp
Garlic	1 clove, peeled and sliced
Vinegar	3 tsp
Green vegetables	30 g, sectioned
Chicken stock	150 ml
Cornflour (cornstarch)	1 Tbsp, mixed with 2 Tbsp water

Method

❉ Heat enough oil for deep-frying in a wok and deep-fry bean curd pieces until golden brown. Drain and leave to cool. When cool, use a small, sharp knife to slice the top off each piece to make a lid. Carefully scoop out some of the bean curd to create an empty box. Set aside.

❉ Leave 2 Tbsp oil in the wok and stir-fry spring onion and ginger. Add minced pork, dried prawns, bamboo shoot, wood ear fungus, 1 Tbsp light soy sauce and ¹/₂ tsp salt.

❉ Fill bean curd boxes with stir-fried ingredients. Replace lid and steam for 5 minutes. Arrange on a serving plate.

❉ In the same wok, heat 2 Tbsp oil and stir-fry garlic slices until fragrant. Add vinegar, green vegetables, remaining light soy sauce and salt and chicken stock. Bring to the boil and thicken with cornflour mixture. Pour sauce over bean curd boxes and serve.

Bean Curd Sandwiches
(Emepror Hongwu's Bean Curd)

The founder of the Ming Dynasty (1368–1644), Zhu Yuanzhang (1328–1398) was born into a poor family in Fengyang, Anhui Province. As a child, Zhu survived by begging. The owner of a roadside stall felt sorry for him and often treated him to a bean curd dish. When he grew up, Zhu became a rebel leader and toppled the Mongol Yuan Dynasty (1279–1368). He proclaimed himself the emperor and took on the reign title of Hongwu.

Even as emperor, Zhu never forgot his humble beginnings. He missed the bean curd he used to eat as a child and he sent for the kind roadside stall owner to recreate the bean curd dish for him. The dish then became known as "Emperor Hongwu's Bean Curd".

Tip: *Handle the bean curd with gentle hands, as they break up easily. Use a flat ladle or spatula to lift the bean curd.*

Ingredients

Spring onions (scallions)	2, finely cut
Ginger	2-cm knob, peeled and chopped
Minced pork	120 g
Prawns (shrimps)	70 g, shelled and minced
Chinese cooking wine (*hua tiao*)	2 tsp
Chicken stock	200 ml
Salt	1 tsp
Cornflour (cornstarch)	1 Tbsp, mixed with 2 Tbsp water
Firm bean curd	250 g, cut into long rectangular pieces
Plain (all-purpose) flour	6 Tbsp
Egg whites	4, whipped and mixed with 3 Tbsp cornflour
Sugar	1 tsp
Vinegar	1 tsp

Method

* Heat 2 Tbsp oil in a wok and stir-fry spring onions and ginger until fragrant. Add pork and prawns and stir-fry to cook.

* Add cooking wine and 100 ml chicken stock, $^1/_2$ tsp salt and half the cornflour mixture. Mix well and bring to the boil. Allow sauce to thicken.

* Arrange half the number of bean curd slices on a serving plate. Place a spoonful of stir-fried ingredients on top of each bean curd slice and sandwich with remaining bean curd slices. Dust with flour and set aside.

* Using the same wok, heat more oil for deep-frying. Coat bean curd sandwiches with egg white-cornflour paste. Deep-fry one at a time, until golden brown. Remove and drain. Arrange on a serving plate.

* Leaving 1 Tbsp oil in the wok, add remaining chicken stock and salt, sugar and vinegar. Cook over low heat for 3 minutes then thicken sauce with remaining cornflour mixture. Pour over bean curd sandwiches to serve.

Shredded Bean Curd and Chicken Soup

When Emperor Qianlong (1711–1799) of the Qing Dynasty (1644–1911) visited Yangzhou, Jiangsu Province, he was treated to the region's finest delicacies. One of the dishes that caught his attention was a fairly simple chicken soup with shredded bean curd and ham. Prawns, chicken and peas were later added to the dish to make it even tastier. Today, the soup remains a famed delicacy of Jiangsu Province.

Ingredients

Chicken stock	800 ml
Chicken breast	30 g, boiled and shredded
Bamboo shoot (*zhu sun*)	30 g, cut into thin strips
Firm bean curd	300 g, cut into thin strips
Salt	1 tsp
Light soy sauce	2 tsp
Ground white pepper	1 tsp
Chinese ham	2 slices, cut into thin strips
Prawns (shrimps)	50 g, shelled
Bean sprouts	30 g, tails plucked and discarded

Method

✤ In a pot, combine chicken stock, chicken breast, bamboo shoot and firm bean curd. Cook over low heat for 8 minutes until soup thickens slightly.

✤ Add salt, light soy sauce and pepper. Cook for another 5 minutes then add ham, prawns and bean sprouts. When prawns turn pink, remove from heat and serve.

Stuffed Vegetables
(Empress Dowager Cixi's Vegetables)

In her later years, Empress Dowager Cixi (1835–1908) practised Buddhism and tried to maintain a strict vegetarian diet during the fasting period. However, she found vegetarian food bland and unappetising and often chided her chefs for their poor culinary skills.

The panicking chefs cracked their heads to come up with more tasty dishes. One day, they stuffed minced chicken into the stems of green vegetables in an effort to hide the meat and presented it to the empress. It was a cunning idea as it allowed the empress to enjoy eating meat without anyone else knowing! The empress liked the dish and it became a regular item on the imperial menu.

Ingredients

Salt	2 tsp
Green vegetables	250 g
Cornflour (cornstarch)	3 Tbsp
Minced chicken	250 g
Chicken stock	250ml
Egg white	1
Chinese cooking wine (hua tiao)	3 tsp
Chinese ham	1 slice, finely cut

Method

❀ Boil some water for blanching vegetables. Stir in 1 tsp salt and blanch vegetables for 45 seconds. Drain and coat stems with cornflour. Place on a plate.

❀ Mix minced chicken with 60 ml chicken stock. Add egg white, 1/4 tsp salt and 1 tsp of cooking wine. Shape into balls and press into vegetable stems. Sprinkle with minced ham and steam 5 minutes.

❀ Bring remaining chicken stock to the boil. Add remaining salt and cooking wine, then thicken with 1 Tbsp cornflour mixed with 2 Tbsp water. Drizzle over vegetables before serving.

Bean Sprout Salad

The 2000 year-old Confucian Mansion where the descendants of Confucius live is a very grand building, second only to the imperial palace. In the past, guests to the mansion were pampered with light and exquisite dishes. This distinctive style of cooking became known as "Confucian cuisine".

Emperor Qianlong (1736–1796) of the Qing Dynasty (1644–1911) once paid a visit to the mansion. During dinner, the host noticed that the emperor was not eating well so he asked the chef to create a light yet tantalising dish. The chef came up with a simple treat of bean sprouts tossed in aromatic peppercorn oil. The dish was well-received by the emperor and it became a perennial favourite at the mansion.

Ingredients

Cooking oil	2 Tbsp
Sichuan peppercorns	1 Tbsp
Bean sprouts	500 g, heads and tails plucked and discarded
Salt	1 tsp

Method

* Heat oil and stir-fry Sichuan peppercorns until fragrant. Remove and discard peppercorns.

* Add bean sprouts and stir-fry briskly. Season with salt. Toss well and serve.

Tip: *This is a very simple and quick recipe. Lightly cook the bean sprouts so they remain juicy and crunchy. Do not overcook as bean sprouts will lose their flavour and nutrition when overdone.*

VEGETABLES & BEAN CURD

Lotus Root Salad

One of the key characters of the Three Kingdoms period (220–280), Cao Cao (155–220) is credited for unifying northern China in 200. He served as prime minister of the Han Dynasty before becoming the first emperor of the Wei Dynasty. The shrewd statesman was one of China's most controversial historical figures. He was both feared for his ruthlessness and admired for his avid support of talented citizens, his wise dictatorship and excellent poetry writing skills.

In 210, Cao Cao led an army to attack Sucheng (today's Qianshan County, Anhui Province). Nine heroic generals of Sucheng helped the people escape to Snow Lake but were themselves captured by Cao Cao's army and drowned in the lake.

According to folklore, right after the generals were drowned, nine hot springs appeared in lake, heating up the water. People changed the name of the lake from "Snow Lake" to "Blood Lake" in memory of the generals. Water from the lake is said to be sweet and lotus roots grown there are known for their sweetness. Cooks in Anhui use these lotus roots to prepare appetisers and desserts.

Ingredients

Lotus root	300 g
Salt	1 tsp
White vinegar	3 Tbsp
Sugar	4 Tbsp
Spring onion (scallion)	1, chopped (optional)

Method

* Wash lotus root well. Peel off skin and cut into thin slices.

* Boil some water and blanch lotus root slices briefly. Remove and cool in cold water. Drain.

* Place lotus root slices in a large bowl. Add salt, white vinegar and sugar. Mix well. Arrange lotus root slices neatly on a serving plate. Garnish with chopped spring onion if desired.

NOODLES

Deep-fried Fish Crisps

Faced with flagging business, a clothing store owner tried to draw customers by treating them to fine food and wine. He hired a cook to whip up a plethora of dishes daily. Once, the cook accidentally dropped some minced fish meat onto a bag of flour. Instead of wasting both the fish and flour, the cook mixed the two ingredients together to make fish crisps. He deep-fried the fish and flour mixture and served it with other fish dishes. The customers enjoyed the dish so much they asked for it on a daily basis. As a result, business boomed at the clothing store and deep-fried fish crisps with fried fish chunks became a much-loved dish for generations.

Tip: Use a pasta machine to make thin noodles if desired. Have a wire strainer on hand to remove the fish crisps quickly once they turn golden brown. Regulate the temperature of the oil as the noodles will burn easily if the oil is too hot.

Ingredients

Freshwater fish	500 g, thickly sliced
Salt	1 1/2 tsp
Chinese cooking wine (hua tiao)	1 1/2 Tbsp
Spring onions (scallions)	2, cut into 2.5-cm lengths
Ginger	2-cm knob, peeled and sliced
Tomato sauce	2 1/2 Tbsp
Sugar	2 Tbsp
Red vinegar	3 Tbsp
Stock or water	6 Tbsp
Cornflour (cornstarch)	2 Tbsp, mixed with 3 Tbsp water
White fish fillet	75 g, minced
Egg white	1
Plain (all-purpose) flour	5 Tbsp
Cornflour (cornstarch)	6 Tbsp
Cooking oil for deep-frying	

Method

* Marinate fish with 1/2 tsp salt, half the cooking wine, spring onions and ginger. Leave for 10 minutes.

* Combine 1 tsp salt, remaining cooking wine, tomato sauce, sugar, red vinegar, stock or water and cornflour mixture for the sauce. Set aside.

* Combine minced fish fillet with egg white and mix until mixture is slightly sticky to the touch. Knead in flour to get pliable dough. Sprinkle some flour on a work surface and roll dough out into a thin sheet. Cut into strips and set aside.

* Heat oil for deep-frying in a wok. Coast fish slices in cornflour and deep-fry until golden brown and crisp. Remove and drain.

* In the same wok, deep-fry fish strips until golden brown and crisp. Remove with a wire strainer and drain.

* Leave 2 Tbsp oil in wok and add sauce ingredients. Bring to the boil, stirring until mixture is slightly thickened. To serve, place fish slices and crisps on a serving plate and pour sauce over.

Yin Mansion's Fried Noodles

Yin Bingshou (1754–1815) was a well-respected governor of Huizhou, Guangdong. On his birthday, some townsfolk presented him with a generous amount of his favourite food—noodles. Pleased, Yin decided to have the noodles cooked and served at his birthday banquet. Back in the kitchen, the mood was a frenzied one as cooks busied themselves with preparation for the banquet. Amid the chaos, the cooks mistakenly put the noodles into hot oil instead of hot water and the noodles were deep-fried. Despite the mistake, the deep-fried noodles tasted good and the cooks served it. The guests enjoyed the noodles thoroughly and the delighted magistrate got his guests to bring home the excess noodles for their families. Everyone who tried the noodles loved it, thus starting the enduring popularity of "Yin Mansion's Fried Noodles".

Tip: *To prevent the noodles from clumping together when fried, loosen them before putting them in the hot oil.*

Ingredients

Cooking oil for deep-frying	
Flat egg noodles	500 g
Prawns (shrimps)	120 g, shelled
Cornflour (cornstarch)	2 Tbsp
Dried Chinese mushrooms	50 g, soaked to soften and sliced
Salt	1 tsp
Stock	150 ml
Oyster sauce	2 tsp
Sugar	1 tsp
Ground white pepper	3/4 tsp
Light soy sauce	1 Tbsp
Chinese chives (jiu cai)	50 g, cut into 3-cm lengths

Method

* Heat oil for deep-frying in a wok. Place noodles in a wire strainer and lower into hot oil for 2 minutes. Drain well.

* Mix prawns and cornflour.

* Leave 2 Tbsp oil in the wok and stir-fry prawns for 1 minute. Add mushrooms, salt, stock, oyster sauce, sugar, pepper and light soy sauce and bring to the boil.

* Add noodles and simmer for 5 minutes. Sprinkle with chives and toss well. Transfer to a serving plate.

Dragon and Phoenix Noodles

Emperor Xizong (1605–1627) of the Ming Dynasty (1368–1644) was an irresponsible ruler who did not care about state affairs and wasted his days away with frivolous pastimes. It was during one of the emperor's dalliances that this dish was created.

Once, when out for a walk in Shaanxi Province, the emperor stopped by a village for a meal of noodles. The emperor fell in love with the noodle dish and also the beautiful young village girl who prepared it for him. He immediately made the girl his concubine and named the noodles "Dragon and Phoenix Noodles".

Serving these noodles has since become a trend at Chinese engagement and wedding parties. It is believed to symbolise good luck and happiness.

Tip: Covering the dough with a damp cloth prevents it from drying out as the dough rests.

Ingredients

Plain (all-purpose) flour	500 g
Salt	2 tsp
Spinach	150 g, stems plucked and discarded
Egg	1, beaten
Cooking oil	2 Tbsp
Spring onions (scallions)	2, shredded
Pork loin	180 g, cut into thin strips
Cucumber	50 g, sliced
Tomatoes	50 g, sliced
Chicken stock	500 ml
Hardboiled quail eggs	8, shelled and cut in half

Method

* In a bowl, combine half the flour, $^1/_2$ tsp salt and enough water to make dough. Cover with a damp cloth for 20 minutes.

* Finely mince spinach leaves and sprinkle with $^1/_2$ tsp salt. Squeeze spinach of its juice using a piece of muslin. Discard fibre. Add remaining flour and beaten egg to spinach juice. Knead into a green dough, adding water if necessary. Cover with a damp cloth for 20 minutes.

* Press plain and green dough together, then flatten. Fold in half and cut into thin noodles. Cook noodles in boiling water for 5 minutes, then rinse in cold water and place into a large serving bowl.

* Heat oil and stir-fry spring onions until fragrant. Add pork and sprinkle in remaining salt. Add cucumber, tomatoes and stock and stir. Ensure pork is well cooked. Allow stock to come to the boil. Pour over noodles and top with quail eggs to serve.

SOUPS

Chicken Soup with Barley
(The First Soup)

Qu Yuan (339–278 B.C.) was a great poet, diplomat and statesman who was deeply loyal to his country. He had a brilliant official career that saw him rise steadily through the ranks. This made some of the other court officials jealous and they spread vicious rumours about him. He was thus expelled from the court in disgrace. Disillusioned, Qu Yuan wandered the countryside, writing the monumental *Li Sao*, a long patriotic poem. At this time, Qu Yuan's country was also in danger of falling into enemy hands. When it was conquered, Qu Yuan could not bear the blow and drowned himself in the Miluo River in Hunan Province.

Qu Yuan once wrote a poem about how the Duke of Peng won the favour of the emperor because he was good at making wild pheasant soup. The duke added barley and plums to the soup, two ingredients that were never before added together in a dish. As this was also the first soup to be ever recorded in historical documents, it became known as "The First Soup".

Ingredients

Chicken	1, about 600 g
Spring onions (scallions)	2, cut into 3-cm lengths
Ginger	2-cm knob, peeled and sliced
Sichuan peppercorns	1 Tbsp
Chinese cooking wine (hua tiao)	3 Tbsp
Barley	120 g
Dried Chinese mushrooms	30 g, soaked to soften and thinly sliced
Bamboo shoot (zhu sun)	30 g, thinly sliced
Chinese ham	30 g, thinly sliced
Green vegetables	30 g, thinly sliced
Salt	1 tsp
Ground white pepper	1 tsp
Vinegar	2 Tbsp

Method

❋ Bring a pot of water to the boil. Carefully place chicken in and allow water to come to the boil again. Remove chicken and rinse.

❋ Place spring onions, ginger and peppercorns into a small muslin bag. Set aside.

❋ Using the same pot, place chicken in and add 1.5 litres water, $1^1/_2$ Tbsp cooking wine, barley and the muslin bag.

❋ Simmer for 2–3 hours until chicken and barley are very soft. Remove chicken, leaving soup to simmer. Leave chicken to cool slightly before shredding meat.

❋ Remove muslin bag and add shredded chicken, mushrooms, bamboo shoot, ham and vegetables. Add remaining cooking wine, salt and pepper. Bring to the boil then stir in vinegar. Serve.

Chicken with Good Fortune Vegetable Soup

During the reign of the Tang Dynasty (618–907), there was a merchant named Wang Yuanbao who lived in Chang'an. He was so fond of eating black hair moss, he would have it at every meal. (The Mandarin name for the moss is *fa cai*, a homonym for "good fortune". Hence it is considered to symbolise good fortune.)

In time, Wang's business prospered and he became one of the richest men in the country. The people believed that Wang's wealth was a direct result of his eating the black hair moss that is associated with good fortune and the demand for it increased. Today, black hair moss is always included in dishes for festive and auspicious occasions with the belief that it will bring prosperity and wealth to the diners.

Ingredients

Minced chicken	180 g
Black hair moss (*fa cai*)	30 g, soaked and torn into pieces
Egg whites	2
Egg yolks	2
Chicken stock	800 ml
Cornflour (cornstarch)	1 Tbsp, mixed with 2 Tbsp water
Eggs	2, fried into a thin omelette
Salt	1 tsp
Chinese cooking wine (*hua tiao*)	3 tsp
Bamboo shoot (*zhu sun*)	30 g, sliced
Spinach	100 g, cut into 5-cm lengths

Method

* Combine minced chicken with black hair moss and egg whites. Set aside.

* Mix egg yolks with 3 Tbsp chicken stock and cornflour mixture. Steam for 8 minutes and leave to cool before cutting into strips. Set aside.

* Spread chicken paste over omelette. Put egg yolk strips on one edge of the omelette, roll up and steam for 10 minutes. Cut into rounds about 1.5-cm thick. Place in a soup bowl.

* Heat remaining chicken stock, then add salt, cooking wine, bamboo shoot and spinach. Bring to the boil and pour into the bowl to serve.

Mandarin Fish Soup

In an effort to defend his kingdom, Emperor Gaozong (1107–1187) of the Song Dynasty (960–1279) sought the military genius of the legendary Yue Fei (1103–1142), a courageous general who fought and won many battles against the barbaric Jin troops. Despite Yue Fei's successful efforts to expel the Jin, the weak-minded emperor carried out negotiations with them and even ceded land to them. He believed foolhardily that these actions would keep the Jin forces at bay.

Once, during a short-lived peaceful period, the emperor went on a cruise down West Lake in Hangzhou. The royal entourage came across an elderly lady known as Fifth Sister Song who sold fish soup for a living. The emperor tried the soup and loved it. Since the royal encounter, Fifth Sister Song's soup has been much sought after and it remains one of Hangzhou's specialties today.

Tip: Ensure that the soup is not boiling when you add the beaten egg. The boiling water will cause the egg to disappear into the soup. Instead, lower the heat and allow the egg to cook gradually in strips.

Ingredients

Mandarin fish	1, about 600 g
Spring onions (scallions)	4, cut into 3-cm lengths
Ginger	2-cm knob, peeled and sliced
Salt	1 tsp
Chicken stock	750 ml
Bamboo shoot (zhu sun)	20 g, thinly sliced
Dried Chinese mushrooms	30 g, soaked to soften and thinly sliced
Light soy sauce	1 Tbsp
Ground white pepper	1 Tbsp
Cornflour (cornstarch)	1 Tbsp, mixed with 2 Tbsp water
Eggs	2, beaten
Vinegar	2 Tbsp

Garnish

Chinese ham	25 g, finely shredded
Spring onions (scallions)	2, finely shredded
Ginger	1-cm knob, peeled and finely shredded

Method

- Gut and wash fish. Discard head.

- Marinate fish with half the spring onions and ginger, and 1/2 tsp salt. Leave for 10 minutes then steam for 5 minutes. Discard spring onions and ginger.

- Peel away skin and flake fish. Discard skin and bones.

- Heat 2 Tbsp oil in a wok and stir-fry remaining spring onions. Add stock and bring to the boil then strain and discard spring onions.

- Add bamboo shoots, mushrooms, light soy sauce, pepper, remaining salt and flaked fish. Thicken with cornflour mixture then slowly stir in beaten eggs. Add vinegar. Garnish with ham, spring onion and ginger to serve.

Sliced Carp and Water Shield Soup

Zhang Han was a high-ranking official in Luoyang during the tumultuous Western Jin Dynasty (265–317). He was in his study one day when a gust of wind swept through the trees and blew into the room. The breeze brought to mind his hometown of Suzhou and he longed for the fish and water shield soup he often had at home. Zhang then realised that being an official had kept him away from home and the people he loved. At that moment, he made the decision to resign from his post and return to his hometown. The name of the soup thus became synonymous with homesickness.

Ingredients

White fish fillet	300 g, thinly sliced
Chinese cooking wine (hua tiao)	1 Tbsp
Salt	1 tsp
Water shield (or other green vegetables)	250 g
Chicken stock	750 ml
Chinese ham	2 slices, thinly sliced

Method

* Marinate fish slices with 2 tsp cooking wine and $1/_2$ tsp salt. Set aside.

* Blanch water shield briefly in a pot of lightly oiled and salted boiling water, until water shield takes on a light green colour. Drain and place in a soup bowl.

* Bring stock to the boil and add fish slices, remaining salt, cooking wine and ham. Pour into the soup bowl and serve.

Pan's Fish Soup

Pan Zuyin (1830–1890) was a linguist and an official of the late Qing Dynasty (1644–1911). He was also a gourmet and introduced numerous dishes to the court, many of which became well-loved classics. This fish soup is one of his gastronomic contributions.

Although Pan liked to eat fish, he did not like it to be cooked with oil. Thus, he made a fish soup using the freshest fish available, top-grade mushrooms, dried prawns and chicken stock. The oil-free soup turned out to be most delicious. He gave the recipe to a friend who owned a famous restaurant in Beijing. The dish was introduced into the restaurant's menu and guests at the restaurant spoke highly of the soup. From then on, it became known as "Pan's Fish Soup".

Tip: *Steam the fish over high heat to cook it fast, but do not overcook as the fish will break up.*

Ingredients

Carp	1, about 600 g
Dried Chinese mushrooms	4–5, soaked to soften and sliced
Spring onions (scallions)	2, cut into 3-cm lengths
Ginger	2-cm knob, peeled and sliced
Dried prawns (shrimps)	1 Tbsp
Chinese cooking wine (hua tiao)	1 Tbsp
Salt	1 tsp
Chicken stock	500 ml

Method

❀ Gut and cut fish in half so there is a head piece and a tail piece.

❀ Blanch fish briefly in boiling water, then drain and place in a large heatproof (flameproof) bowl. Add mushrooms, spring onions, ginger, dried prawns, cooking wine, salt and chicken stock.

❀ Steam over high heat for 15 minutes. If preferred, carefully remove and discard spring onions and ginger before serving.

SOUPS

Longevity Fish Strip Soup

During the Qing Dynasty (1644–1911), the food-loving governor of Chaozhou in Guangdong Province housed a large team of cooks in his residence.

To celebrate his mother's 60th birthday, he organised a big party. As his mother loved to eat fish, the governor requested that the cooks prepare fish dishes for the party. The chief cook cleverly thought of a recipe for fish soup. He flattened fish fillets, then boiled and cut them into long, narrow strips. (In Mandarin, the words for "long" and "narrow" sound like the term for "longevity".) The governor was impressed with the taste of the dish and its significance. He then rewarded the chief cook handsomely.

When the chief cook retired to his home province of Jiangxi a few years later, he brought the recipe with him and the soup became the region's traditional delicacy.

Ingredients

Carp fillet	150 g, finely minced
Cornflour (cornstarch)	150 g
Chicken breast	60 g, boiled and shredded
Dried Chinese mushrooms	3, soaked to soften and sliced
Chicken stock	750 ml
Salt	1 tsp
Ground white pepper	1 tsp
Spring onion (scallion)	1, finely cut

Method

* Divide minced fish into two portions and roll them into balls. Dust with cornflour and roll balls out into thin sheets.

* Boil water in a large pan and gently lower in the sheets of fish paste. Boil for 5 minutes then remove and rinse with cold water. Cut into thin and long strips. Set aside.

* Heat chicken stock and add shredded chicken, mushrooms, fish strips, salt and pepper. Bring to the boil and transfer to soup bowls. Garnish with spring onion and serve.

Wild Vegetable Soup
(Soup for Protecting the Nation)

Zhao Bing (1278–1279) was the last emperor of the Song Dynasty (960–1279). When the Mongols conquered it, Zhao sought refuge in an old temple in Chaozhou. He was visibly weakened and shaken by his ordeal and the monks showed him great hospitality. As they were short of food themselves, the monks went out of their way to pick wild vegetables to make a soup that would nourish Zhao.

The emperor was so moved by the monks' efforts to keep up his health he named the dish "Soup for Protecting the Nation". However, this boost of morale was short-lived and the dispirited emperor drowned himself soon after.

Ingredients

Baking soda	$^1/_2$ tsp
Spinach or other tender green vegetables	500 g, cut into 5-cm lengths
Dried Chinese mushrooms	100 g, soaked to soften, stems discarded
Chicken stock	500 ml
Salt	1 tsp
Cooking oil	2 Tbsp
Chinese ham	1 slice, finely cut

Method

✽ Boil some water and stir in baking soda. Blanch vegetables briefly in the hot water. Drain and rinse in cold water, then drain again. Set aside.

✽ Mix mushrooms with 100 ml chicken stock and $^1/_2$ tsp salt. Steam for 20 minutes.

✽ Heat oil in a wok and stir-fry vegetables for about 3 minutes. Add in remaining stock and salt.

✽ Add steamed mushrooms and bring to the boil. Spoon soup into serving bowls and sprinkle ham over to serve.

SNACKS & DESSERTS

Chestnut Soup with Osmanthus Blossoms

The story behind this dish can be traced back to tale involving the legendary Chang Er and Wu Gang. Chang Er was the wife of Hou Yi, a divine archer who had a magic bow and magic arrows. Long ago, it was believed that there were 10 suns circling earth, each taking turns to light and warm earth. One day, all 10 suns appeared at the same time, causing a great drought. Hou Yi saved the earth by shooting nine of the suns down with his arrows. The Queen Mother of the West rewarded him with an elixir pill for immortality. In one of the many versions of this tale, Chang Er stole the pill and was transformed into a fairy to live on the moon. Wu Gang was a woodcutter who was banished to the moon. He became Chang Er's friend and servant.

One mid-autumn evening, Chang Er looked down from Guanghan Palace (the lunar palace) at the reflection of the moon on West Lake in Hangzhou, Zhejiang Province and was enraptured by its beauty. She began to dance while Wu Gang made music by tapping the trunk of the osmanthus tree. His tapping caused osmanthus blossoms to fall from the tree and into a pot of chestnut soup a monk was preparing in a temple near West Lake. The soup was particularly tasty that night and from then, it became a popular dish in Zhejiang as well as in neighbouring Jiangsu Province.

Ingredients

Lotus root powder	2 Tbsp
Chestnuts	120 g, shelled, boiled and thinly sliced
Sugar	120 g
Candied plums	5, sliced
Sweetened osmanthus blossoms (gui hua)	2 tsp
Dried rose petals	3, shredded

Method

❀ Add ¹/₂ Tbsp of water to lotus root powder to form a paste. Adjust the amount of water as necessary.

❀ In a pot, boil 500 ml water. Add chestnut slices and sugar. Stir in lotus root paste and bring to the boil again. Turn off heat and add sliced plums, sweetened osmanthus blossoms and rose petals. Serve.

SNACKS & DESSERTS

Barley Soup (Emperor Liu Xiu's Soup)

Wang Mang (45 B.C.–23 A.D.) was a commander in the army under the Han Dynasty (206 B.C.–220 A.D.). He seized the throne and established his own Xin Dynasty (9–23). Many rebellions broke out throughout China during his reign and one of the factions was led by Liu Xiu (6–57), a Han prince.

Once, when Liu Xiu was out in battle, he passed by a small village where an old man invited him in for a bowl of soup made with barley. The soup tasted good and it energised Liu Xiu. Thereafter, Liu Xiu overthrew Wang Mang and restored the Han Dynasty. Liu Xiu's reign was marked by numerous socio-economic reforms, including the lowering of taxes, abolishment of inapproppriate laws and the liberation of slaves.

Despite the luxury of palace life, Liu Xiu did not forget the kindness of the old man who offered him soup. He had the soup listed on the imperial menu. The recipe was then improved upon and it spread throughout the country as "Emperor Liu Xiu's Soup".

Ingredients

Barley	120 g
Dried lily buds (golden needles)	50 g, soaked to soften
Gingko nuts	50 g, soaked to soften
Lotus seeds	50 g, soaked to soften
Dried red dates (hong zao)	50 g, pitted
Preserved plums	5
Dried longan pulp	2 Tbsp
Haw candy	10 g
Sugar	80 g
Rock sugar	80 g
Cornflour (cornstarch)	1 Tbsp , mixed with 2 Tbsp water

Method

❈ Steam barley for 25 minutes. Steam soaked dried lily buds, gingko nuts and lotus for 15 minutes.

❈ Boil 1.5 litres of and place all the ingredients except the cornflour mixture in. Allow the water to come to the boil again then lower heat. Remove from heat and stir in cornflour mixture to thicken soup slightly before serving.

White Fungus and Wolfberry Soup

Brilliant military strategist Zhang Liang (?–186 B.C.) helped Liu Bang (256–195 B.C.) topple the Qin kingdom. Following the fall of Qin, Liu Bang founded the Han Dynasty (206 B.C.–220 A.D.) and Zhang was made a duke.

Zhang, however, remained acutely aware of the treacheries of court life. (Past emperors have been known to persecute those who helped them ascend the throne.) He soon retired and lived in seclusion, immersing himself in the study of Taoism. As an indication of his pure political intentions, Zhang often cooked and ate white fungus soup.

Years later, during the Tang Dynasty (618–907), high-ranking officials like Fan Xuanling (570–648) and Du Ruhui (585–630) re-examined Zhang Liang's white fungus political metaphor. They added Chinese wolfberries to the white fungus soup to symbolise drops of blood. They believed that good statesmen should be prepared to die for a just cause represented by the Chinese wolfberries, and not be simply concerned with pursuing a clean reputation, as represented by the white fungus.

Ingredients

Dried white fungus (*bai mu er*)	60 g, soaked to soften
Rock sugar	120 g
Chinese wolfberries (*gou qi zi*)	3 Tbsp
Egg whites	2

Method

❀ In a pot, add 1 litre water and white fungus. Bring to the boil then lower heat and simmer for 1 hour.

❀ Add rock sugar and Chinese wolfberries. Cook for 10 minutes then stir in egg whites. Skim off any foam that appears. Serve.

Stir-fried Egg Yolks

The story behind this unique dish concerns a magistrate in Xiangzhou (today's Anyang), Henan Province. He was a filial son and aware that his elderly father's teeth were getting weaker, he asked the family cook to make a special dish with his father's favourite food, egg yolks. The dish was both soft and tasty and his father enjoyed it very much.

To celebrate his father's birthday, the magistrate invited many friends to his home for a banquet. He also requested that the cook prepare the special egg yolk dish for his father.

As he was preparing the dish in a much larger quantity than usual, the cook lost account of the amount of ingredients used. When he realised that he had poured too much water into the pot with the egg yolks, he tried to thicken the mixture by adding cornflour and then stirring in some cooking oil. The result was a brightly coloured dish of tasty egg yolks which all the guests enjoyed.

Ingredients

Egg yolks	6
Eggs	2
Sugar	120 g
Water	3 Tbsp
Cornflour (cornstarch)	2 Tbsp, mixed with 3 Tbsp water
Cooking oil	6 Tbsp

Method

❀ Combine egg yolks, eggs, sugar, water and cornflour mixture. Mix well.

❀ Heat half the oil and gradually pour in egg paste. Stir-fry over low heat while stirring and sprinkling in remaining oil until egg paste no longer sticks to wok. Serve.

Tip: The secret to cooking this dish well is the low heat used. Monitor the egg paste as it cooks. If it becomes too hard, sprinkle in some water. If it becomes too soft, stir in a little cornflour.

Glutinous Rice with Fruit

Behind this sweet dish is a story of true courage. During the Western Zhou period (1027–771 B.C.), eight righteous gentlemen banded together to bring down the despotic emperor. Before heading out to kill him, they took a pledge of unity while holding glutinous rice in their hands. The stickiness of the rice was used as a symbol of their resolve and steadfastness in carrying out this act. In honour of their valour, this dish was created.

Tip: Arrange the ingredients inside the bowl nicely so it makes an attractive pattern when the rice is turned out onto a plate. Slice it as you would a cake when serving.

Ingredients

Lotus seeds	60 g
Barley	60 g
Glutinous rice	280 g
Sugar	120 g
Cooking oil	1 1/2 Tbsp
Dried red dates (hong zao)	25 g, cut into small cubes
Preserved white melon strips	25 g, cut into small cubes
Dried longan pulp	25 g, cut into small cubes
Shelled melon seeds	25 g
Cornflour (cornstarch)	1 Tbsp, mixed with 2 Tbsp water

Method

* Place lotus seeds and barley in separate bowls. Add 2 Tbsp water into each bowl and steam for 30 minutes.

* Place glutinous rice in a bowl with half the sugar and 4 Tbsp water. Steam for 25 minutes.

* Lightly oil the inside of a rice bowl. Arrange lotus seeds, barley, red dates, preserved white melon strips, longan flesh and melon seeds all around the inside of the bowl. Spoon glutinous rice into the prepared bowl and steam for 10 minutes. Remove from heat and turn glutinous rice out onto a plate.

* Boil about 200 ml water and stir in remaining sugar. Stir in cornflour mixture to thicken syrup then pour over glutinous rice. Serve.

GLOSSARY

Dried Chillies
Crisp, dried chillies are more aromatic than fresh chillies.

Dried Chinese Mushrooms
The best quality dried Chinese mushrooms have unparalleled fragrance and a meaty texture. Before using, soak in warm water till softened and squeeze out the excess water.

Dried Wood Ear Fungus
Also known as black fungus and *hei mu er*, this ingredient is prized for its crunchy bite. Soak in warm water for about 15 minutes to soften before using.

Dried White Fungus
Also known as *bai mu er*. A close relative of wood ear fungus, but rare and hence more expensive. Popularly used in Chinese sweet soups.

Black Hair Moss
Known by its auspicious nickname, *fa cai*, (which means "prosperity" in Mandarin), this hair-like ingredient is popularly used in Chinese New Year dishes. Soften in water before adding to stir-fries or stews.

Dried Lily Buds
Also known as golden needles, lily buds lend an earthy flavour to many vegetarian dishes. They are usually softened by soaking in warm water, then tied into a knot before use.

Pickled Mustard Leaves/ Stems
Also known as *mei cai*, these are cooked mustard vegetables that are steeped in a salt and sugar solution, and then dried. Wash well to get rid of excess salt and any sand before cooking.

Preserved Prunes
Available from Chinese grocery stores or medical shops, preserved prunes make healthy snacks and are also commonly used in sweet soups or cakes.

Dried Longan Pulp
Sweet with a mild smoky flavour, dried longans are popularly used in Chinese sweet desserts.

Preserved Winter Melon Strips
Winter melon cut into strips, given a thick sugary coat and dried. Favoured for its cooling health properties and used in sweet soups. Can be eaten as a snack on its own.

Red Dates
Also known as *hong zao*, dried red dates add a subtle sweetness to both savoury soups and sweet desserts. Red dates are believed to be good for the blood and nervous system.

Chinese Wolfberries
Also called boxthorne berries or *gou qi zi*, they have a pleasant, mild sweet taste. Mainly used for tonic purposes, they are combined with other seasonings and lend colour to a dish. They are sold dried in packets.

Osmanthus Blossoms
Also known as *gui hua*, these dried flowers are sprinkled over sweet soups and cakes for their fragrance. Osmanthus flowers are said to improve digestion.

Barley
Hulled barley, sold at Chinese grocery stores or medical shops as Chinese barley is more nutritious than pearl barley. However it takes a longer time to cook and is far chewier than the latter. Use either one in these recipes, but cook for a longer period of time if using Chinese barley.

Red Yeast Rice
Also known as *hong qu*, red yeast rice is made by fermenting red yeast on rice. It imparts a reddish colour to foods it is cooked with. Red yeast rice is believed to be effective in improving cholesterol levels.

Dried Bean Curd Skin
These thin yellow bean curd sheets are also known as *fu pi*. They are commonly used as edible food wrappers. Wipe with a damp cloth before use.

Firm Bean Curd
This pressed bean curd is the hardest of all the bean curd types.

Preserved Red Bean Curd

It is red in colour and thick in consistency. Made from fermented red rice that has been added into a pickling. It has a pungent smell and salty taste, and is usually used to season meat dishes.

Preserved Black Beans

Fermented black beans with a pungent and salty flavour. They are preserved in salt and commonly used in Asian cooking.

Spicy Bean Paste

This reddish-brown paste consists of a mixture of fermented soy beans and red chillies.

Cinnamon Sticks

A dark brown spice that comprises thick, woody, brittle-layered curls of bark. They have a warm and sweet taste. They are commonly used to flavour stews and sauces in Asian cooking.

Cloves

These are the dried , unopened flower buds of the clove tree. They have a sweet, aromatic flavour and can be used ground or whole in savoury dishes.

Sichuan Peppercorns

Different from black or white peppercorns. These are reddish in colour and have a rather mild taste.

Fennel Seeds
Similar to anise seeds, but sweeter and milder. They are like a larger, paler version of cumin seeds.

Rock Sugar
Large irregularly shaped crystals of sugar used mainly in Chinese sweet soups or desserts.

Lotus Seeds
Seeds from the lotus pod. They are often boiled whole in soups and desserts, but may also be cooked and mashed, and used as a sweet filling in Chinese pastries.

Gingko Nuts
These are seeds of the gingko tree. They are pale yellow when raw and turn slightly translucent when cooked. Available fresh or canned. Pluck the young green shoots out before use, as they are bitter.

Haw Candy
Also known as haw jelly, these popular sweet-sour sweets are made from hawthorn berries, believed to aid weight loss. Sold in Chinese medical shops or in the snack section of Chinese grocery stores.

Quail Eggs
Though small in size, these eggs are said to be high in cholesterol. They are usually hardboiled before use in dishes.

GLOSSARY

Spring Onions (Scallions)

These are onions that have small bulbs and long green stalks. They are usually added raw into dishes for flavour or used as garnish.

Chinese Chives

Also known as garlic chives or *jiu cai*, these flat deep green leaves have a much stronger flavour compared to Western chives. Trim off and discard the white stems before use.

Coriander (Cilantro)

This strong-flavoured herb is also known as Chinese parsley. It is very different from Western parsley and should never be used as a substitute or vice versa. In Chinese cooking, it is favoured for its aroma and is commonly used as a garnish.

Bamboo Shoot

Also known as *zhu sun*. These are the pale yellow spear-shaped young shoots of bamboo trees. Available fresh or canned. The canned shoots are less tough and do not need to be boiled like fresh bamboo shoots before using. Winter bamboo shoots are more tender and less fibrous.

Bean Sprouts
These white bean sprouts are commonly stir-fried and enjoyed for their juicy, crunchy texture. Choose plump, crisp-looking sprouts with buds at the tip. Avoid browned and limp ones.

Chinese Cabbage
This has a mild flavour and a rather large and long shape. Its closely-packed broad leaves have a white centre section and light green edges.

Lotus Roots
The gourd-shaped root of the lotus plant. When the root is cut across, distinctive holes can be seen. The texture is light, slightly dry and crisp. Commonly used in Chinese soups.

Water Chestnuts
The edible tuber of a water plant, water chestnuts have thin dark brown skin and sweet, white and crisp flesh. They are available fresh or canned.

Dried Prawns (Shrimps)
Sun-dried, steamed prawns with a mild flavour. Soak in water for 15 minutes to get rid of excess salt before using. Use whole or pound for extra flavour.

WEIGHTS & MEASURES

Quantities for this book are given in Metric and American (spoon) measures. Standard spoon measurements used are: 1 tsp = 5 ml and 1 Tbsp = 15 ml. All measures are level unless otherwise stated.

LIQUID AND VOLUME MEASURES

Metric	Imperial	American
5 ml	1/6 fl oz	1 teaspoon
10 ml	1/3 fl oz	1 dessertspoon
15 ml	1/2 fl oz	1 tablespoon
60 ml	2 fl oz	1/4 cup (4 tablespoons)
85 ml	2 1/2 fl oz	1/3 cup
90 ml	3 fl oz	3/8 cup (6 tablespoons)
125 ml	4 fl oz	1/2 cup
180 ml	6 fl oz	3/4 cup
250 ml	8 fl oz	1 cup
300 ml	10 fl oz (1/2 pint)	1 1/4 cups
375 ml	12 fl oz	1 1/2 cups
435 ml	14 fl oz	1 3/4 cups
500 ml	16 fl oz	2 cups
625 ml	20 fl oz (1 pint)	2 1/2 cups
750 ml	24 fl oz (1 1/5 pints)	3 cups
1 litre	32 fl oz (1 3/5 pints)	4 cups
1.25 litres	40 fl oz (2 pints)	5 cups
1.5 litres	48 fl oz (2 2/5 pints)	6 cups
2.5 litres	80 fl oz (4 pints)	10 cups

OVEN TEMPERATURE

	°C	°F	Gas Regulo
Very slow	120	250	1
Slow	150	300	2
Moderately slow	160	325	3
Moderate	180	350	4
Moderately hot	190/200	370/400	5/6
Hot	210/220	410/440	6/7
Very hot	230	450	8
Super hot	250/290	475/550	9/10

LENGTH

Metric	Imperial
0.5 cm	1/4 inch
1 cm	1/2 inch
1.5 cm	3/4 inch
2.5 cm	1 inch

DRY MEASURES

Metric	Imperial
30 grams	1 ounce
45 grams	1 1/2 ounces
55 grams	2 ounces
70 grams	2 1/2 ounces
85 grams	3 ounces
100 grams	3 1/2 ounces
110 grams	4 ounces
125 grams	4 1/2 ounces
140 grams	5 ounces
280 grams	10 ounces
450 grams	16 ounces (1 pound)
500 grams	1 pound, 1 1/2 ounces
700 grams	1 1/2 pounds
800 grams	1 3/4 pounds
1 kilogram	2 pounds, 3 ounces
1.5 kilograms	3 pounds, 4 1/2 ounces
2 kilograms	4 pounds, 6 ounces

ABBREVIATION

tsp	teaspoon
Tbsp	tablespoon
g	gram
kg	kilogram
ml	millilitre